Bo E. Sernelius

Surface Modes in Physics

Bo E. Sernelius

Surface Modes in Physics

Berlin · Weinheim · New York · Chichester
Brisbane · Singapore · Toronto

Author:
Professor Bo E. Sernelius
Department of Physics and Measurement Technology
Linköping University
Linköping
Sweden
http://www.ifm.liu.se/~boser/
bos@ifm.liu.se

> This book was carefully produced. Nevertheless, author and publisher do not warrant the information contained therein to be free of errors. Readers are advised to keep in mind that statements, data, illustrations, procedural details or other items may inadvertently be inaccurate.

With 193 figures and 6 tables

1st edition

Library of Congress Card No.: applied for

A catalogue record for this book is available from the British Library.

Die Deutsche Bibliothek - CIP Cataloguing-in-Publication-Data
A catalogue record for this publication is available from Die Deutsche Bibliothek

ISBN 3-527-40313-2

© WILEY-VCH Verlag Berlin GmbH, Berlin (Bundesrepublik Deutschland) 2001
Printed on acid-free paper.
All rights reserved (including those of translation in other languages). No part of this book may be reproduced in any form - by photoprinting, microfilm, or any other means - nor transmitted or translated into machine language without written permission from the publishers. Registered names, trademarks, etc. used in this book, even when not specifically marked as such, are not to be considered unprotected by law.
Printing: Strauss Offsetdruck GmbH, 69509 Mörlenbach.
Bookbinding: J. Schäffer GmbH & Co. KG, 67269 Grünstadt.
Printed in the Federal Republic of Germany.
WILEY-VCH Verlag Berlin GmbH
Bühringstrasse 10, D-13086 Berlin
Federal Republic of Germany

PREFACE

This book is aimed at MSc and PhD students, last year MSci students, Postdocs, teachers, and professional scientists. The book is suitable as the main text of a complimentary course to more general courses in electromagnetism, surface physics and condensed matter. It may also be useful for physicists wanting to broaden their fields in the direction of biology and chemistry; this is in line with the general trend that the scientific focus shifts away from physics towards these other disciplines and especially towards interdisciplinary work.

The reason I wrote this book is the following: At many occasions I came in contact with publications referring to modes at surfaces and interfaces and in slab geometries. Also the dispersion of the modes, with and without retardation effects, were given. I didn't really understand how these modes and their dispersions came about but I felt that I should know about the subject even though it is outside my field of research. So, I put these problems into the pile of problems that I would look into later if and when I got time. Then suddenly I was, more or less, forced to deal with these modes in connection with a work in which I tried to explain the anomalous behavior of the mean-free-path in beryllium. One of the many possible explanations to the anomalies is the excitation of surface plasmons by the outgoing photoelectron. My experience is that the best way to understand a subject is to try to derive everything by yourself. This I did for the surface plasmons. In doing so I realized that the derivation and origin of other modes were very similar. Furthermore, when working on these problems, I found that these modes were very important and had effects on numerous physical quantities like the surface energy, the Casimir effect (the attraction between two flat, neutral, metallic surfaces), the capillary force, the van der Waals interaction, step formation and crystal growth at crystal surfaces, the tip-effect, adhesion, and so forth. I realized that many students, teachers and scientists were unaware of how these effects come about. This made me experience an urge to share these insights with a broad audience. During the development of this book I included this subject as a part of my own research and it gradually, by its own force, shifted towards the focus of my research. New experimental high accuracy results for the force between metal surfaces have brought the van der Waals and Casimir forces into focus. These forces are important parts of this book.

The modes and the effects of these modes are what a solid-state theorist would call many-body effects. Many-body theory is a very beautiful, but also very complicated, theory. There is a big step one has to pass before one masters the many-body theory and can use it in practical applications. However, as we will see in the text, we may get away without having to use this theory. We only need to be able to determine the electromagnetic fields that are solutions to Maxwell's equations with the proper boundary conditions at the surfaces and interfaces. Thus the problem is reduced into one of classical fields.

At our university, Linköping university, the research is focused on the physics of condensed matter, and solids in particular. An ordinary piece of solid has a dimension of 1 cm cubed or larger. Most of the atoms and electrons in such a sample belong to the interior or bulk and are not influenced by the presence of the surfaces. Just the atoms in a few atomic layers closest the surface are affected by the surface. The larger the sample the smaller is the fraction of surface to bulk atoms. In most cases the bulk properties dominate, but in many cases the presence of surfaces and interfaces plays an important role.

In the majority of experiments one perturbs the system and studies the response to the perturbation. Let us choose photoemission as an example. Here light is used as an external probe. The response to the external perturbation is that electrons are excited and some of them leave the sample. The energy and momentum distributions of these escaping photoelectrons are studied. The presence of the surface has effect on the experimental result in several ways. The light has to pass the surface and the surface potential gives rise to an additional excitation contribution – we have bulk and surface photoemission. The photoelectrons must pass the surface before they reach the detector and may in doing so excite surface modes. If the energy of the photons in the experiment is chosen such that the kinetic energy of the photoelectrons is close to the plasmon energy of the solid the mean-free-path of the electrons is very short, just a few Ångströms. This means that only electrons excited very close to the surface can escape and be detected. These electrons do not represent an average electron of the solid and measured properties are not the bulk properties but properties very much influenced by the presence of the surface.

In the example above I discussed how the experimental result may be modified by the presence of the surface of the sample although the overall properties of the sample are determined by the bulk. If the sample is small or thin enough there is no real bulk and the properties are very much determined by the surfaces. This is the case for thin films and in some artificial quantum structures.

In other situations one is interested in surface physics. Examples are crystal growth, interaction between an external particle and the sample, and so forth.

In all these cases the presence of surface modes may be very important and everyone should at least be aware of them. They may turn up as unexpected effects in experiments and they may also be utilized to gain information about changes at the surface.

I have for several years successfully used the material in this book as the main text in a course for students with varied backgrounds: graduate students in physics or in cross disciplinary subjects; last year undergraduate students in physics and in civil engineering; master students in physics; colleagues. Since I know that equations and other mathematical expressions or relations might scare some people, especially outside the field of physics, I first tried to keep the number of equations at a minimum. However, when I consulted the students they urged me to include all equations and all steps in the derivations. Consequently, the book contains a large number of derivations that should be of help in really understanding the background of many relations that can be found in other books. The detailed derivations should furthermore be useful for researchers wanting to get into this field.

I use CGS units and not SI units in the text. This is because CGS units are better suited for electromagnetic problems and also for applied solid-state theory. In Appendix 1, I have included a conversion table for symbols and formulas which makes it easy to convert each equation or expression into the corresponding SI units form.

Many persons have contributed to the development of the field covered by this book and

I cannot give just credit to them all. I just list the names of a handful of pioneers in the same order as their contributions appeared, historically. They are: J. D. van der Waals for the empirical discovery of the van der Waals force between atoms; F. London for the theoretical explanation of the van der Waals force between atoms; H. B. G. Casimir for the explanation of the Casimir force between plates of perfect metals in terms of zero-point energies; E. M. Lifshitz for the formulation of the forces between macroscopic objects in terms of the solutions to Maxwell's equations.

Finally, I want to thank Professor A. A. Maradudin for carefully reading the manuscript and for providing many useful suggestions. I am furthermore indebted to my wife Ulla and my children Natalie and David for bearing with me during the preparation of the manuscript; at times I walked around with empty eyes looking like a zombie, completely absorbed by the task at hand. I also thank my wife for reading the final version of the manuscript trying to find missed typing errors. I furthermore apologize to my family for spoiling our evenings and weekends here in Oak Ridge, Tennessee, where I am just now on a sabbatical. Instead of enjoying the change of milieu as we had planned to do I have been editing the manuscript, transforming it into camera ready format.

Bo E. Sernelius

Oak Ridge, in December, 2000

CONTENTS

Introduction		13
1 Bulk modes		17
1.1	Bulk modes in terms of fields	17
1.2	Bulk modes in terms of potentials	24
2 Model dielectric functions		31
2.1	Lorentz' classical model for the dielectric function of insulators	32
2.2	Drude's classical model for the dielectric function of metals	36
2.3	Modelling	36
2.4	Dielectric function of a plasma	51
2.5	Static dielectric function for a dilute gas of permanent dipoles	54
2.6	Debye rotational relaxation	56
2.7	Dielectric properties of water	60
2.8	Superluminal speeds	64
	2.8.1 Speed of light in vacuum	64
	2.8.2 Einstein's special theory of relativity	65
	2.8.3 Tachyons	66
	2.8.4 Trivial examples	67
	2.8.5 EPR paradox	68
	2.8.6 Phase velocity versus group velocity	68
	2.8.7 Surpassing the sonic speed barrier	71
	2.8.8 Faster than the speed of light in a medium	73
	2.8.9 Superluminal speeds caused by changes in the vacuum	74
	2.8.10 Tunneling	74
	2.8.11 What do we mean by signals, information and message?	76
	2.8.12 Conclusions	76
3 Zero-point energy of modes		79

4 Modes at flat interfaces — 99
- 4.1 Modes at a single interface — 104
  - 4.1.1 Metal-vacuum interface — 107
  - 4.1.2 Semiconductor-vacuum interface — 111
- 4.2 Modes in slab geometry — 117
  - 4.2.1 Metal slab in vacuum — 124
  - 4.2.2 Semiconductor slab in vacuum — 128
  - 4.2.3 Vacuum gap in a metal — 132
  - 4.2.4 Vacuum gap in a semiconductor — 136
- 4.3 The Casimir effect — 138
  - 4.3.1 Casimir effect at zero temperature — 139
  - 4.3.2 Casimir effect at finite temperature — 142
- 4.4 Metal surfaces — 145
  - 4.4.1 Surface energy of metals — 145
  - 4.4.2 Optical properties of mercury — 149
  - 4.4.3 Surface tension of mercury — 153
- 4.5 Quantum wells — 154
  - 4.5.1 Casimir and van der Waals forces between two 2D metallic sheets — 154
  - 4.5.2 Plasmon-pole approximation — 161

5 Forces — 167
- 5.1 Two molecules with permanent dipole moments — 168
- 5.2 One ion and one molecule with permanent dipole moment — 170
- 5.3 Two molecules one with and one without permanent dipole moment — 171
- 5.4 Two molecules without permanent dipole moments — 173
- 5.5 Two ions — 177
- 5.6 Three or more polarizable atoms — 177
- 5.7 Interaction between macroscopic objects — 180
- 5.8 Interaction between two spheres: limiting results — 181
- 5.9 Interaction between two spheres: general results — 182
  - 5.9.1 Radially varying dielectric functions — 185
- 5.10 General expression for small separations — 186
- 5.11 Cylinders and half-spaces — 186
- 5.12 Summation of pair interactions — 188
- 5.13 Derivation of the van der Waals equation of state — 192

6 Energy and force — 197
- 6.1 Interaction energy at zero temperature — 198
  - 6.1.1 Interaction between two polarizable atoms revisited: no retardation — 199
  - 6.1.2 Interaction between two polarizable atoms revisited: retardation — 202
- 6.2 Interaction energy at finite temperature — 209
- 6.3 Surface energy, method 1: no retardation — 213
- 6.4 Surface energy, method 1: retardation — 215
- 6.5 Surface energy, method 2: no retardation — 216
- 6.6 Surface energy, method 2: retardation — 218
- 6.7 Finite temperatures — 221
  - 6.7.1 Retarded interaction energy — 226
- 6.8 Recent results for metals — 227
- 6.9 Adhesion, cohesion, and wetting — 232
  - 6.9.1 Work of adhesion and cohesion — 232
  - 6.9.2 Wetting — 234
  - 6.9.3 Model calculations — 236
    - 6.9.3.1 Modelling of adhesion, cohesion and wetting — 236
    - 6.9.3.2 Birds of a feather flock together — 242
  - 6.9.4 Capillary rise — 244
- 6.10 Finding the pair interactions — 253
  - 6.10.1 Non-retarded limit — 254
  - 6.10.2 Retarded limit — 255

7 Modes at non-planar interfaces — 257
- 7.1 Modes at the surface of a sphere — 257
  - 7.1.1 Metal sphere in vacuum — 261
  - 7.1.2 Dielectric sphere in vacuum — 262
  - 7.1.3 Spherical void in a metal — 263
  - 7.1.4 Spherical void in a dielectric — 264
  - 7.1.5 Modes in a layered sphere — 265
    - 7.1.5.1 Metallic spherical shell in vacuum — 268
  - 7.1.6 When liquids stay dry — 269
- 7.2 Modes at the surface of a cylinder — 271
  - 7.2.1 Metal cylinder in vacuum — 274
  - 7.2.2 Cylindrical void in a metal — 275
- 7.3 Modes at an edge — 277
  - 7.3.1 Metallic wedge in vacuum — 280
  - 7.3.2 Wedge void in a metal — 281
- 7.4 Modes in a needle (a paraboloid of revolution) — 282

8 Different mode types — 285
- 8.1 Polar semiconductors or ionic insulators — 285
- 8.2 Metallic systems — 293
- 8.3 Characterization of different surface mode types — 298
- 8.4 Spatial dispersion — 302
- 8.5 Surface roughness — 306
- 8.6 The ATR method — 306
- 8.7 Earthquakes, rainbow and optical glory — 314

9 Colloids — 317
- 9.1 Milk — 318
- 9.2 Stability of colloids — 319
- 9.3 Formation of the double layer — 323
  - 9.3.1 Flat double layer — 324
  - 9.3.2 Spherical double layer — 326
- 9.4 Gouy and Chapman theory — 328
  - 9.4.1 Gouy and Chapman theory of a flat double layer — 329
  - 9.4.2 Gouy and Chapman theory of a spherical double layer — 336
- 9.5 Stern's theory of a flat double layer — 337
- 9.6 The ζ-potential — 338
- 9.7 Interaction energy and force between objects with double layers — 339
  - 9.7.1 Interaction between two flat double layers — 342
    - 9.7.1.1 Total potential between two layers — 347
    - 9.7.1.2 Stability conditions — 349
  - 9.7.2 Interaction between spherical particles — 354

Appendix 1 Conversion table from CGS to SI units — 361

Appendix 2 Fourier-transform conventions — 363

Index — 365

INTRODUCTION

All objects in the universe, living bodies as well as dead objects, are surrounded by electromagnetic fields or modes. Also within the objects or bodies, if there are interfaces between materials of different dielectric or magnetic properties, there are modes.

These modes are solutions to Maxwell's equations with the proper boundary conditions. They are localized to the surfaces and interfaces. They play an important role in many situations. One set of effects derives from the fact that these modes contain energy. As a result they contribute to the surface energy and surface tension of the objects, which means that they have effect on the stability and shape of the object itself. They also give rise to forces between objects. These forces are always there and are of particular importance if the objects are small.

The modes are responsible for a diversity of, often spectacular, effects like that liquid helium climbs up the walls of a beaker, that rain is associated with thunderstorms and that drops of coffee can survive for seconds on top of the coffee surface in a drip coffee maker. The modes are important for wetting, capillary forces, adhesion and cohesion. They are important for catalytic reactions at surfaces. Another effect is that their presence modifies the optical properties of the objects; with the term optical properties we do not mean to limit ourselves to the visible range of the spectrum. In this book we will cover these two aspects, the energy and forces on the one hand and the optical properties on the other; the emphasis will be on the first of the two. The energy of each mode is sensitive to small changes at the surface like the presence of foreign atoms. These changes can be detected by optical means or other. Thus, the modes can be utilized, and are so, in sensors like gas sensors. The forces between objects are used in the atomic force microscope and similar instruments.

Many of the materials around us are colloids. We ourselves are to a large extent made up from colloids and the nutrients in our food are often on colloidal form. The effective surface area is very large for colloids which means that the effects studied here are of great importance. We will spend some time discussing colloids and their stability. The forces derived from the surface modes are one component determining their stability. Milk is an example of a colloid. We know how to prepare milk in order to enhance its stability and we also know how to quickly destroy the stability; white tea or tea with lemon is O.K. but putting both milk and lemon in the tea is not such a good idea. Another example from the kitchen is gravy. The gravy has to be prepared with care and a little mistake can easily break the stability and ruin the dinner. Many of the ingredients used in the food industry, cosmetic industry, and pharmaceutical industry, like stabilizers of various kinds, utilize these forces determining the colloidal stability to improve the properties of the materials. In composites one makes use of the large effective surface area and hence surface energy to strengthen the material; the thinner the fibers in the composite the stronger the material.

Many of the important processes in biological systems occur at surfaces of cells and membranes. One may say that life itself in the early stages developed, and life still develops, at surfaces and interfaces. The interactions are important when, for example, viruses attach themselves to the surfaces of the cells. Thus, the modes are also important in biological systems. Furthermore, the energetics of surfaces and interfaces are important for biocompatibility which is an important property for biological implants.

The modes at the surface of an object are determined by the dielectric and magnetic properties of the materials on both sides of the surface and also by the shape of the object. We will learn how to determine the modes on flat surfaces, at wedges, at tips and on objects of various shapes, like spheres and cylinders. We will not treat magnetic materials, only dielectric.

Before we confront the surface-modes we discuss what modes we have in the bulk of a material and how these are related to the dielectric properties of the material. This we do in Chapter 1. We find that the dielectric function of the bulk material is important for the dispersion of normal modes and hence devote Chapter 2 to model dielectric functions representing materials of different types. We describe the general behavior of the dielectric properties of different types of material, like metals, semiconductors, polar semiconductors and ionic insulators, heavily-doped semiconductors, rare-gas solids, liquids and gases. We also introduce model dielectric functions for these materials and demonstrate the different optical properties that films from these materials attain. In Chapter 3 we demonstrate that the description of the effects from interactions in a many-particle system can be cast in a form where the energy is expressed in terms of the zero-point energies of the normal modes – the modes we deal with in this book. This chapter is intended for readers with some experience in many-body theory. It demonstrates that the more relaxed formalism in this book really treats the same thing as the formally more strict many-body formalism. The chapter may safely be passed by without any harm to the remainder.

Flat or planar interfaces are of special importance. These are treated in Chapter 4 where we discuss modes at such interfaces and for systems in slab geometry. In slab geometry there are two parallel interfaces. In this case there is an additional type of mode present, not localized to each interface but bounded by the two interfaces. These modes give rise to the Casimir force at large separations. In Section 4.5, where we treat the forces between two quantum wells, we demonstrate that the formalism in this book produces the same results as the many-body formalism does.

In Chapter 5 we discuss forces in general and we study forces between atoms, molecules and larger objects and derive the van der Waals force between these objects. In particular we derive the expression for the van der Waals force between two polarizable atoms using the coupled equations-of-motion for the electrons in the atoms.

Chapter 6 is devoted to the energy and force caused by the modes, both at zero and finite temperatures. We make a more general derivation of the interaction energy between two polarizable atoms, both neglecting and taking the retardation effects into account. Also here we use the coupled equations-of-motion. At the end of the chapter we rederive these expressions in an alternative way. We furthermore calculate the surface energy of planar surfaces and the force between plates of different materials. We reobtain the result for the Casimir force between metal plates and we make contact with the research front by presenting theoretical and experimental results for the force between two gold surfaces at

Introduction

finite temperature. We furthermore discuss adhesion, cohesion, wetting, and capillary rise.

Chapter 7 is devoted to the modes of non-planar interfaces on objects like spheres, cylinders, wedges and needles.

In Chapter 8 we discuss the modes in different systems and how they can be excited and studied. We treat in some detail the *ATR* method and present modelling of different types of *ATR* spectra. Throughout the book we treat the interface as ideal. We assume that the interface is perfectly smooth, that the material and its dielectric function on each side of the interface is identical all the way up to the interface. A real interface is not perfectly smooth; there is some *surface roughness*. There is a varying extent of relaxation of the crystal structure near a surface or an interface. The separation between the atomic layers are not the same as in the bulk and there may even be surface reconstruction where the crystal structure is modified at the surface. Besides, the treatment where the dielectric functions are the same all the way up to interface is an assumption that is in conflict with the concept of the dielectric function itself. The dielectric function is a result of an averaging of fields over regions in space, large compared to the volume per polarizable entity making up the material. This means that the dielectric function near the interface should have contributions from the polarizable entities on the other side of the interface. Furthermore, *spatial dispersion* may in some situations have the effect that there are more than one transmitted wave when a wave impinges on an interface. These complications are at least briefly touched upon at the end of Chapter 8. They seem to suggest that the treatment of the interface as ideal is of no use at all but, on the contrary, it seems to work quite well. However, one should still always keep in mind what the basic assumptions are.

Finally in Chapter 9 we treat colloids. We define the term colloid and give examples from all different types of colloid. We demonstrate how the so-called double layer is formed and its importance for the stability.

Each chapter starts with some thought-provoking questions in italics. The answers to these are to be found somewhere in the chapter.

1 Bulk modes

What determines the normal modes in the bulk of a material?

1.1 Bulk modes in terms of fields

When Maxwell in 1865 published [1] his equations, which were a formalization of the accumulated research on electromagnetism, he was not aware of that these equations also were applicable to light; that light was electromagnetic waves. It was not until Hertz [2] in 1889 discovered electromagnetic waves of longer wavelengths that one realized that the different forms of electromagnetic waves were all of the same origin.

We can divide the treatment of physical problems according to Figure 1.1, where the most exact and general formalism is used in the upper right-hand corner. In our every-day life we are mostly concerned with macroscopic objects that move with speeds small compared with the speed of light. Then we can safely use the theory in the lower left corner. This treatment is the simplest of the four. The results from the formalisms in the other three corners turn into the classical results when the speeds and the sizes of the objects get low and large, respectively. What is surprising is that Maxwell's equations (MEs) look the same in all four corners.

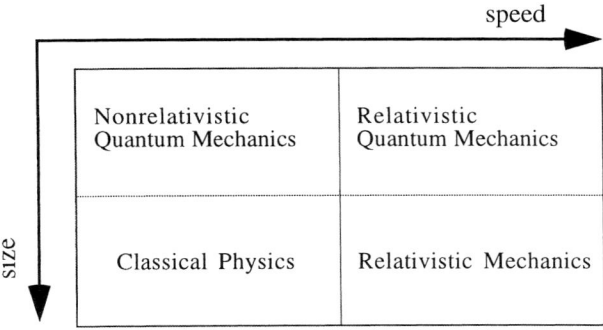

Figure 1.1: Physics in different domains.

The MEs read

$$\nabla \cdot \mathbf{D} = 4\pi\rho$$
$$\nabla \cdot \mathbf{B} = 0$$
$$\nabla \times \mathbf{E} = -\frac{1}{c}\frac{\partial \mathbf{B}}{\partial t} \qquad (1.1)$$
$$\nabla \times \mathbf{H} = \frac{4\pi}{c}\mathbf{J} + \frac{1}{c}\frac{\partial \mathbf{D}}{\partial t}$$

where the functions **E**, **D**, **B** and **H** are the electric field, the electric displacement, the magnetic induction and the magnetic field, respectively. We refer the reader to Appendix 1 for conversion of all equations and quantities into SI-units.

We also have the constitutive equations:

$$\mathbf{D} = \varepsilon \mathbf{E}$$
$$\mathbf{B} = \mu \mathbf{H} \qquad (1.2)$$

where ε and μ are the dielectric function and magnetic permeability of the medium, respectively. We assume, throughout, that the medium is isotropic or has cubic symmetry so that these material characteristics can be considered scalar quantities. The functions ρ and **J** often denote the total free charge and current densities — external and internal. The effects on the fields from induced charge and current densities originating from the bound charges are included in the dielectric function. We will here treat metals on the same footing as dielectrics. Then it is better to let ρ and **J** represent external charge and current densities only. We let the effect from the free charges in the metal be included in the dielectric function. This is achieved by using Ohm's law:

$$\mathbf{J}^{ind} = \sigma \mathbf{E} \qquad (1.3)$$

where σ is the conductivity, and the equation of continuity:

$$\frac{\partial \rho^{ind}}{\partial t} + \nabla \cdot \mathbf{J}^{ind} = 0 \qquad (1.4)$$

Now, in writing the constitutive equations the way we have done in Equation (1.2) we have assumed that the dielectric function and magnetic permeability of the medium are local, that is, the displacement and magnetic induction in a point at a certain time do only depend on the values of the electric and magnetic fields in that same point and at that same instant of time. This means that the two material parameters are delta functions in the space and time variables. This also means that their Fourier transforms with respect to space and time variables are constants, that is, are independent of frequency and momentum (see Appendix 2 for our conventions regarding the Fourier transforms.) This is sometimes a good approximation, but we need to be more general. The general form of the constitutive

1.1 Bulk Modes in Terms of Fields

relations for homogeneous and isotropic systems is

$$\mathbf{D}(\mathbf{r},t) = \iint d^3r'\, dt'\, \varepsilon(\mathbf{r}-\mathbf{r}', t-t')\mathbf{E}(\mathbf{r}',t')$$
$$\mathbf{B}(\mathbf{r},t) = \iint d^3r'\, dt'\, \mu(\mathbf{r}-\mathbf{r}', t-t')\mathbf{H}(\mathbf{r}',t')$$
(1.5)

Similarly we have

$$\mathbf{J}^{ind}(\mathbf{r},t) = \iint d^3r'\, dt'\, \sigma(\mathbf{r}-\mathbf{r}', t-t')\mathbf{E}(\mathbf{r}',t')$$
$$\mathbf{P}(\mathbf{r},t) = \iint d^3r'\, dt'\, \alpha(\mathbf{r}-\mathbf{r}', t-t')\mathbf{E}(\mathbf{r}',t')$$
$$\mathbf{M}(\mathbf{r},t) = \iint d^3r'\, dt'\, \chi(\mathbf{r}-\mathbf{r}', t-t')\mathbf{H}(\mathbf{r}',t')$$
(1.6)

where the first equation is Ohm's law and the other two give the polarization and magnetization of the system in the linear response regime in terms of the dielectric and magnetic polarizabilities, α and χ, respectively. The polarization and magnetization are related to the fields according to:

$$\mathbf{D} = \mathbf{E} + 4\pi\mathbf{P}$$
$$\mathbf{B} = \mathbf{H} + 4\pi\mathbf{M}$$
(1.7)

The integral relations above are so-called convolution integrals. They have the nice property that their Fourier transforms will be just the product of the two factors in the integrand. Thus, we have

$$\mathbf{D}(\mathbf{q},\omega) = \varepsilon(\mathbf{q},\omega)\mathbf{E}(\mathbf{q},\omega)$$
$$\mathbf{B}(\mathbf{q},\omega) = \mu(\mathbf{q},\omega)\mathbf{H}(\mathbf{q},\omega)$$
(1.8)

and

$$\mathbf{J}^{ind}(\mathbf{q},\omega) = \sigma(\mathbf{q},\omega)\mathbf{E}(\mathbf{q},\omega)$$
$$\mathbf{P}(\mathbf{q},\omega) = \alpha(\mathbf{q},\omega)\mathbf{E}(\mathbf{q},\omega)$$
$$\mathbf{M}(\mathbf{q},\omega) = \chi(\mathbf{q},\omega)\mathbf{H}(\mathbf{q},\omega)$$
(1.9)

These last two equations mean that

$$\varepsilon(\mathbf{q},\omega) = 1 + 4\pi\alpha(\mathbf{q},\omega)$$
$$\mu(\mathbf{q},\omega) = 1 + 4\pi\chi(\mathbf{q},\omega)$$
(1.10)

We now take the Fourier transform of the MEs and the equation of continuity. In Appendix 2 we have listed the effects this has on the different differential operators. To demonstrate how this comes about we explicitly go through one example. We have chosen the third of MEs. The left hand side can be written as

$$
\begin{aligned}
\nabla \times \mathbf{E} &= \nabla \times \int \frac{d^3q}{(2\pi)^3} \int_{-\infty}^{\infty} \frac{d\omega}{(2\pi)} e^{i(\mathbf{q}\cdot\mathbf{r}-\omega t)} \mathbf{E}(\mathbf{q},\omega) \\
&= \int \frac{d^3q}{(2\pi)^3} \int_{-\infty}^{\infty} \frac{d\omega}{(2\pi)} \nabla \times e^{i(\mathbf{q}\cdot\mathbf{r}-\omega t)} \mathbf{E}(\mathbf{q},\omega) \\
&= \int \frac{d^3q}{(2\pi)^3} \int_{-\infty}^{\infty} \frac{d\omega}{(2\pi)} i\mathbf{q} \times e^{i(\mathbf{q}\cdot\mathbf{r}-\omega t)} \mathbf{E}(\mathbf{q},\omega) \\
&= \int \frac{d^3q}{(2\pi)^3} \int_{-\infty}^{\infty} \frac{d\omega}{(2\pi)} e^{i(\mathbf{q}\cdot\mathbf{r}-\omega t)} [i\mathbf{q} \times \mathbf{E}(\mathbf{q},\omega)]
\end{aligned}
\tag{1.11}
$$

The first step is to express $\mathbf{E}(\mathbf{r},t)$ in terms of its Fourier transform, $\mathbf{E}(\mathbf{q},\omega)$; in the second step we move the rotational operator or curl, $\nabla \times$, inside the integral; in the third step we let it operate on the only factor that contains the spatial coordinate, the exponential factor; this produces the operator $i\mathbf{q}\times$ which operates on the vector $\mathbf{E}$. We make the corresponding steps for the right hand side of the equation:

$$
\begin{aligned}
-\frac{1}{c}\frac{\partial \mathbf{B}}{\partial t} &= -\frac{1}{c}\frac{\partial}{\partial t} \int \frac{d^3q}{(2\pi)^3} \int_{-\infty}^{\infty} \frac{d\omega}{(2\pi)} e^{i(\mathbf{q}\cdot\mathbf{r}-\omega t)} \mathbf{B}(\mathbf{q},\omega) \\
&= \int \frac{d^3q}{(2\pi)^3} \int_{-\infty}^{\infty} \frac{d\omega}{(2\pi)} \left(-\frac{1}{c}\right) \frac{\partial}{\partial t} e^{i(\mathbf{q}\cdot\mathbf{r}-\omega t)} \mathbf{B}(\mathbf{q},\omega) \\
&= \int \frac{d^3q}{(2\pi)^3} \int_{-\infty}^{\infty} \frac{d\omega}{(2\pi)} \left(-\frac{1}{c}\right) (-i\omega) e^{i(\mathbf{q}\cdot\mathbf{r}-\omega t)} \mathbf{B}(\mathbf{q},\omega) \\
&= \int \frac{d^3q}{(2\pi)^3} \int_{-\infty}^{\infty} \frac{d\omega}{(2\pi)} e^{i(\mathbf{q}\cdot\mathbf{r}-\omega t)} \left[i\omega \frac{1}{c} \mathbf{B}(\mathbf{q},\omega) \right]
\end{aligned}
\tag{1.12}
$$

Thus we end up with

$$
\begin{aligned}
&\int \frac{d^3q}{(2\pi)^3} \int_{-\infty}^{\infty} \frac{d\omega}{(2\pi)} e^{i(\mathbf{q}\cdot\mathbf{r}-\omega t)} [i\mathbf{q} \times \mathbf{E}(\mathbf{q},\omega)] \\
&= \int \frac{d^3q}{(2\pi)^3} \int_{-\infty}^{\infty} \frac{d\omega}{(2\pi)} e^{i(\mathbf{q}\cdot\mathbf{r}-\omega t)} \left[i\omega \frac{1}{c} \mathbf{B}(\mathbf{q},\omega) \right]
\end{aligned}
\tag{1.13}
$$

Since the Fourier transform and its inverse are unique transformations what is inside the square brackets on the two sides must be equal. Thus we have

1.1 Bulk Modes in Terms of Fields

$$i\mathbf{q} \times \mathbf{E}(\mathbf{q},\omega) = i\omega \frac{1}{c} \mathbf{B}(\mathbf{q},\omega) \tag{1.14}$$

or

$$i\mathbf{q} \times \mathbf{E}(\mathbf{q},\omega) = i\omega \frac{1}{c} \mu(\mathbf{q},\omega) \mathbf{H}(\mathbf{q},\omega) \tag{1.15}$$

where we have used one of the constitutive relations relating the magnetic field to the magnetic induction.

Going through the same procedure, or using the shortcuts given in Appendix 2, for the rest of the equations we find for the MEs:

$$\begin{aligned}
&i\varepsilon(\mathbf{q},\omega)\mathbf{q} \cdot \mathbf{E}(\mathbf{q},\omega) = 4\pi\rho^{ext}(\mathbf{q},\omega) + 4\pi\rho^{ind}(\mathbf{q},\omega) \\
&i\mu(\mathbf{q},\omega)\mathbf{q} \cdot \mathbf{H}(\mathbf{q},\omega) = 0 \\
&i\mathbf{q} \times \mathbf{E} = i\omega \frac{1}{c} \mu(\mathbf{q},\omega) \mathbf{H}(\mathbf{q},\omega) \\
&i\mathbf{q} \times \mathbf{H} = \frac{4\pi}{c} \mathbf{J}^{ind}(\mathbf{q},\omega) + \frac{4\pi}{c} \mathbf{J}^{ext}(\mathbf{q},\omega) - i\omega\varepsilon(\mathbf{q},\omega) \frac{1}{c} \mathbf{E}(\mathbf{q},\omega)
\end{aligned} \tag{1.16}$$

and for the equation of continuity:

$$-i\omega\rho^{ind}(\mathbf{q},\omega) = -i\mathbf{q} \cdot \mathbf{J}^{ind}(\mathbf{q},\omega) \tag{1.17}$$

We may now use Ohm's law to substitute the induced current density in favor of the electric field and with the help from the equation of continuity we may also substitute the induced charge density in favor of the electric field. Making these substitutions leads to

$$\begin{aligned}
&\tilde{\varepsilon}(\mathbf{q},\omega)\mathbf{q} \cdot \mathbf{E}(\mathbf{q},\omega) = -4\pi i\rho^{ext}(\mathbf{q},\omega) \\
&\mu(\mathbf{q},\omega)\mathbf{q} \cdot \mathbf{H}(\mathbf{q},\omega) = 0 \\
&\mathbf{q} \times \mathbf{E} = \frac{\omega}{c} \mu(\mathbf{q},\omega) \mathbf{H}(\mathbf{q},\omega) \\
&\mathbf{q} \times \mathbf{H} = -\frac{4\pi i}{c} \mathbf{J}^{ext}(\mathbf{q},\omega) - \frac{\omega}{c} \tilde{\varepsilon}(\mathbf{q},\omega) \mathbf{E}(\mathbf{q},\omega)
\end{aligned} \tag{1.18}$$

where we have introduced

$$\tilde{\varepsilon}(\mathbf{q},\omega) = \varepsilon(\mathbf{q},\omega) + i\frac{4\pi\sigma(\mathbf{q},\omega)}{\omega} \tag{1.19}$$

The last term of this equation is the contribution to the dielectric function from the free charges. The MEs separate out the transverse and longitudinal components of the fields. This separation into longitudinal and transverse modes is in general only possible for a homogeneous system. In the presence of interfaces the boundary conditions give rise to additional equations leading to mixing of the longitudinal and transverse modes and the final result is other types of normal modes. Figure 1.2 (a) illustrates a transverse field. The wave is moving vertically and the Electric-field vector is always pointing perpendicular to this direction, in the horizontal direction.

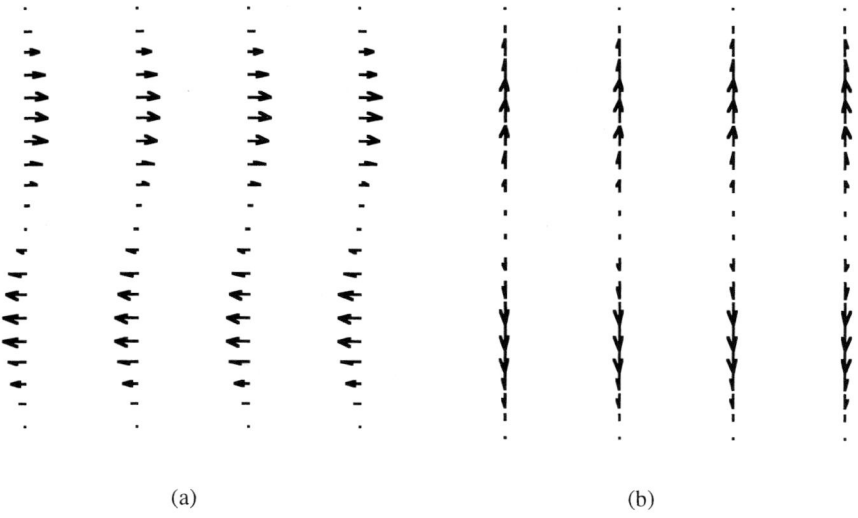

(a) (b)

Figure 1.2: Transverse field (a) and longitudinal field (b).

An example of longitudinal waves is given in Figure 1.2 (b). Also this wave is moving vertically but the field is here always parallel to the direction of motion.

We get two sets of equations, namely

$$\tilde{\varepsilon}_L(\mathbf{q},\omega)qE_L(\mathbf{q},\omega) = -i4\pi\rho^{ext}(\mathbf{q},\omega)$$
$$\mu_L(\mathbf{q},\omega)qH_L(\mathbf{q},\omega) = 0$$
$$0 = \frac{\omega}{c}\mu_L(\mathbf{q},\omega)H_L(\mathbf{q},\omega) \qquad (1.20)$$
$$0 = -\frac{4\pi i}{c}J_L^{ext}(\mathbf{q},\omega) - \frac{\omega}{c}\tilde{\varepsilon}_L(\mathbf{q},\omega)E_L(\mathbf{q},\omega)$$

and

1.1 Bulk Modes in Terms of Fields

$$qE_{\perp,x} = \frac{\omega}{c}\mu_\perp(\mathbf{q},\omega)H_{\perp,y}(\mathbf{q},\omega)$$

$$qE_{\perp,y} = -\frac{\omega}{c}\mu_\perp(\mathbf{q},\omega)H_{\perp,x}(\mathbf{q},\omega)$$

$$qH_{\perp,x} = -\frac{4\pi i}{c}J^{ext}_{\perp,y}(\mathbf{q},\omega) - \frac{\omega}{c}\tilde{\varepsilon}_\perp(\mathbf{q},\omega)E_{\perp,y}(\mathbf{q},\omega) \quad (1.21)$$

$$qH_{\perp,y} = \frac{4\pi i}{c}J^{ext}_{\perp,x}(\mathbf{q},\omega) + \frac{\omega}{c}\tilde{\varepsilon}_\perp(\mathbf{q},\omega)E_{\perp,x}(\mathbf{q},\omega)$$

Equations (1.20) are relations between longitudinal fields only and Equations (1.21) relations between transverse fields. This last system of equations can easily be solved and the result is

$$E_{\perp,x} = i\frac{4\pi\omega\mu_\perp(\mathbf{q},\omega)J^{ext}_{\perp,x}(\mathbf{q},\omega)}{(cq)^2 - \omega^2\mu_\perp(\mathbf{q},\omega)\tilde{\varepsilon}_\perp(\mathbf{q},\omega)}$$

$$E_{\perp,y} = i\frac{4\pi\omega\mu_\perp(\mathbf{q},\omega)J^{ext}_{\perp,y}(\mathbf{q},\omega)}{(cq)^2 - \omega^2\mu_\perp(\mathbf{q},\omega)\tilde{\varepsilon}_\perp(\mathbf{q},\omega)}$$

$$H_{\perp,x} = -i\frac{4\pi cqJ^{ext}_{\perp,y}(\mathbf{q},\omega)}{(cq)^2 - \omega^2\mu_\perp(\mathbf{q},\omega)\tilde{\varepsilon}_\perp(\mathbf{q},\omega)} \quad (1.22)$$

$$H_{\perp,y} = i\frac{4\pi cqJ^{ext}_{\perp,x}(\mathbf{q},\omega)}{(cq)^2 - \omega^2\mu_\perp(\mathbf{q},\omega)\tilde{\varepsilon}_\perp(\mathbf{q},\omega)}$$

From Equations (1.20) and (1.22) we find that there will be solutions in the absence of an external current or charge density if one of the following relations are fulfilled:

$$\omega^2 = \frac{(cq)^2}{\mu_\perp(\mathbf{q},\omega)\tilde{\varepsilon}_\perp(\mathbf{q},\omega)}$$

$$\tilde{\varepsilon}_L(\mathbf{q},\omega) = 0 \quad (1.23)$$

$$\mu_L(\mathbf{q},\omega) = 0$$

These equations give the normal modes of the system. They are the conditions for self-sustained fields. The first condition of (1.23) arises from the requirement that if, for example, we wish to have a nonzero result for $E_{\perp,y}(\mathbf{q},\omega)$ in the limit $J^{ext}_{\perp,y}(\mathbf{q},\omega)$ tends to zero, the denominator of the first of Equations (1.22) has to vanish. The second condition arises from the requirement that if we wish to have a nonzero result for $E_L(\mathbf{q},\omega)$ in the limit $J^{ext}_L(\mathbf{q},\omega)$ (or equivalently $\rho^{ext}(\mathbf{q},\omega)$) tends to zero $\tilde{\varepsilon}_L(\mathbf{q},\omega)$ has to vanish for the two inhomogeneous of Equations (1.20) to be fulfilled. The last condition is needed for a finite

$H_L(\mathbf{q},\omega)$. Otherwise the two homogeneous of Equations (1.20) are not fulfilled. Equations (1.23) look the same in SI units if the relative dielectric functions and permeabilities are used.

The first of Equations (1.23) leads to the dispersion of light inside the medium. The second gives the dispersion of longitudinal modes, like bulk plasmons and phonons. The third gives the dispersion of magnetic modes in a magnetic system. We will not treat magnetic systems in this book. We let the magnetic permeability be unity. This means that **H** and **B** are equal and it is enough to keep one of them. We will however, throughout the book keep them both so that it is easy for the reader to extend the material to magnetic systems if needed.

To summarize, we have in this section derived the generating equations for the two types of electromagnetic mode that are present in the bulk of materials. One type is transverse and the other is longitudinal. We wanted to treat all materials, metals and non-metals, on the same footing. Therefore we treated the free charges in metallic systems in the same way as we did the bound polarization charges. This is not the conventional treatment and to indicate this we put a tilde over the generalized dielectric function. From Chapter 2 and onwards we will drop this tilde but it is implicitly assumed that when we deal with metals it is this generalized form of dielectric function we use.

1.2 Bulk modes in terms of potentials

Often it is useful to introduce potentials and work with them instead of with the fields. These potentials are auxiliary functions, "help-functions." They are not real, measurable quantities as are the fields. Sometimes one gets so used to the potentials that one really thinks that they are real. The scalar potential, for example, may seem very much real. This section has been included to alert the reader to the fact that potentials can produce spurious modes that are not really there and goes away when the actual fields are derived from the potentials.

The potentials can be chosen in many different ways. Two very common ways to choose the potential are the Coulomb and Lorentz gauges. We will go through their derivation now. Let **A** and Φ denote the vector and scalar potentials, respectively. We will need the following three relations from vector analysis:

$$\begin{aligned}
&\nabla \times (\nabla \varphi) = \mathbf{0} \\
&\nabla \cdot (\nabla \times \mathbf{C}) = \mathbf{0} \\
&\nabla \times (\nabla \times \mathbf{C}) = \nabla(\nabla \cdot \mathbf{C}) - \nabla^2 \mathbf{C}
\end{aligned} \quad (1.24)$$

(The last equation assumes a Cartesian coordinate system.)

1.2 Bulk Modes in Terms of Potentials

Let us now start from the two homogeneous MEs. The first equation,

$$\nabla \cdot \mathbf{B} = 0 \tag{1.25}$$

is fulfilled if we let

$$\mathbf{B} = \nabla \times \mathbf{A} \tag{1.26}$$

This is seen from the second of Equations (1.24). Let us now make use of this relation in the second of the homogeneous MEs (1.1). This gives

$$\nabla \times \left(\mathbf{E} + \frac{1}{c} \frac{\partial \mathbf{A}}{\partial t} \right) = 0 \tag{1.27}$$

From the first of the Equations (1.24) we see that this is fulfilled if what is inside the parentheses is the gradient of a scalar. Thus let

$$\mathbf{E} + \frac{1}{c} \frac{\partial \mathbf{A}}{\partial t} = -\nabla \Phi \tag{1.28}$$

Thus for a choice of potentials the fields are obtained from the relations

$$\mathbf{E} = -\nabla \Phi - \frac{1}{c} \frac{\partial \mathbf{A}}{\partial t}$$
$$\mathbf{B} = \nabla \times \mathbf{A} \tag{1.29}$$

We still have a rather large flexibility to choose the potentials to fit our particular problem, but a change in the scalar potential has to be accompanied by a change in the vector potential and vice versa. Let us assume that we have made a choice of $\mathbf{A}$ and Φ. We see from the first of Equations (1.24) that adding a gradient of a scalar function to $\mathbf{A}$ will not change $\mathbf{B}$. We are allowed to change the potentials in the following way:

$$\mathbf{A} \rightarrow \mathbf{A}' = \mathbf{A} + \nabla \Lambda$$
$$\Phi \rightarrow \Phi' = \Phi - \frac{1}{c} \frac{\partial \Lambda}{\partial t} \tag{1.30}$$

These transformations are called gauge transformations. The fields, the real quantities, are not changed with such transformations; only the potentials, the auxiliary functions, are.

Let us now turn to the remaining MEs, the inhomogeneous ones. We have

$$\nabla \cdot \left(\tilde{\varepsilon}_L \nabla \Phi + \frac{\tilde{\varepsilon}_L}{c} \frac{\partial \mathbf{A}}{\partial t} \right) = -4\pi \rho^{ext}$$

$$\nabla \times \left(\frac{1}{\mu_\perp} \nabla \times \mathbf{A} \right) + \frac{1}{c} \frac{\tilde{\varepsilon}_L \nabla \partial \Phi}{\partial t} + \frac{\tilde{\varepsilon}}{c^2} \frac{\partial^2 \mathbf{A}}{\partial t^2} = \frac{4\pi}{c} \mathbf{J}^{ext}$$

(1.31)

which may be rewritten as

$$\tilde{\varepsilon}_L \nabla^2 \Phi + \frac{\tilde{\varepsilon}_L}{c} \frac{\partial}{\partial t} (\nabla \cdot \mathbf{A}) = -4\pi \rho^{ext}$$

$$\frac{1}{\mu_\perp} \nabla (\nabla \cdot \mathbf{A}) - \frac{1}{\mu_\perp} \nabla^2 \mathbf{A} + \frac{1}{c} \frac{\tilde{\varepsilon}_L \nabla \partial \Phi}{\partial t} + \frac{\tilde{\varepsilon}}{c^2} \frac{\partial^2 \mathbf{A}}{\partial t^2} = \frac{4\pi}{c} \mathbf{J}^{ext}$$

(1.32)

and after rearrangements in the second one we have

$$\tilde{\varepsilon}_L \nabla^2 \Phi + \frac{\tilde{\varepsilon}_L}{c} \frac{\partial}{\partial t} (\nabla \cdot \mathbf{A}) = -4\pi \rho^{ext}$$

$$\nabla^2 \mathbf{A} - \frac{\mu_\perp \tilde{\varepsilon}}{c^2} \frac{\partial^2 \mathbf{A}}{\partial t^2} - \nabla \left(\nabla \cdot \mathbf{A} + \frac{\mu_\perp \tilde{\varepsilon}_L}{c} \frac{\partial \Phi}{\partial t} \right) = -\frac{4\pi \mu_\perp}{c} \mathbf{J}^{ext}$$

(1.33)

These are two coupled differential equations; each of them contains both $\mathbf{A}$ and Φ. They can be decoupled by proper gauge transformations. In *Coulomb gauge* or transverse gauge we choose

$$\nabla \cdot \mathbf{A} = 0 \tag{1.34}$$

With this choice the vector potential is purely transverse and the scalar potential satisfies Poisson's equation:

$$\nabla^2 \Phi = -4\pi \rho / \tilde{\varepsilon}_L \tag{1.35}$$

This means that the scalar potential is instantaneous. There are no retardation effects for the scalar potential. So if for example someone on the moon were to play around with some charges, the potential would here on earth change according to this immediately, without any time delay as if information could be transferred faster than with the speed of light.

We will now get rid of the scalar potential from the equation for the vector potential, the second of Equations (1.33). Let us take the time derivative of Equation (1.35). We have

$$\nabla \cdot \left(\nabla \frac{\partial \Phi}{\partial t} \right) = -\frac{4\pi}{\tilde{\varepsilon}_L} \frac{\partial \rho^{ext}}{\partial t} \tag{1.36}$$

The equation of continuity, Equation (1.4), is valid for the external charge densities and

1.2 Bulk Modes in Terms of Potentials

currents as well as for the induced quantities, discussed earlier. Thus we have

$$\nabla \cdot \left(\nabla \frac{\partial \Phi}{\partial t} \right) = \frac{4\pi}{\tilde{\varepsilon}_L} \nabla \cdot \mathbf{J}^{ext} \tag{1.37}$$

This means that the longitudinal parts of $\nabla \partial \Phi / \partial t$ and $(4\pi/\tilde{\varepsilon}_L) \mathbf{J}^{ext}$ are the same. Since the first is purely longitudinal, in the present choice of gauge, we have

$$\nabla \frac{\partial \Phi}{\partial t} = \frac{4\pi}{\tilde{\varepsilon}_L} \mathbf{J}_L^{ext} \tag{1.38}$$

We have divided the external current into transverse and longitudinal parts:

$$\begin{aligned} \mathbf{J}^{ext} &= \mathbf{J}_\perp^{ext} + \mathbf{J}_L^{ext} \\ \nabla \cdot \mathbf{J}_\perp^{ext} &= 0 \\ \nabla \times \mathbf{J}_L^{ext} &= 0 \end{aligned} \tag{1.39}$$

Thus the decoupled differential equations for the potentials are

$$\begin{aligned} \nabla^2 \Phi &= -4\pi \frac{\rho^{ext}}{\tilde{\varepsilon}_L} \\ \nabla^2 \mathbf{A} - \frac{\mu_\perp \tilde{\varepsilon}_\perp}{c^2} \frac{\partial^2 \mathbf{A}}{\partial t^2} &= -\frac{4\pi \mu_\perp}{c} \mathbf{J}_\perp^{ext} \end{aligned} \tag{1.40}$$

The Fourier transformed versions are (see Appendix 2)

$$\begin{aligned} -q^2 \Phi &= -4\pi \frac{\rho^{ext}}{\tilde{\varepsilon}_L} \\ -q^2 \mathbf{A} + \frac{\mu_\perp \tilde{\varepsilon}_\perp}{c^2} \omega^2 \mathbf{A} &= -\frac{4\pi \mu_\perp}{c} \mathbf{J}_\perp^{ext} \end{aligned} \tag{1.41}$$

The conditions for self-sustained potentials are

$$\begin{aligned} \tilde{\varepsilon}_L(\mathbf{q}, \omega) &= 0 \\ \omega^2 &= \frac{(cq)^2}{\mu_\perp(\mathbf{q}, \omega) \tilde{\varepsilon}_\perp(\mathbf{q}, \omega)} \end{aligned} \tag{1.42}$$

These are the same as for self-sustained fields, obtained in the previous section. Nothing strange appeared with this choice of gauge.

Another common gauge is the Lorentz gauge or in the presence of matter the *generalized Lorentz gauge*. In this case we put

$$\nabla \cdot \mathbf{A} + \frac{\mu_\perp \tilde{\varepsilon}_L}{c} \frac{\partial \Phi}{\partial t} = 0 \tag{1.43}$$

and get

$$\tilde{\varepsilon}_L \nabla^2 \Phi - \left(\frac{\mu_\perp \tilde{\varepsilon}_L^2}{c^2} \frac{\partial^2 \Phi}{\partial t^2} \right) = -4\pi \rho^{ext}$$

$$\nabla^2 \mathbf{A} - \frac{\mu_\perp \tilde{\varepsilon}}{c^2} \frac{\partial^2 \mathbf{A}}{\partial t^2} = -\frac{4\pi \mu_\perp}{c} \mathbf{J}^{ext} \tag{1.44}$$

The Fourier transformed versions are

$$\tilde{\varepsilon}_L \left(q^2 - \frac{\omega^2 \mu_\perp \tilde{\varepsilon}_L}{c^2} \right) \Phi = 4\pi \rho^{ext}$$

$$\left(q^2 - \frac{\omega^2 \mu_\perp \tilde{\varepsilon}_\perp}{c^2} \right) \mathbf{A}_\perp = \frac{4\pi \mu_\perp}{c} \mathbf{J}_\perp^{ext} \tag{1.45}$$

$$\left(q^2 - \frac{\omega^2 \mu_\perp \tilde{\varepsilon}_L}{c^2} \right) \mathbf{A}_L = \frac{4\pi \mu_\perp}{c} \mathbf{J}_L^{ext}$$

Thus it appears to be a longitudinal "light mode" apart from the longitudinal collective mode and the transverse "light mode". This is a spurious solution. To see that this is so we have to return to the real quantities, the fields. We have

$$\Phi = \frac{4\pi \rho^{ext}}{\tilde{\varepsilon}_L \left(q^2 - \frac{\omega^2 \mu_\perp \tilde{\varepsilon}_L}{c^2} \right)}$$

$$\nabla \cdot \mathbf{A} + \frac{\mu_\perp \tilde{\varepsilon}_L}{c} \frac{\partial \Phi}{\partial t} = 0 \tag{1.46}$$

$$\mathbf{E} = -\nabla \Phi - \frac{1}{c} \frac{\partial \mathbf{A}}{\partial t}$$

or

1.2 Bulk Modes in Terms of Potentials

$$\Phi = \frac{4\pi\rho^{ext}}{\tilde{\varepsilon}_L \left(q^2 - \frac{\omega^2 \mu_\perp \tilde{\varepsilon}_L}{c^2} \right)}$$

$$i\mathbf{q}\cdot\mathbf{A} - i\frac{\omega\mu_\perp \tilde{\varepsilon}_L}{c}\Phi = 0 \qquad (1.47)$$

$$\mathbf{E} = -i\mathbf{q}\Phi + i\frac{\omega}{c}\mathbf{A}$$

where the second equation is just the Lorentz gauge condition and the last is the field expressed in terms of the potentials.

Thus, we have

$$\begin{aligned} E_L &= -iq\Phi + i\frac{\omega^2\mu_\perp \tilde{\varepsilon}_L}{qc^2}\Phi \\ &= i\left(-q + \frac{\omega^2\mu_\perp \tilde{\varepsilon}_L}{qc^2}\right)\frac{4\pi\rho^{ext}}{\tilde{\varepsilon}_L \left(q^2 - \frac{\omega^2 \mu_\perp \tilde{\varepsilon}_L}{c^2} \right)} \\ &= -i\frac{4\pi\rho^{ext}}{q\tilde{\varepsilon}_L} \end{aligned} \qquad (1.48)$$

The spurious longitudinal mode has disappeared. We will sometimes use the potentials to find the normal modes. Then it is important to keep in mind that spurious modes might appear.

References

[1] J. C. Maxwell, Phil. Trans., **155**, 459 (1865).
[2] H. Hertz, Ann. Physik, **36**, 1 (1889).

2 Model dielectric functions

Why is blue and violet light always refracted more than yellow and red? Why is copper and gold red and yellow, respectively, while silver and aluminum are no color? Why is the dielectric constant of water as high as 80? Have superluminal speeds been achieved?

We saw in the previous chapter that the material properties determining the bulk modes are the dielectric properties. We will later find that the modes at interfaces and surfaces are determined by the dielectric properties of the materials making up the system. Thus, the dielectric function is very important for the effects treated in this book.

We will represent the dielectric properties of the media treated here by simple model dielectric functions. We describe the construction of these in some detail since it leads to interesting insights.

The imaginary part of the dielectric function is non-zero in one or several frequency regions. These are the regions where light is absorbed; at the transverse optical phonon energy in a polar or ionic system; at the region of interband transitions in a solid; just below this region of interband transitions in semiconductors and insulators where there may be exciton lines; at the position of dipolar excitations of the atoms in gases. The list could be made very long.

The real and imaginary parts of the dielectric function are not independent. They are intimately connected. If, for example, the imaginary part is known for all frequencies the real part can be completely determined and vice verse. These relationships are given by the very useful *Kramers Kronig dispersion relations*:

$$B_1(\omega) = \frac{2}{\pi} \int_0^\infty d\omega' \, B_2(\omega') P \frac{\omega'}{\omega'^2 - \omega^2}$$
$$B_2(\omega) = \frac{-2}{\pi} \int_0^\infty d\omega' \, B_1(\omega') P \frac{\omega}{\omega'^2 - \omega^2},$$
(2.1)

where the indices *1* and *2* indicate the real and imaginary parts, respectively, of the correlation function *B*; the dielectric function, ε, or rather (ε-1) is such a correlation function that obeys these relations. The *P* denotes the principle part. These relations follow from the analytic properties of retarded correlation functions; a retarded correlation function describes how the system responds to an external perturbation.

We will present models for each of the regions of non-vanishing imaginary part. These models are intended to be used in the book or be saved for later use. Before going into these models we present some classical models.

2.1 Lorentz' classical model for the dielectric function of insulators

In this model the electrons are assumed to be bound to the nucleus with forces obeying Hooke's law. The forces are assumed to be isotropic and damping can be included through frictional forces proportional to the electron velocity.

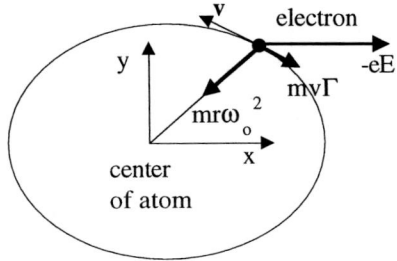

Figure 2.1: Schematic illustration of the electron orbit around the center of the atom. Indicated by thick arrows are the forces acting on the electron: the centripetal force towards the center; the frictional force in the direction opposite to that of the velocity **v**; the external force in the direction opposite to that of the external electric field **E**.

The equation of motion for an electron is

$$m\ddot{\mathbf{r}} + m\Gamma\dot{\mathbf{r}} + m\omega_0^2\mathbf{r} = -e\mathbf{E} \tag{2.2}$$

This is just the Newton's second law, stating that mass times acceleration equals the force exerted on the particle; we have collected all terms except the external force on the left hand side of the equation. The first term is just the mass times the acceleration; the second the frictional force; the third the restoring force binding the electron to the nucleus, placed at the origin. The term on the right hand side of the equation is the force due to the external field, **E**.

Fourier transforming this differential equation gives (see Appendix 2)

$$m\mathbf{r}(\omega)\left(-\omega^2 - i\omega\Gamma + \omega_0^2\right) = -e\mathbf{E}(\omega) \tag{2.3}$$

Thus

$$\mathbf{r}(\omega) = \frac{e\mathbf{E}(\omega)}{m\left(\omega^2 - \omega_0^2 + i\omega\Gamma\right)} \tag{2.4}$$

2.1 Lorentz' Classical Model ...

The induced dipole moment per electron is **p** = -e**r**, that is,

$$\mathbf{p}(\omega) = -\frac{e^2}{m} \frac{1}{\left(\omega^2 - \omega_0^2 + i\omega\Gamma\right)} \mathbf{E}(\omega) \tag{2.5}$$

Thus if we assume that there is one electron per atom and n atoms per unit volume we have for the atomic polarizability

$$\alpha^{at}(\omega) = -\frac{e^2}{m} \frac{1}{\left(\omega^2 - \omega_0^2 + i\omega\Gamma\right)} \tag{2.6}$$

for the polarizability

$$\alpha(\omega) = n\alpha^{at}(\omega) \tag{2.7}$$

and for the dielectric function, according to Equation (1.10),

$$\varepsilon(\omega) = 1 + 4\pi n \alpha^{at}(\omega)$$

$$= 1 - \frac{4\pi n e^2}{m} \frac{1}{\left(\omega^2 - \omega_0^2 + i\omega\Gamma\right)} \tag{2.8}$$

$$= 1 - \frac{\omega_{pl}^2}{\left(\omega^2 - \omega_0^2 + i\omega\Gamma\right)}$$

where ω_{pl} is the plasma frequency.
The real and imaginary parts are

$$\varepsilon_1(\omega) = 1 - \frac{\omega_{pl}^2\left(\omega^2 - \omega_0^2\right)}{\left(\omega^2 - \omega_0^2\right)^2 + (\omega\Gamma)^2} \tag{2.9}$$

and

$$\varepsilon_2(\omega) = \frac{\omega_{pl}^2 \omega\Gamma}{\left(\omega^2 - \omega_0^2\right)^2 + (\omega\Gamma)^2} \tag{2.10}$$

respectively. This treatment can be generalized to insulators with more than one electron per atom. We then have

$$\varepsilon(\omega) = 1 - \frac{4\pi e^2}{m} \sum_j \frac{n_j}{\left(\omega^2 - \omega_j^2 + i\omega\Gamma_j\right)}$$

$$= 1 - \sum_j \frac{\omega_{pl,j}^2}{\left(\omega^2 - \omega_j^2 + i\omega\Gamma_j\right)} \quad ; \quad \sum_j \omega_{pl,j}^2 = \omega_{pl}^2 = \frac{4\pi n e^2}{m} \tag{2.11}$$

where n_j is the density of electrons with resonance frequency ω_j and n is the total density of electrons.

Let us study an example using this dielectric function with the values:

$$\begin{cases} \hbar\omega_0 = 4eV \\ \left(\hbar\omega_{pl}\right)^2 = 60(eV)^2 \\ \Gamma = 1eV \end{cases} \tag{2.12}$$

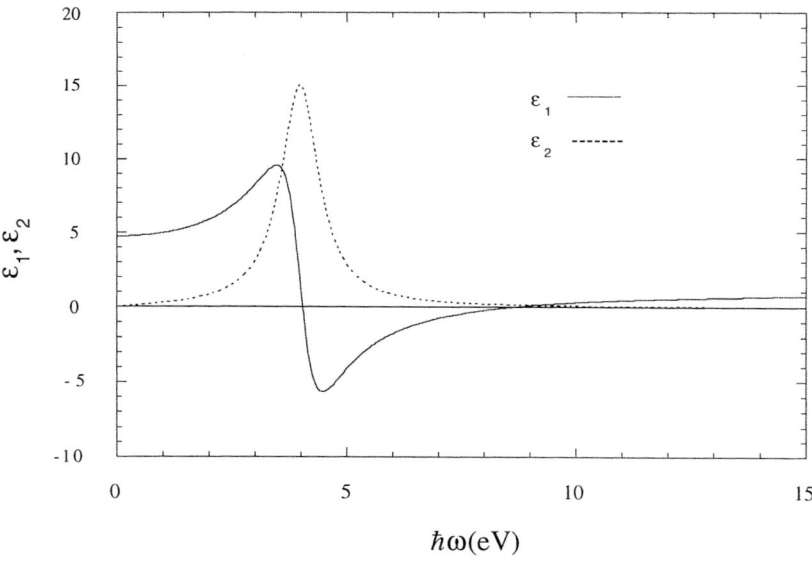

Figure 2.2: Real and imaginary parts of a dielectric function in the Lorentz model.

The real and imaginary parts of the dielectric function for the system with this choice of parameters are shown in Figure 2.2. The dashed curve represents the imaginary part; it also represents the absorption of the system; the larger the frictional force the wider this peak will be; in the limit of no friction this peak becomes a δ-function.

Let us use this dielectric function to find the light dispersion in the medium. The equation giving the light-dispersion curve was given in Section 1.1:

$$\omega^2 = \frac{(cq)^2}{\varepsilon(\mathbf{q},\omega)} \tag{2.13}$$

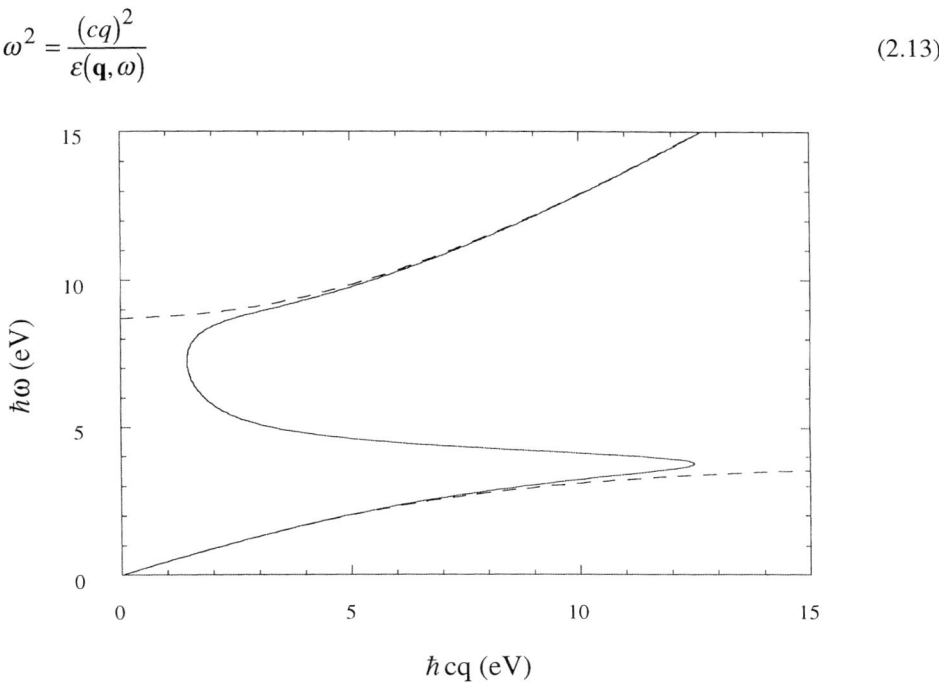

Figure 2.3: Light dispersion curve for a system with the dielectric function in Figure 2.2. The dashed curve is the result in absence of damping.

The result is shown in Figure 2.3. As comparison we have also shown the result in absence of damping (dashed curve). In the absence of damping there is a gap in energy where there is no mode. Light with energy in this range cannot travel through the medium; it is totally reflected if impinging on the surface of the material from outside. If we have damping, on the other hand, the light may enter. The light is not totally reflected, but the reflectivity is large and the fraction of light that enters is to a large extent absorbed. For a complex valued dielectric function the solution of Equation (2.13) gives a complex valued energy. We have plotted the real part of this above. The complex valued energy means that the excitation dies out with time.

The speed of light in the medium, in units of c (the speed of light in vacuum), is given by the slope of the curve. In absence of damping this slope is always smaller than unity. In the case of damping there are regions where it exceeds unity! Is the speed of light exceeded? Yes, if we define it as the the speed of a wave packet moving through the system! However the amplitude of this wave packet decreases all the time and the speed of the wave front will probably not exceed c. It should not be possible to transmit information with a speed exceeding c. There is at present an active research in the field of superluminal speeds so we devote Section 2.8 to this topic.

2.2 Drude's classical model for the dielectric function of metals

The Drude model for metals is obtained from the Lorentz model by letting the electrons be free and not bound to the atoms. This is obtained by letting ω_0 be zero:

$$\varepsilon(\omega) = 1 - \frac{\omega_{pl}^2}{\left(\omega^2 + i\omega\Gamma\right)} = 1 - \frac{\omega_{pl}^2}{\omega(\omega + i\Gamma)} \qquad (2.14)$$

This dielectric function works really well. In the generalized Drude model one lets Γ be frequency dependent. In doing so one has to let Γ be complex valued for ε to obey the Kramers Kronig dispersion relations. It turns out that the real part of Γ then stays rather constant all the way up to the plasma energy and the imaginary part is of less importance. Thus the simple Drude model is valid all the way up to the plasma energy.

2.3 Modelling

In this section we introduce simplified models for the contributions to the dielectric function from the various excitation regions. We use a model expression with three parameters to represent each of these regions of non-vanishing imaginary part. These parameters are the center of the region, ω_0, the width of the region, Δ, and the area of the peak (see Figure 2.4). We assume that the overall behavior of the system is not so sensitive to the detailed shape of the peaks. In Figure 2.4 we show the simplest case with only one region of finite imaginary part for positive frequencies. There is always a corresponding peak for negative frequencies; there, the sign is opposite. We have in this model symmetric peaks, with sharp edges. This is not generally the case in a real system.

2.3 Modelling

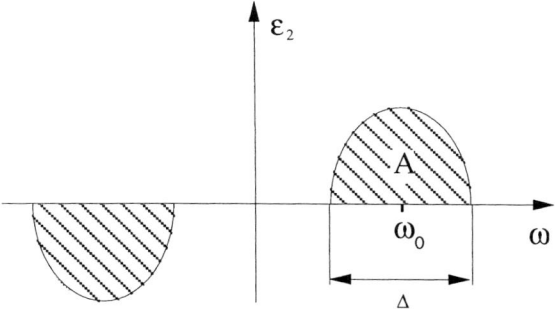

Figure 2.4: The imaginary part of the model dielectric function.

This dielectric function is for example what one obtains for a metal, at zero temperature, in the Random Phase Approximation (RPA) for $q>2k_F$, where k_F is the Fermi wave vector. For smaller momentum the whole peak is no longer on the positive frequency side. The peaks overlap and partly cancel each other as in Figure 2.5.

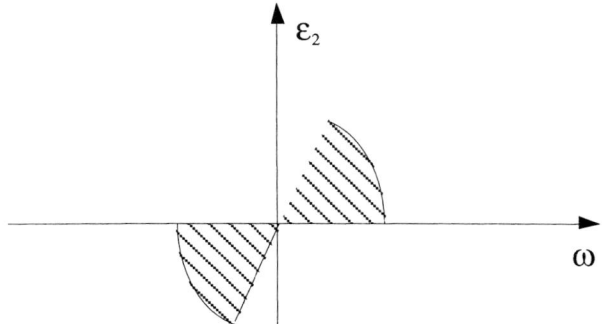

Figure 2.5: The imaginary part for the dielectric function of a metal at small momentum.

Now, these three parameters are not independent. The *f-sum rule* says that

$$\int_0^\infty d\omega \omega \varepsilon_2(\omega) = C = 2\pi^2 \sum_i \frac{n_i q_i^2}{m_i} \tag{2.15}$$

where the sum runs over all the particle types involved in the screening. The density, charge and mass of the particles are represented by n, q, and m, respectively. Thus we have

$$A\omega_0 = 2\pi^2 \sum_i \frac{n_i q_i^2}{m_i} \tag{2.16}$$

for the simple example here. In the more general case we have

$$\sum_j A^j \omega_0^j = 2\pi^2 \sum_i \frac{n_i q_i^2}{m_i} \tag{2.17}$$

This shows that if there are several possible interband transitions for the same carriers, that is, there are several bands to which the carriers can be excited, there is a redistribution of oscillator strength from the original peaks to the new ones. No additional oscillator strength is generated. If we add an extra peak the size of the earlier peaks are reduced. This fact might seem a bit odd. Why should additional possibilities for excitation lead to a reduction in the absorption via the original channels? The reason is that the matrix elements for transitions to the original bands are different in the system with additional channels. Everything is intimately connected. The *f-sum rule* is just a result of this.

The analytical expression we will use is the following:

$$\varepsilon_2(\omega) = \frac{6A}{\Delta^3}\left\{\left[\left(\frac{\Delta}{2}\right)^2 - (\omega-\omega_0)^2\right]\theta\left[\omega-\left(\omega_0-\frac{\Delta}{2}\right)\right]\theta\left[\left(\omega_0+\frac{\Delta}{2}\right)-\omega\right]\right.$$
$$\left. -\left[\left(\frac{\Delta}{2}\right)^2 - (\omega+\omega_0)^2\right]\theta\left[\omega-\left(-\omega_0-\frac{\Delta}{2}\right)\right]\theta\left[\left(-\omega_0+\frac{\Delta}{2}\right)-\omega\right]\right\} \tag{2.18}$$

where $\theta(\omega)$ represents the Heaviside unit step function.
This implies the following real part:

$$\varepsilon_1(\omega) = 1 + \frac{6A}{\pi\Delta^3}\left\{2\Delta\omega_0 + \left[\left(\frac{\Delta}{2}\right)^2 - (\omega-\omega_0)^2\right]\ln\left|\frac{\left(\frac{\Delta}{2}\right)-(\omega-\omega_0)}{\left(\frac{\Delta}{2}\right)+(\omega-\omega_0)}\right|\right.$$
$$\left. -\left[\left(\frac{\Delta}{2}\right)^2 - (\omega+\omega_0)^2\right]\ln\left|\frac{\left(\frac{\Delta}{2}\right)-(\omega+\omega_0)}{\left(\frac{\Delta}{2}\right)+(\omega+\omega_0)}\right|\right\} \tag{2.19}$$

It could be useful to specify some limits for these functions. We do this in Equations (2.20)-(2.22). Equation (2.20) gives the result for zero frequency at a finite Δ-value:

$$\varepsilon_1(\Delta,0) = 1 + \frac{6A}{\pi\Delta^3}\left\{2\Delta\omega_0 + 2\left[\left(\frac{\Delta}{2}\right)^2 - (\omega_0)^2\right]\ln\left|\frac{\left(\frac{\Delta}{2}\right)+\omega_0}{\left(\frac{\Delta}{2}\right)-\omega_0}\right|\right\} \tag{2.20}$$

2.3 Modelling

Equation (2.21) is the result when also the Δ-value is zero:

$$\varepsilon_1(0,0) = 1 + \frac{2}{\pi} A \frac{\omega_0}{\omega_0^2} = \frac{\omega_0^2 + \omega_{pl}^2}{\omega_0^2} \quad (2.21)$$

and Equation (2.22) is for finite frequencies and zero Δ-value:

$$\varepsilon_1(0,\omega) = 1 - \frac{2}{\pi} A \frac{\omega_0}{\omega^2 - \omega_0^2} = 1 - \frac{\omega_{pl}^2}{\omega^2 - \omega_0^2} \quad (2.22)$$

For a metal in RPA we have

$$\begin{aligned} \omega_0 &= \frac{\hbar q^2}{2m} \\ \Delta &= 2v_F q \\ A &= \frac{4\pi^2 n e^2}{\hbar q^2} \end{aligned} \quad (2.23)$$

In the metal case we use the dimensionless variables:

$$W = \frac{\hbar \omega}{4E_F} \quad ; \quad Q = \frac{q}{2k_F} \quad ; \quad y = \frac{me^2}{\hbar^2 k_F} \quad (2.24)$$

This gives for the model parameters

$$\begin{aligned} \omega_0 &= Q^2 \\ \Delta &= 2Q \\ A &= \frac{y}{6Q^2} \end{aligned} \quad (2.25)$$

and the *f-sum rule* becomes

$$\int_0^\infty dW W \varepsilon_2(W) = \frac{y}{6} \quad (2.26)$$

To illustrate these results we chose aluminum as an example. Its density parameter, y, has the value 1.08 and the region of absorption is shown in Figure 2.6. In Figures 2.7 and 2.8 we show the real and imaginary parts, respectively, of the dielectric function along the four vertical lines in Figure 2.6.

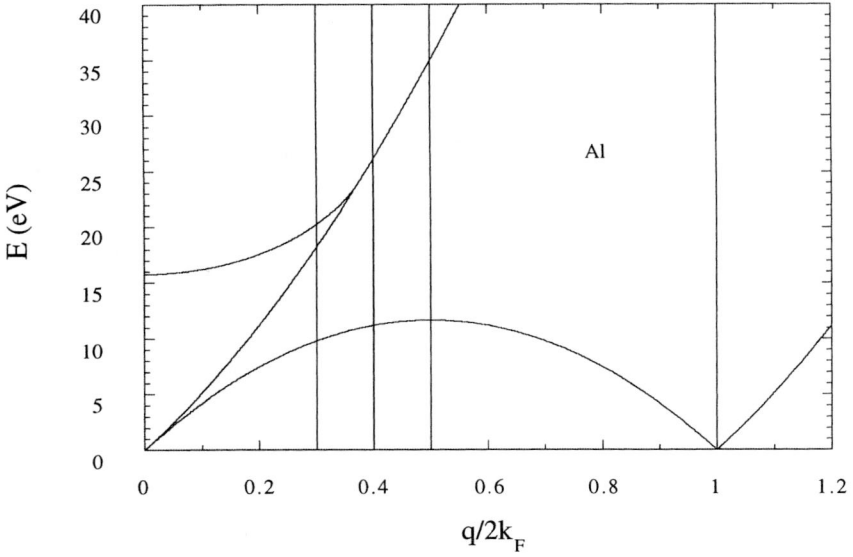

Figure 2.6: The boundaries of the region of absorption in the ωq-plane for aluminum. To the left is also shown the plasmon dispersion. The four vertical lines indicate where we have made the calculations of the dielectric function reported in Figures 2.7 and 2.8.

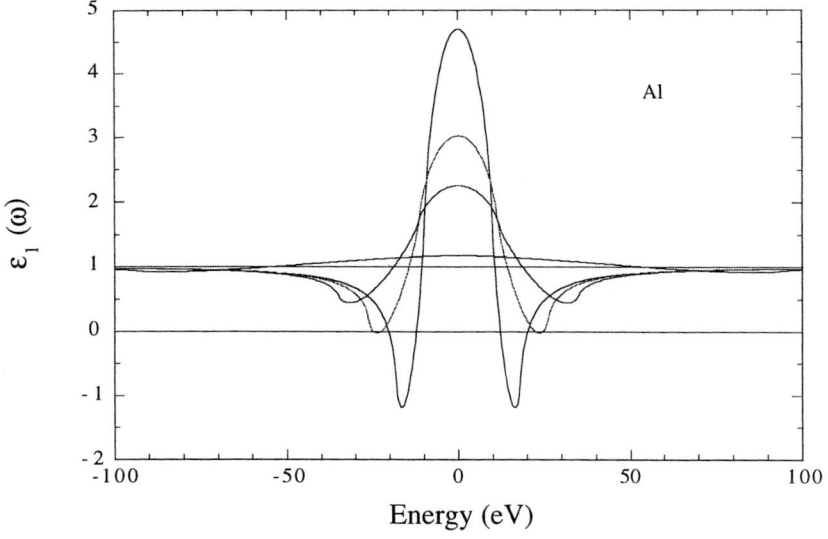

Figure 2.7: The real part of the dielectric function of aluminum for four constant q-values. The maxima of the curves at zero energy decrease monotonically with momentum.

2.3 Modelling

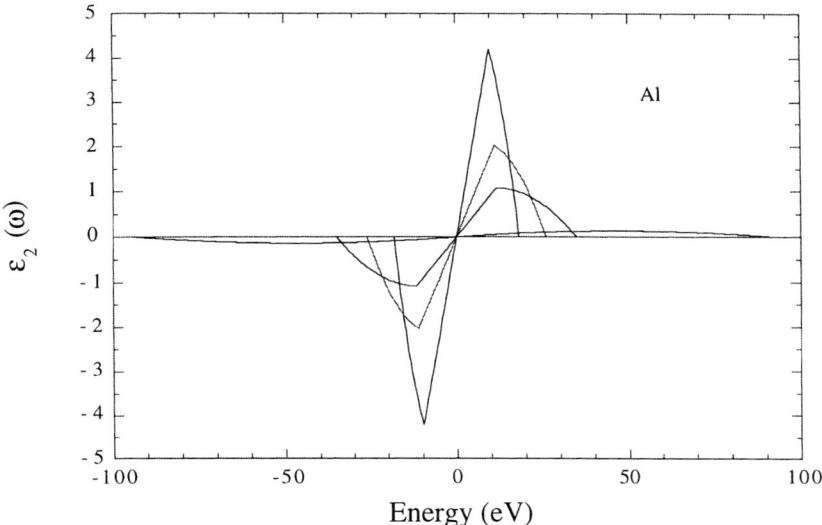

Figure 2.8: The imaginary part of the dielectric function of aluminum for four constant q-values. The maximum values of the curves decrease monotonically with momentum.

Of the four q-values studied only the two at 0.3 and 0.4 have a region where the real part of the dielectric function is negative. For the lowest of the two q-values part of this region is outside the single-particle continuum; for the other this region is completely within the continuum.

We have demonstrated that with the rather simple analytical expressions for the real and imaginary parts of the dielectric function in Equations (2.18)-(2.19) we can extract much information about a metal. It can also be interesting to treat the limit of these functions when the width of the peaks goes to zero, that is, when the imaginary part consists of δ-functions. Then we have

$$\varepsilon_2(\omega) = A[\delta(\omega - \omega_0) - \delta(\omega + \omega_0)]$$
$$\varepsilon_1(\omega) = 1 + \frac{2}{\pi} A \frac{\omega_0}{\omega_0^2 - \omega^2}$$
(2.27)

and

$$\omega_L = \omega|_{\varepsilon_1(\omega)=0} = \sqrt{\omega_0^2 + \frac{2A\omega_0}{\pi}} = \sqrt{\omega_0^2 + \omega_{pl}^2}$$
$$\varepsilon_1(0) = 1 + \frac{2}{\pi} A \frac{\omega_0}{\omega_0^2} = \frac{\omega_L^2}{\omega_0^2}$$
(2.28)

If we are studying a system where the excitations at low frequencies are well separated from excitations higher up we can describe the low frequency region with a similar dielectric function but where the constant unity in the above equations is replaced by another, larger constant. For metals core excitations, that is, excitations involving the core electrons, occur at high frequencies, and contribute in making the constant somewhat larger than unity. The effect is more important for polar semiconductors and ionic insulators and we use ε_∞ for the constant which is standard for these systems. Thus, the more general form is

$$\varepsilon_2(\omega) = A[\delta(\omega - \omega_0) - \delta(\omega + \omega_0)]$$

$$\varepsilon_1(\omega) = \varepsilon_\infty + \frac{2}{\pi} A \frac{\omega_0}{\omega_0^2 - \omega^2}$$

$$\omega_L = \omega|_{\varepsilon_1(\omega)=0} = \sqrt{\omega_0^2 + \frac{2A\omega_0}{\varepsilon_\infty \pi}} = \sqrt{\omega_0^2 + \omega_{pl}^2}$$

$$\frac{\varepsilon_0}{\varepsilon_\infty} = 1 + \frac{2}{\pi \varepsilon_\infty} A \frac{\omega_0}{\omega_0^2} = \frac{\omega_L^2}{\omega_0^2}$$

(2.29)

The last relation is the so-called *Lyddane-Sachs-Teller* relation. If this dielectric function is representing a simple polar crystal, cubic with two ions in the primitive unit cell, the frequencies ω_L and ω_0 are the long-wavelength–limiting frequencies of the longitudinal optical and transverse optical vibration modes of the crystal, respectively.

Let us now study a model dielectric function for a polar semiconductor. At low frequencies there are excitations involving the optical phonons and at higher frequencies interband-transitions, transitions across the band-gap, contribute. We neglect the width of the imaginary part for both the phonon- and interband-regions. The real part of the dielectric function then becomes

$$\varepsilon(\omega) = 1 + \frac{\omega_T^2(\varepsilon_0 - \varepsilon_\infty)}{\omega_T^2 - \omega^2} + \frac{\omega_0^2(\varepsilon_\infty - 1)}{\omega_0^2 - \omega^2}; \quad \omega_0 = \frac{E_g}{\hbar}$$

(2.30)

This is illustrated in Figure 2.9.

2.3 Modelling

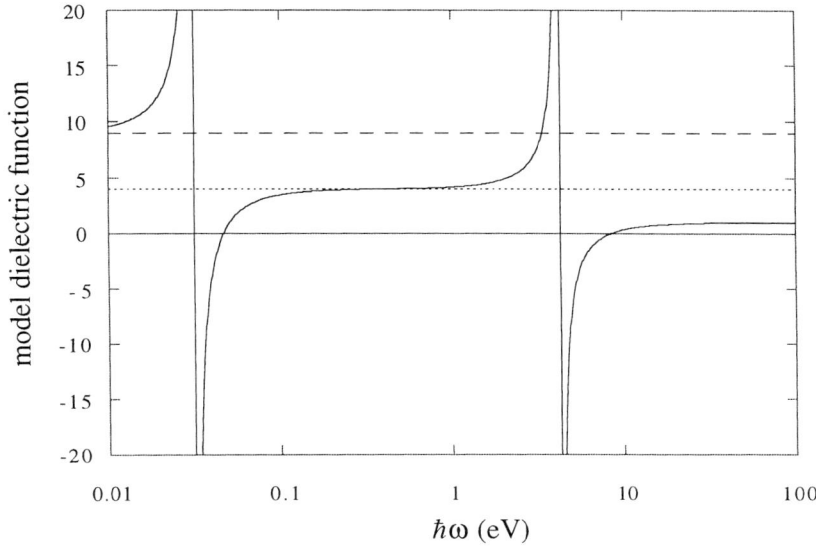

Figure 2.9: The real part of a model dielectric function for a polar semiconductor. The imaginary part has been approximated by two δ-functions; one at the transverse optical phonon energy and one at the band-gap energy. The horizontal dashed and dotted lines are indicating the ε_0 and ε_∞ values, respectively.

To get a good view of both regions we have given the results on a logarithmic energy axis. The light dispersion curve, which is obtained as the solution to the equation

$$\omega = \frac{cq}{\sqrt{\varepsilon(\omega)}} \quad (2.31)$$

in such a system, that is, the relation between frequency and momentum for freely propagating light in the medium, is demonstrated in Figure 2.10. As a reference we show the dispersion curve for light in vacuum, the diagonal solid line. In the present system there are three branches. They are separated by two forbidden frequency regions (hatched in the figure); one between the two phonon frequencies; one between the interband energy and the shifted plasma energy.

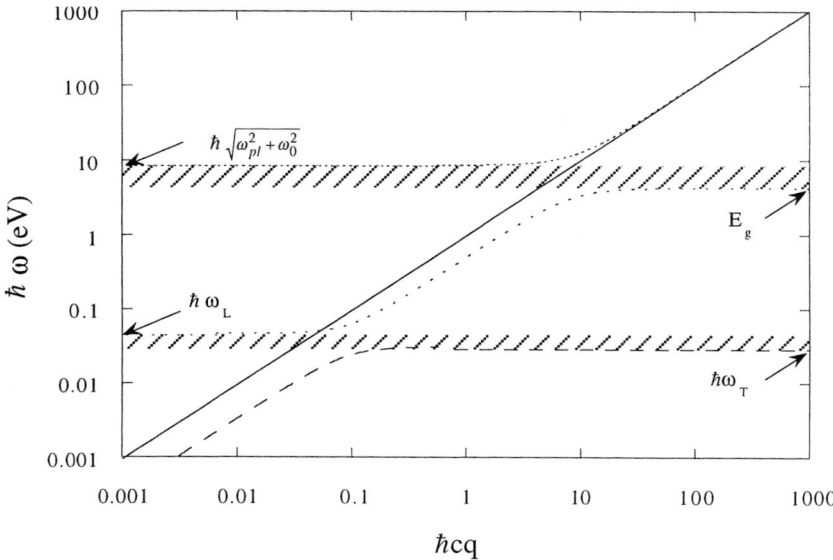

Figure 2.10: The dispersion curve for light in the medium characterized by the dielectric function in Figure 2.9.

The forbidden regions correspond to the frequencies where the dielectric function is negative. Light, in these frequency regions, cannot penetrate the interior of the material but is reflected off. We will see later that these regions are the interesting ones from our point of view since it is here that there may be localized modes at the surface. As we said there are three regions or branches of photon modes. The speed of light for the first branch starts out with the speed of light in vacuum, c, reduced by $1/\sqrt{\varepsilon_0}$, then when the frequency saturates below that of the transverse phonon mode the speed drops to zero. The second mode starts out with zero speed of light. The speed increases towards c reduced by $1/\sqrt{\varepsilon_\infty}$, then when the frequency saturates below that of the interband excitation mode the speed drops to zero again. The third mode starts out with zero speed of light. The speed increases towards c. We have here referred to the group velocity, which is always lower than that of speed of light in vacuum. The phase velocity, on the other hand, may exceed this value. It does for the parts of the branches that are to the left of the light dispersion curve

We note that the refractive index is monotonically increasing with frequency in each of the branches. This is a general property for all transparent materials. This answers the first question put forward in this chapter namely why the blue and violet light is always refracted more than the red and yellow. This is also why the sky is blue and the sunset red! The atoms in colorless gases have excitation energies in the ultraviolet. The scattering cross section increases monotonically with frequency in the visible range. However, there are regions of anomalous dispersion where the real part of the dielectric function decreases with frequency. This may occur in regions of absorption, in so-called absorption bands. An example of this is water in the radio-frequency range. We will come back to this example later in this chapter, in Section 2.7.

2.3 Modelling

Like in our example above the modes always come in pairs, one longitudinal and one transverse excitation; the plasmon and the interband excitation; the longitudinal optical phonon and the transverse one. If there is a finite width of the region with finite imaginary part of the dielectric function the longitudinal mode might fall inside this region. Then the mode is damped. It can also disappear completely; the real part of the dielectric function never turns negative. For a metal the plasmon mode is undamped for small momentum. For larger momentum it enters the single particle continuum and is damped. For even larger momentum it disappears completely.

The situation for a simple metal is the following: The transverse mode has zero frequency. This corresponds to having the band gap zero in Figure 2.10. Thus there is no photon mode below the the transverse frequency. The dielectric function is

$$\varepsilon_2(\omega) = A[\delta(\omega - \delta) - \delta(\omega + \delta)] = \frac{\pi}{2\delta} \omega_{pl}^2 [\delta(\omega - \delta) - \delta(\omega + \delta)]$$
$$\varepsilon_1(\omega) = 1 + \frac{2}{\pi} A \frac{\delta}{\delta^2 - \omega^2} = 1 - \frac{\omega_{pl}^2}{\omega^2}$$
(2.32)

The last step may seem odd. We take the limit when δ goes to zero, but at the same time A increases and $A\delta$ is constant, all according to the *f-sum rule*.

If there are no interband transitions below the plasma energy the metal reflects all light below this energy. The plasma energy for all metals is above the visible range of the spectrum. This means that all visible light is reflected. This gives the metallic appearance in common for all metals. However for copper and gold, for example, there are interband transitions in the blue and *UV*-range. These wave-lengths are absorbed. Thus the reflected light contains a reduced amount of blue light and hence appears slightly yellow or red.

The reflectivity is given by

$$R = \frac{(n-1)^2 + k^2}{(n+1)^2 + k^2}$$
(2.33)

where n and k are the real and imaginary parts, respectively, of the refractive index. For our simple model system for a semiconductor we get the result displayed in Figure 2.11.

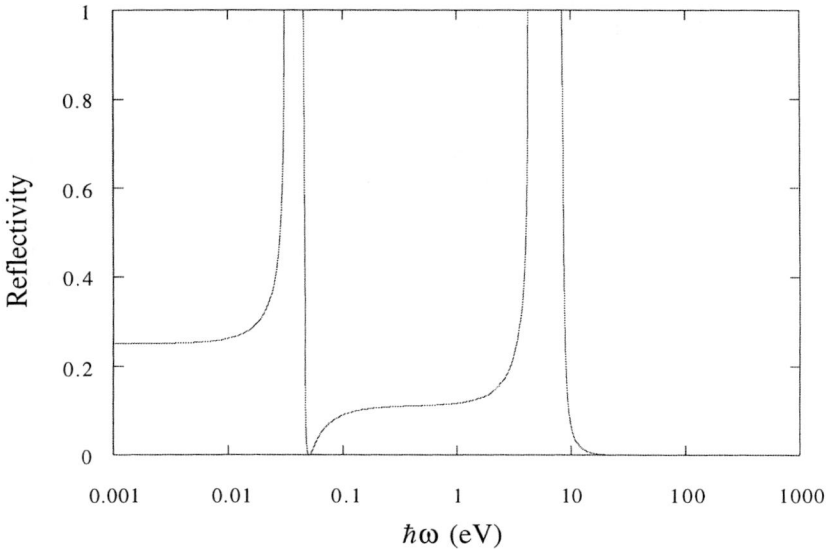

Figure 2.11: The reflectivity for the system described in Figures 2.9 and 2.10.

The system is totally reflecting in the two regions with negative dielectric function. The reflectivity for silver is given in Figure 2.12:

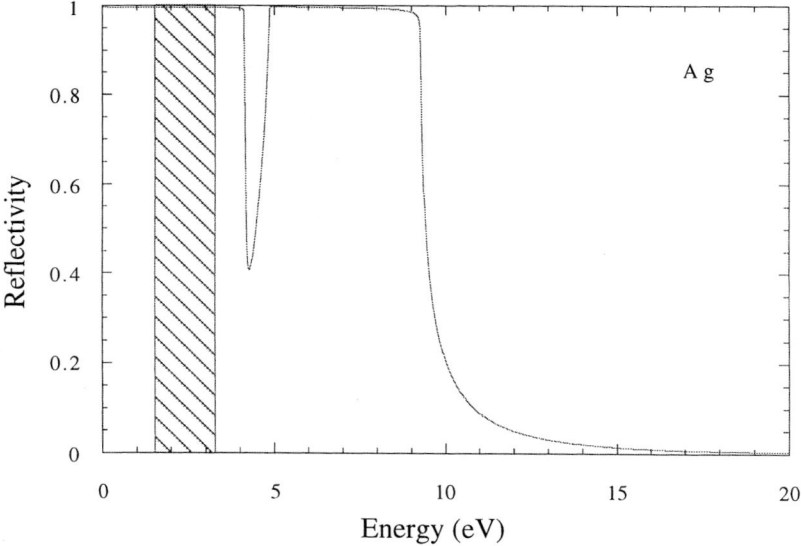

Figure 2.12: The reflectivity of silver with our model.

2.3 Modelling

The visible region is in the range 1.6 eV (0.77 μm) - 3.2 eV (0.39 μm). This region is indicated by the hatched area in Figure (2.12). We see that silver is totally reflecting in the whole visible range. The situation is different for copper. Long–wave-length light is reflected more than short–wave-length light and the metal gives a red appearance. The situation is similar in gold which attains a yellowish appearance.

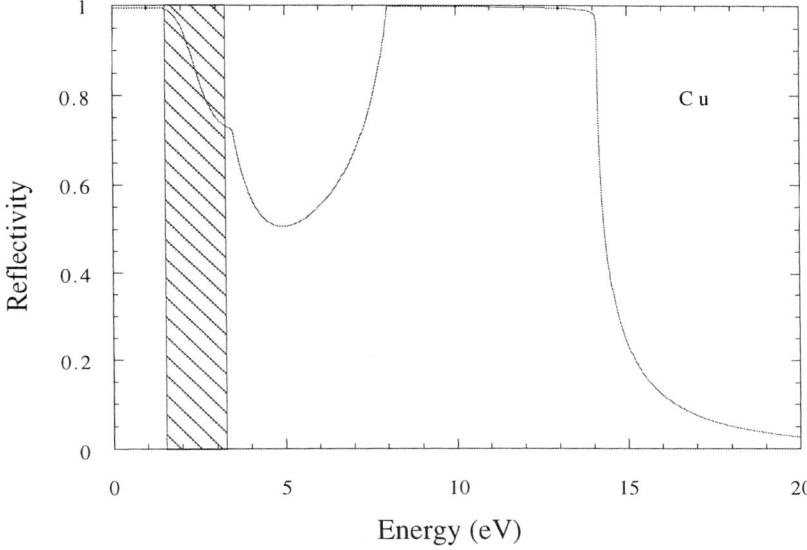

Figure 2.13: The reflectivity of copper with our model.

The dip in the reflectivity is due to interband transitions leading to absorption in this range. Also silver has interband transitions below the plasma energy but these are above the visible range. Now if we make very thin sheets of copper it is possible to see through it. The question is what color is the light that passes through? We investigate this in the next series of figures.

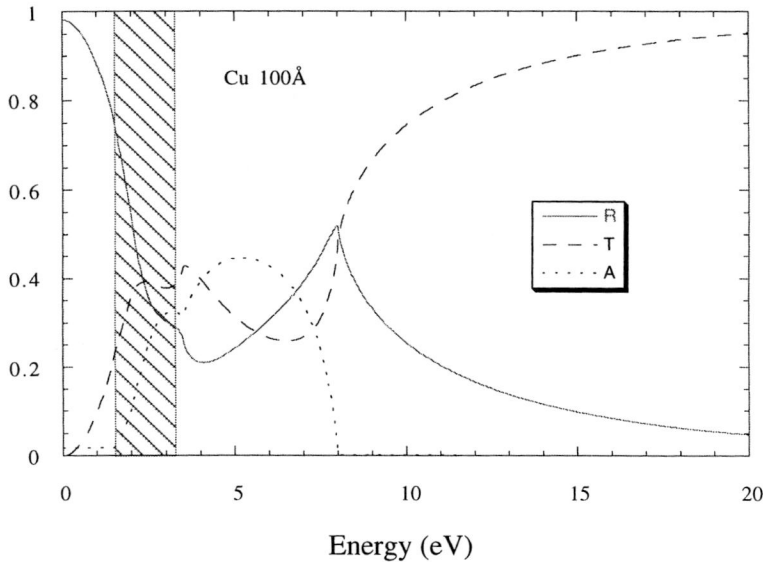

Figure 2.14: The reflectance, transmittance and absorption of a 100 Å thick film of copper with our model.

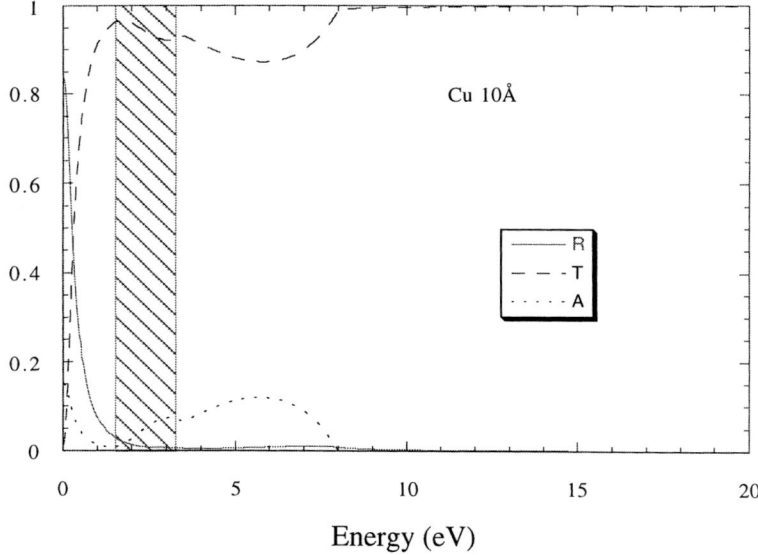

Figure 2.15: The reflectance, transmittance and absorption of a 10 Å thick film of copper with our model.

If the thickness of the film is 100 Å the film will look greenish, but if the film is only

2.3 Modelling

10 Å thick most of the light passes through and the light is white. For a 1000 Å thick film basically no light passes through.

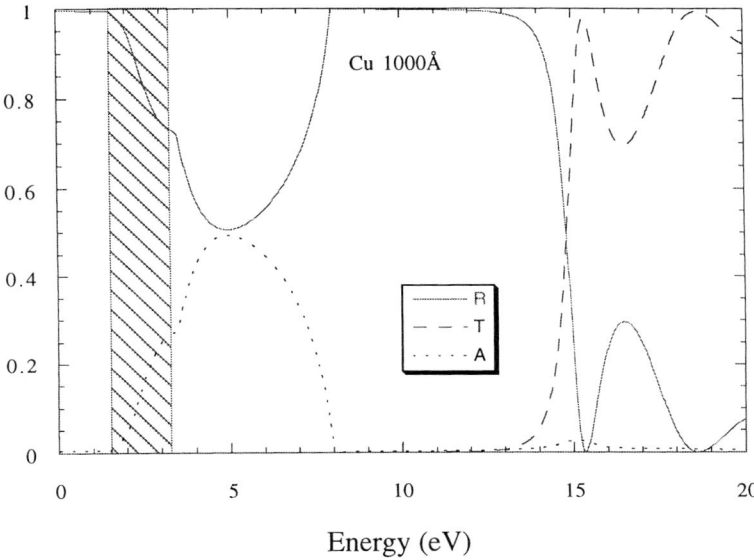

Figure 2.16: The reflectance, transmittance and absorption of a 1000 Å thick film of copper with our model.

The oscillations on the high-energy side of Figure 2.16 is due to interference effects from multiple scattering in the film. We have assumed that the coherence length for the light is large compared to the film thickness. If not these oscillations disappear. We have used the form, demonstrated in Figure 2.17, for the imaginary part of the dielectric function for Cu. The real part is obtained from this through the *Kramers Kronig dispersion relations*.

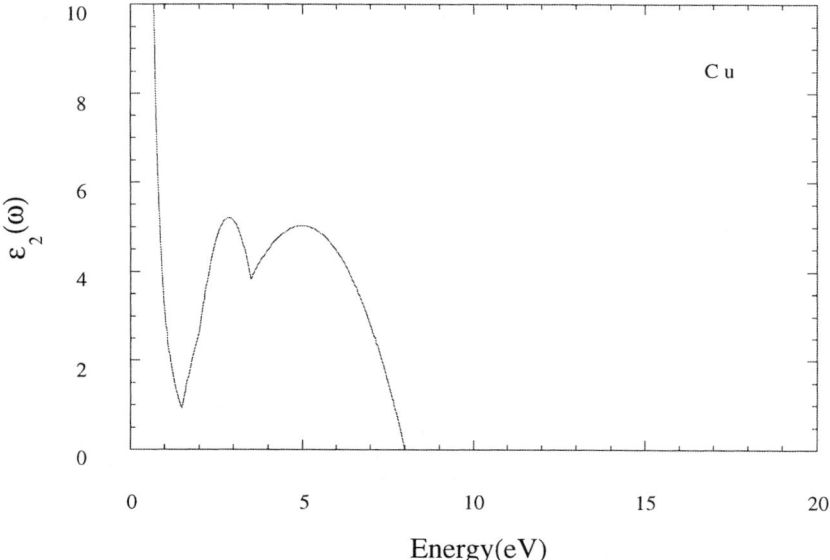

Figure 2.17: The imaginary part of the dielectric function used for copper.

Two interband transitions are included and we have used the same shape of the absorption as discussed before. The positions, widths and strengths have been chosen to mimic the experimentally measured dielectric function [1]. On the low-energy side there is also a contribution from the Drude tail (intraband contributions). This contribution is found from the experimental resistivity in the following way:

$$\varepsilon(\omega) = 1 + \frac{4\pi i}{\omega} \sigma(\omega)$$

$$\sigma(\omega) = \frac{ne^2}{m(1/\tau - i\omega)} = \frac{1}{\rho_0 - i\omega m/(ne^2)} = \frac{1}{\rho_0 - i4\pi\omega/\omega_{pl}^2} \quad (2.34)$$

For metals the frequency dependent resistivity caused by scattering against impurities and phonons is almost constant up to the plasma energy. We find

$$\varepsilon_2(\omega) = \frac{4\pi}{\omega} \operatorname{Re} \sigma(\omega) = \frac{4\pi}{\omega} \frac{\rho_0}{(\rho_0)^2 + \left(4\pi\omega/\omega_{pl}^2\right)^2}$$

$$\approx \frac{1}{\omega^3} \frac{\rho_0 \omega_{pl}^4}{4\pi} \quad ; \text{ for large } \omega. \quad (2.35)$$

In the calculations we have used the room temperature value 1.7 $\mu\Omega$cm for the

2.4 Dielectric function of a plasma

In several situations it can be useful to have a dielectric function for a plasma; to describe an electron-hole plasma in a highly excited semiconductor; the dielectric properties of an electrolyte; the dielectric properties of a metallic system with carriers in different pockets in the Brillouin zone, for example, a heavily doped many-valley semiconductor.

We may get a qualitative expression for the dielectric function of a plasma by using classical arguments just as in the Lorentz model. We get the results expressed in terms of some material parameters, coefficients of friction or relaxation times. These parameters can then be determined from more strict many-body calculations. We start from the equations of motion for the two plasma components. Let the average velocity for component *1* and *2* be v and u, respectively. Let q, γ, n be the charge, the coefficient of friction against impurities of concentration n_i, and density, of component *1*, respectively. The corresponding parameters for component *2* is given by capital letters. The coefficient of mutual friction is denoted by η. The equations of motion read

$$m\dot{v} = qE - \gamma n_i v - \eta N(v-u)$$
$$M\dot{u} = QE - \Gamma n_i u - \eta n(u-v)$$
(2.36)

Fourier-transforming (see Appendix 2) these with respect to time gives

$$-i\omega m v = qE - \gamma n_i v - \eta N(v-u)$$
$$-i\omega M u = QE - \Gamma n_i u - \eta n(u-v)$$
(2.37)

From this system of equations we may extract the velocities of the two components and obtain the current. The current is proportional to the electric field and the constant of proportionality is identified as the conductivity:

$$nv = \frac{Nu}{\eta}(-i\omega_2 + \Gamma n_i / N + \eta n / N) - \frac{1}{\eta}QE \quad ; \omega_2 = \frac{\omega M}{N}$$
$$Nu = \frac{nv}{\eta}(-i\omega_1 + \gamma n_i / n + \eta N / n) - \frac{1}{\eta}qE \quad ; \omega_1 = \frac{\omega m}{n}$$
(2.38)

$$Nu = \frac{Q(-i\omega_1 + \gamma_i/n + \eta N/n) + \eta q}{\left[(-i\omega_1 + \gamma_i/n + \eta N/n)(-i\omega_2 + \Gamma n_i/N + \eta n/N) - \eta^2\right]} E$$

$$nv = \frac{q(-i\omega_2 + \Gamma n_i/N + \eta n/N) + \eta Q}{\left[(-i\omega_1 + \gamma_i/n + \eta N/n)(-i\omega_2 + \Gamma n_i/N + \eta n/N) - \eta^2\right]} E$$

(2.39)

We arrive at

$$\sigma = (NuQ + nvq)/E$$
$$= \frac{Q^2(-i\omega_1 + \gamma_i/n + \eta N/n) + q^2(-i\omega_2 + \Gamma n_i/N + \eta n/N) + 2\eta q Q}{\left[(-i\omega_1 + \gamma_i/n + \eta N/n)(-i\omega_2 + \Gamma n_i/N + \eta n/N) - \eta^2\right]} \quad (2.40)$$

This is the general expression. If we restrict ourselves to the simple case where the two components have the same mass, the same charge except for the sign, and the same impurity relaxation time we get

$$\sigma = \begin{bmatrix} N = n \\ Q = -q \\ \Gamma = \gamma \end{bmatrix} = \frac{q^2(-i\omega_1 - i\omega_2 + 2\gamma_i/n)}{\left[(-i\omega_1 + \gamma_i/n + \eta)(-i\omega_2 + \gamma_i/n + \eta n) - \eta^2\right]}$$

$$= [M = m] = \frac{2nq^2(-im\omega + \gamma_i)}{\left[(-im\omega + \gamma_i + n\eta)^2 - (n\eta)^2\right]} = \frac{2nq^2}{(-im\omega + \gamma_i + 2n\eta)}$$

(2.41)

From this result we find the dielectric function as

$$\varepsilon = \kappa + \frac{4\pi i \sigma}{\omega} = \kappa - \frac{4\pi 2nq^2 m}{\left[(\gamma_i + 2n\eta)^2 + (m\omega)^2\right]} + i\frac{4\pi 2nq^2(\gamma_i + 2n\eta)}{\omega\left[(\gamma_i + 2n\eta)^2 + (m\omega)^2\right]}$$

$$= \kappa - \frac{\omega_{pl}^2}{\left[\omega^2 + (\gamma_i/m + 2n\eta/m)^2\right]} + i\frac{\omega_{pl}^2(\gamma_i/m + 2n\eta/m)}{\omega\left[\omega^2 + (\gamma_i/m + 2n\eta/m)^2\right]} ;$$

(2.42)

$$\omega_{pl}^2 = \frac{4\pi nq^2}{m} + \frac{4\pi NQ^2}{M} = \frac{8\pi nq^2}{m}$$

For small ω and zero temperature we have $\gamma(\omega) \propto const$ and $\eta(\omega) \propto \omega^2$. For finite temperatures also the inter-particle friction tends to a constant.

2.4 Dielectric Function of a Plasma

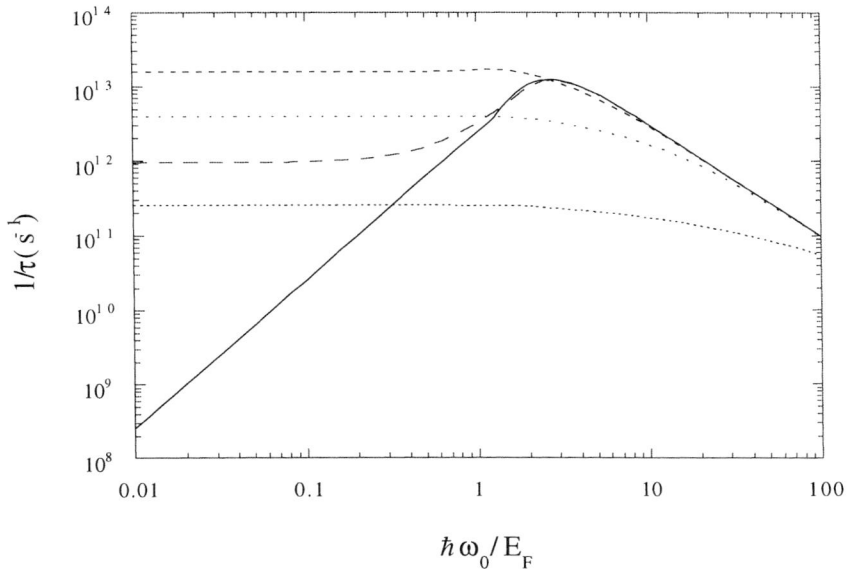

Figure 2.18: The inverse relaxation time for a plasma for different temperatures, given in the text.

This is demonstrated [2] in Figure 2.18. The curves are for the T/T_F values 0, 0.1, 1, 10, and 100 for the solid, long dashed, short dashed, sparsely dotted, and fully dotted curves, respectively. The Fermi temperature, T_F, is the temperature for which the thermal energy, $k_B T$ equals the Fermi energy. The constant k_B is the Boltzmann constant. The results are for a plasma of fermions of density 1×10^{20} cm^{-3} with masses of 0.3 m_e and in a medium with a dielectric constant of 10. These are reasonable values for an electron-hole plasma in a typical semiconductor. The relaxation time is related to the coefficient of friction: $1/\tau = 2n\eta/m$.

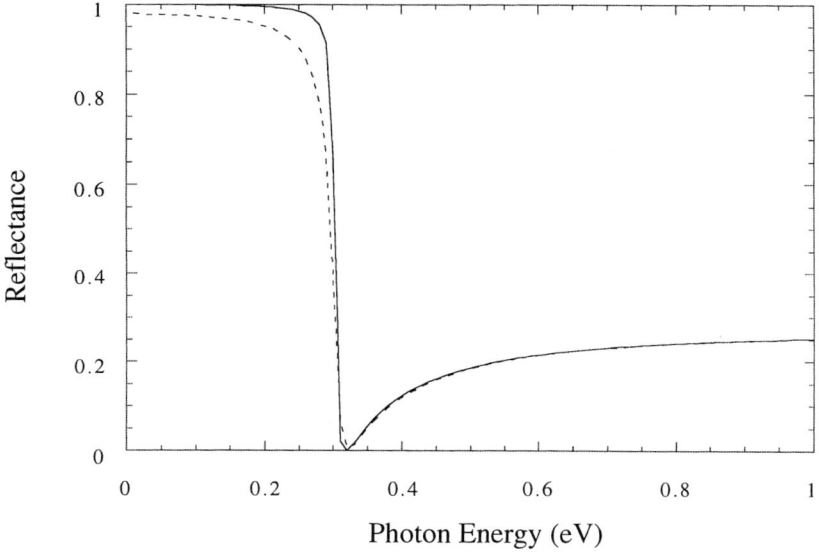

Figure 2.19: The reflectance for a plasma at zero and finite temperature.

The reflectance for the system with the plasma is depicted in Figure 2.19. The solid curve is for zero temperature and the dashed one for the Fermi temperature. The system is almost totally reflecting for energies below the plasma energy. This behavior is typical for metallic systems.

2.5 Static dielectric function for a dilute gas of permanent dipoles

A system of permanent dipoles with freedom to rotate and adjust the direction of their dipoles to an applied field behaves dielectrically a little differently than the systems treated so far. The system can, for example, represent a cloud of water molecules. The dielectric function is

$$\varepsilon(0) = 1 + 4\pi n \alpha^{mol} \tag{2.43}$$

The polarizability is obtained in the following way. The energy of a dipole in an electric field **E** is

2.5 Static Dielectric Function for a Dilute Gas of Permanent Dipoles

$$U = -\mathbf{p} \cdot \mathbf{E} = -pE\cos(\theta) = -pEx \tag{2.44}$$

The energy depends on the orientation of the dipole relative the electric field. For finite temperatures this orientation, expressed in the parameter x, will on the average have a value obtained from the following averaging procedure:

$$\langle x \rangle = \int_{-1}^{1} dx\, x e^{-\beta U(x)} \bigg/ \int_{-1}^{1} dx\, e^{-\beta U(x)} \tag{2.45}$$

This gives

$$\langle x \rangle = \coth(\beta pE) - 1/\beta pE \approx \frac{1}{3}(\beta pE) \tag{2.46}$$

where the last result is valid to linear order in the electric field.
Now,

$$\alpha^{mol} = \frac{p\langle x \rangle}{E} = \frac{p^2}{3}\beta \tag{2.47}$$

and

$$\varepsilon(0) = 1 + \frac{4\pi n}{3} p^2 \beta \tag{2.48}$$

The result is valid for high temperature where $\beta pE \ll 1$. In the other extreme limit where the field is very strong or the temperature low $\langle x \rangle = 1$ and the dipole moment is aligned with the field. Thus a strong static electric field may quench the orientation-polarizability. The dipole moment for the water molecule is 1.87×10^{-18} esu-cm or simply 1.87 Debye units. This means that the electric field must be much smaller than 10^7 Vcm^{-1} for the result to be valid. This is a very high field strength so the high-temperature limit is valid in most cases. Water is a very important substance so we address its dielectric properties in special sections, below.

2.6 Debye rotational relaxation

Water is a very important substance and we will pay it a little extra attention. Water has an extremely large static dielectric function of around 80. The low frequency contribution to the dielectric function, responsible for this large static value, is due to relaxation of the permanent dipoles of the water molecules. It is very nicely described by the simple Debye [3] rotational relaxation, as illustrated in Figure 2.20.

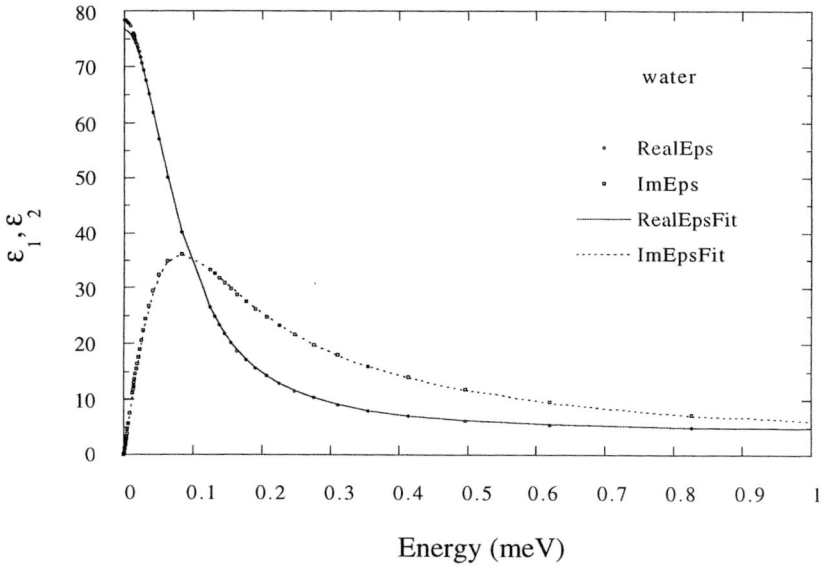

Figure 2.20: The real and imaginary parts of the dielectric function of water compared to the theoretical fits, given in Equation (2.50).

This contribution to the polarizability can be written as

$$\alpha(\omega) = \frac{C_{rot}}{1 - i\omega/\omega_{rot}} \; ; \; \omega_{rot}^{-1} = 4\pi\beta\eta a^3 \tag{2.49}$$

where η (≈ 0.01 poise) and a are the viscosity and the radius of the molecule, respectively. The theoretical fit used in the figure is given in Equation (2.50):

2.6 Debye Rotational Relaxation

$$\begin{cases} \varepsilon_1(\omega) = 4.35 + \dfrac{72.24}{1+\left(\hbar\omega/8.27\cdot 10^{-2}\,meV\right)^2} \\ \varepsilon_2(\omega) = \dfrac{72.24\cdot\left(\hbar\omega/8.27\cdot 10^{-2}\,meV\right)}{1+\left(\hbar\omega/8.27\cdot 10^{-2}\,meV\right)^2} \end{cases} \quad (2.50)$$

The fit is very good for both the real and imaginary parts.

We note that in this region of the spectrum, the radio-frequency range, water shows anomalous dispersion. The real parts of the dielectric function and refractive index (see Equation (2.21)) decrease with frequency or energy. We have earlier claimed that these functions always increase with frequency in regions free of absorption. Since the imaginary part of the dielectric function is finite here we have absorption. Thus the *anomalous dispersion* does not contradict our claim. We should note, though, that the real and imaginary parts of the dielectric function behave a bit unusual. The real part looks more like an imaginary part and the imaginary part looks like a real part with reverted sign.

We will now very briefly sketch the derivation of the dielectric function for this system. In Section 2.5 we studied the polarization of a system of permanent dipoles in the presence of a static electric field, at finite temperature. The Brownian motion prevented the complete alignment of the dipoles. Now we need to extend the treatment to time varying electric fields. If we start without an applied field the dipoles are pointing in random directions. If we suddenly turn on a constant field the dipoles will eventually get a distribution of directions according to the results we found before. This does not occur momentarily. It takes some time. There are two reasons: One is that the dipoles have moments of inertia, which means that they can not momentarily change direction. The other is that there is an internal friction in the liquid related to the viscosity. This friction will also give rise to an absorption and dissipation of energy for a similar reason as impurity friction gives rise to absorption and resistivity in a metal. Debye [3] generalized the Boltzmann equation to the case of rotational movements instead of translational. We will not go into the derivation of this differential equation here. We will just give the result and continue from there.

The generalized Boltzmann equation for the distribution function is as obtained by Debye

$$\zeta \frac{\partial f(\theta,t)}{\partial t} = \frac{1}{\sin\theta}\frac{\partial}{\partial \theta}\left[\sin\theta\left(\frac{1}{\beta}\frac{\partial f(\theta,t)}{\partial \theta} + f(\theta,t)pE\sin\theta\right)\right] \quad (2.51)$$

where the inner friction, ζ, is defined from the relation

$$M = \zeta\dot{\theta} \quad (2.52)$$

If we force a molecule to rotate with angular velocity, $\dot{\theta}$, the friction gives rise to a moment **M** working against the rotation. The inner friction is related to the viscosity of the liquid. If the molecule may be represented by a sphere of radius a the relation is

$$\zeta = 8\pi \eta a^3 \tag{2.53}$$

For a static field the distribution function is

$$f(\theta) = \exp(\beta p E \cos\theta) \approx 1 + \beta p E \cos\theta \tag{2.54}$$

according to Equations (2.44) and (2.45), where we in the last step have linearized the distribution function, that is, assumed that the field is weak.

In the case of a time dependent electric field, $E = E_0 e^{-i\omega t}$, we try the solution

$$f(\theta,t) = A\left[1 + B\beta p E_0 e^{-i\omega t} \cos\theta\right] \tag{2.55}$$

where A and B are constants. If we put this solution into the differential equation we find

$$i\omega\zeta A B\beta p E_0 e^{-i\omega t} \cos\theta = \frac{1}{\sin\theta}\frac{\partial}{\partial\theta}\left\{\sin\theta\left[\frac{-1}{\beta} A B\beta p E_0 e^{-i\omega t} \sin\theta + \right.\right. \\ \left.\left. + A\left(1 + B\beta p E_0 e^{-i\omega t} \cos\theta\right) p E_0 e^{-i\omega t} \sin\theta\right]\right\} \tag{2.56}$$

Tidying up and keeping only terms linear in E_0 we arrive at

$$B = \frac{1}{1 - i\omega\zeta\beta/2} = \frac{1}{1 - i\omega\tau} \; ; \; \tau = \zeta\beta/2 \tag{2.57}$$

Thus we have

$$f(\theta,t) = A\left[1 + \frac{\beta p \cos\theta}{1 - i\omega\tau} E_0 e^{-i\omega t}\right] \tag{2.58}$$

With this distribution function we get the following molecular polarizability:

$$\alpha^{mol} = \frac{p\langle x\rangle}{E}$$

$$= \frac{p}{E_0 e^{-i\omega t}} \int_{-1}^{1} dx\, xA\left[1 + \frac{\beta p x}{1 - i\omega\tau} E_0 e^{-i\omega t}\right] \bigg/ \int_{-1}^{1} dx A\left[1 + \frac{\beta p x}{1 - i\omega\tau} E_0 e^{-i\omega t}\right]$$

$$= \frac{p}{E_0 e^{-i\omega t}}\left[A\frac{\beta p}{1 - i\omega\tau} E_0 e^{-i\omega t}\frac{2}{3}\right] \bigg/ 2A = \frac{\beta p^2}{3(1 - i\omega\tau)}$$

$$\tag{2.59}$$

2.6 Debye Rotational Relaxation

Introducing the frequency

$$\omega_{rot}^{-1} = \tau = \frac{\zeta\beta}{2} = \frac{8\pi\eta a^3 \beta}{2} = 4\pi\beta\eta a^3 \qquad (2.60)$$

we may rewrite the molecular polarizability as

$$\alpha^{mol} = \frac{\beta p^2}{3(1 - i\omega/\omega_{rot})} \qquad (2.61)$$

and the polarizability becomes

$$\alpha_{rot}(\omega) = 4\pi n \frac{\beta p^2}{3(1 - i\omega/\omega_{rot})} \qquad (2.62)$$

The total dielectric function is

$$\varepsilon(\omega) = 1 + \alpha_{pol}(\omega) + \alpha_{rot}(\omega) \qquad (2.63)$$

where the first part of the polarizability comes from the polarizability of the water molecules and the second from the rotational alignment of the dipoles in the electric field. The two first terms would be the total result were water not polar. In the radio frequency range this part of the dielectric function has saturated at the value 4.35, according to Equation (2.50). This is a typical value for the static dielectric function for a non-metallic and non-polar substance.

2.7 Dielectric properties of water

The optical constants for water as found in the *Handbook of Optical Constants of Solids II* [4] are presented in Figure 2.21. These data cover many energy scales and are the results of several different types of measurement.

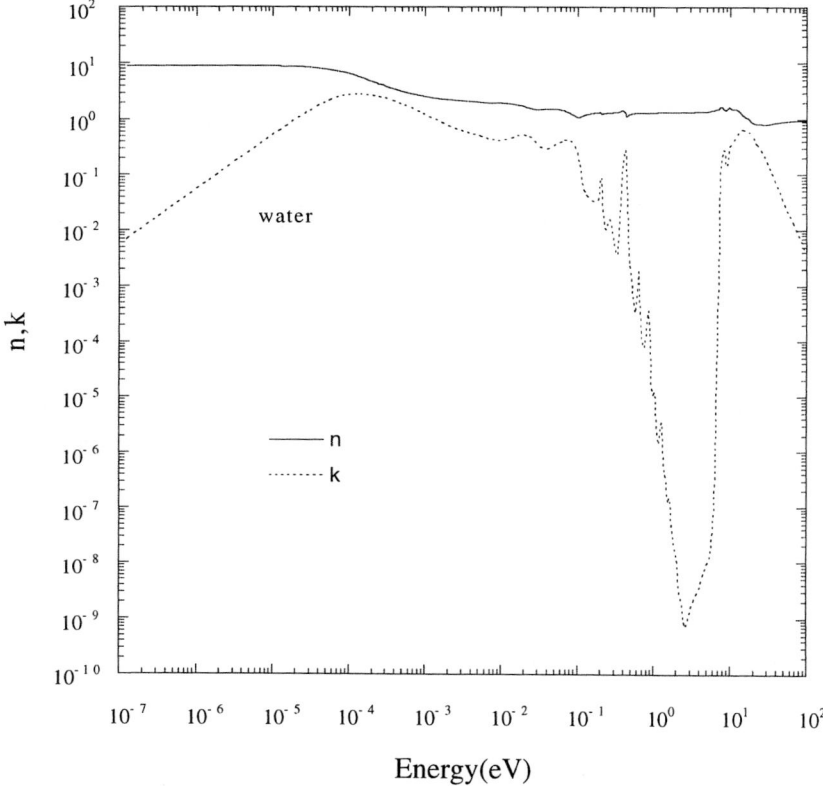

Figure 2.21: The optical properties of water in terms of *n* and *k*, the real and imaginary parts, respectively, of the complex refractive index.

From these data it is easy to get the dielectric function. It is given in Figure 2.22.

2.7 Dielectric Properties of Water

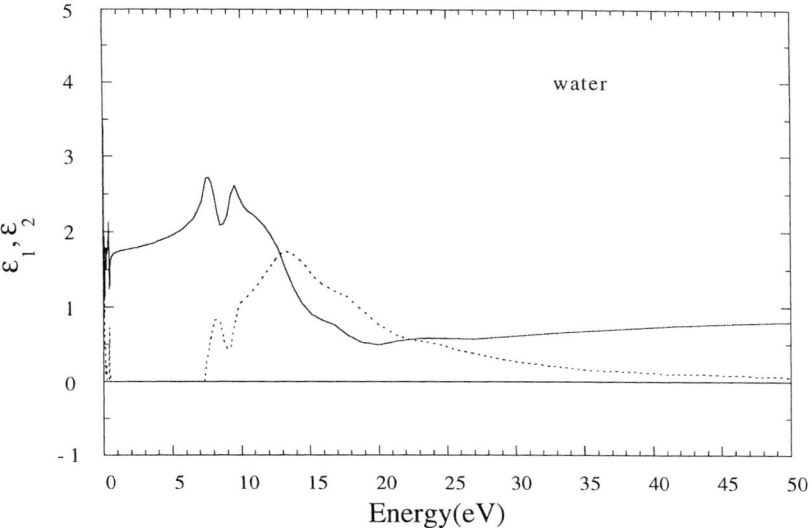

Figure 2.22: The real (solid curve) and imaginary (dotted curve) parts of the dielectric function of water as obtained from the optical data in Figure 2.21.

We see that there are "interband" absorption or electronic absorption in the region from 7.2 eV and up to 50 eV. There is no absorption in the visible range. Let us expand this region. It is done in Figure 2.23.

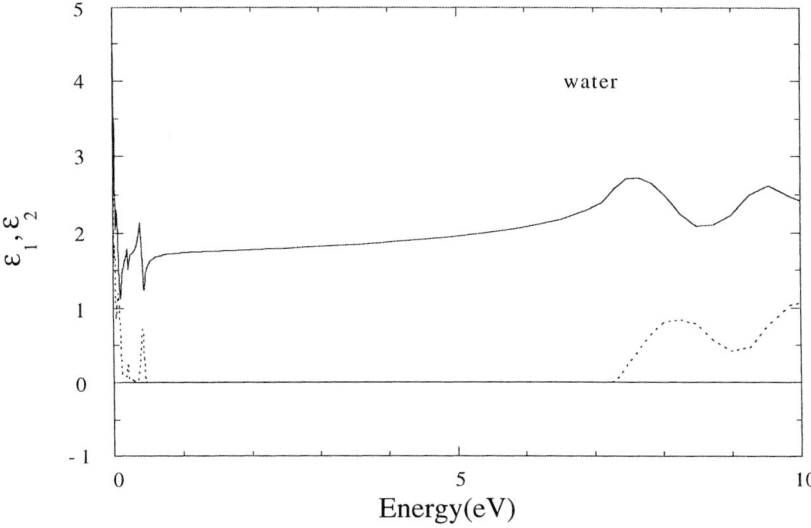

Figure 2.23: Expanded view of Figure 2.22 in the visible range.

The real part and the refractive index increase in the whole visible range, leading to the usual separation of colors. If we now go further down in energy, Figure 2.24, we find absorption bands around 400 meV and 200 meV.

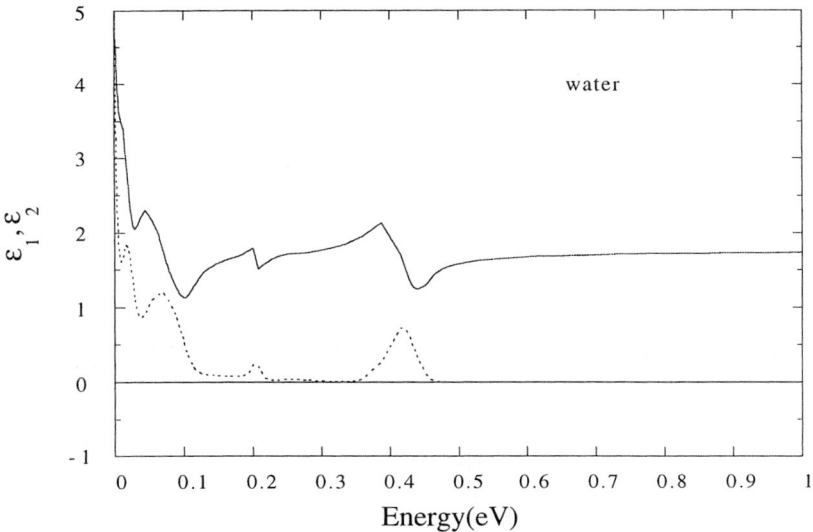

Figure 2.24: Expanded view of Figure 2.22 in near infrared range.

These are the known vibrational energies for the water molecule in Figure 2.25.

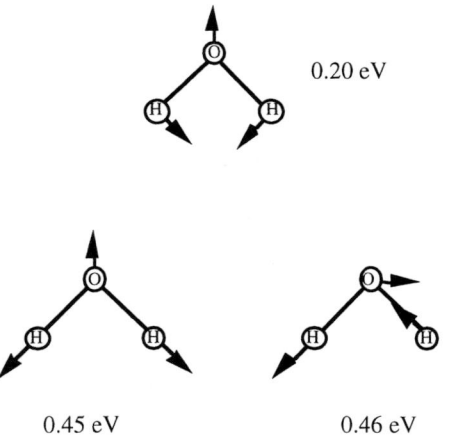

Figure 2.25: Different vibration modes of water molecules.

Further down in energy, Figure 2.26, we have more absorption bands.

2.7 Dielectric Properties of Water

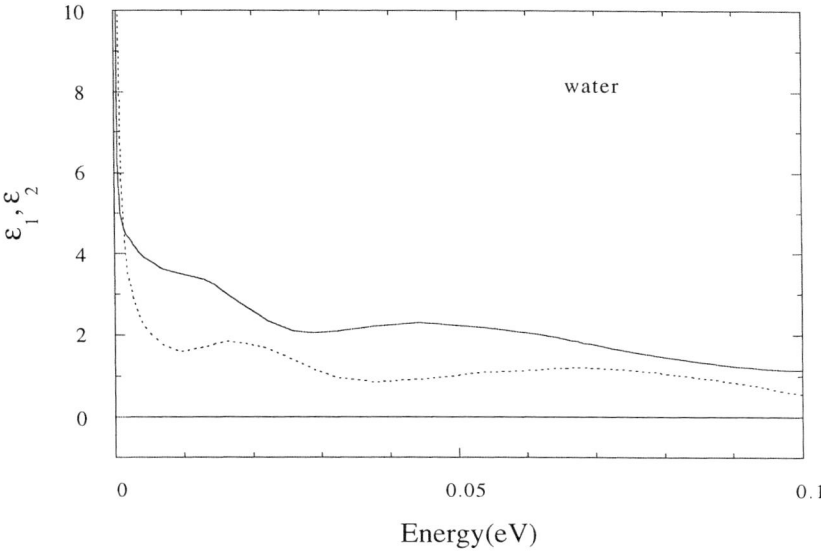

Figure 2.26: Expanded view of Figure 2.22 in the infrared range.

There is one very broad band centered around 70 meV and one around 20 meV. These are probably phonon like modes.

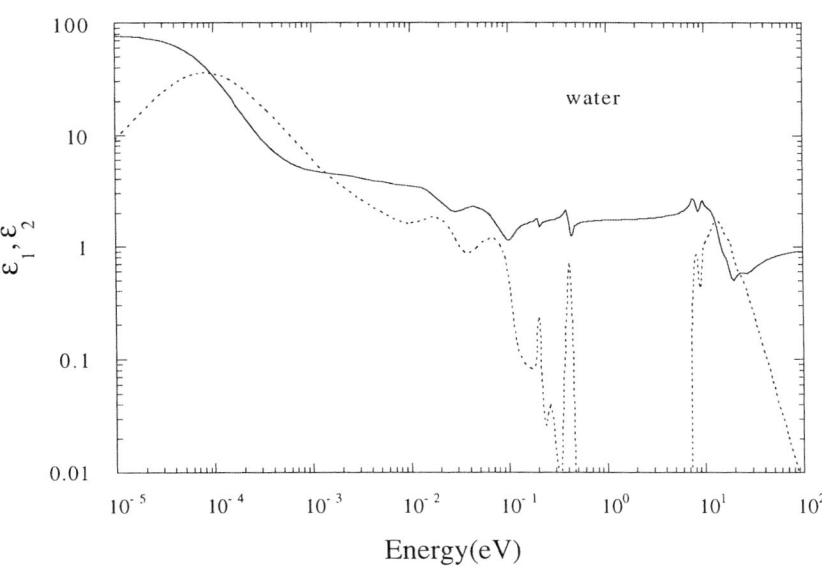

Figure 2.27: The real (solid curve) and imaginary (dotted curve) parts of the dielectric function of water on a logarithmic plot.

Even further down in energy there is a large contribution coming from rotations. In Figure 2.27 we have plotted the result on a logarithmic energy scale. We find that the rotational contributions are centered around 0.1 meV. This low energy contribution results in a large static dielectric constant and high index of refraction. This contribution was discussed in Section 2.6.

2.8 Superluminal speeds

In Section 2.1 we found that the group velocity for electromagnetic waves in an absorbing system may exceed the speed of light in vacuum. This fact prompts us to discuss in what situations we may attain superluminal speeds and what this implies.

2.8.1 Speed of light in vacuum

Maxwell's equations for electromagnetic waves in vacuum are

$$\nabla \cdot \mathbf{E} = 0$$
$$\nabla \cdot \mathbf{B} = 0$$
$$\nabla \times \mathbf{E} = -\frac{1}{c}\frac{\partial \mathbf{B}}{\partial t} \quad (2.64)$$
$$\nabla \times \mathbf{B} = \frac{1}{c}\frac{\partial \mathbf{E}}{\partial t}$$

Making use of the relation, Equation (1.24):

$$\nabla \times (\nabla \times \mathbf{C}) = \nabla(\nabla \cdot \mathbf{C}) - \nabla^2 \mathbf{C} \quad (2.65)$$

gives the two decoupled differential equations:

$$\nabla^2 \mathbf{E} = \frac{1}{c^2}\frac{\partial^2 \mathbf{E}}{\partial t^2}$$
$$\nabla^2 \mathbf{B} = \frac{1}{c^2}\frac{\partial^2 \mathbf{B}}{\partial t^2} \quad (2.66)$$

2.8 Superluminal Speeds

whose solutions are plane waves with phase velocity c. This constant c ($=\sqrt{1/\mu_0\varepsilon_0}$) happens to be the speed of light in vacuum (remember that Maxwell, at the time he wrote down his equations in 1865, did not know that these equations were valid also for light, that light was an electromagnetic wave.) After Fourier transforming these equations we find

$$q^2 \mathbf{E} = \frac{\omega^2}{c^2} \mathbf{E}$$
$$q^2 \mathbf{B} = \frac{\omega^2}{c^2} \mathbf{B}$$
(2.67)

and thus $\omega^2 = c^2 q^2$, or $\omega = cq$.

This means that the phase and the group velocities both equal c. Since this velocity is finite it will take a finite time for any electromagnetic signal to travel between two points in space — there are retardation effects.

2.8.2 Einstein's special theory of relativity

The speed of light in vacuum plays a central role in Einstein's special theory of relativity. In one of the postulates it is stated that light travels through vacuum with the same speed, c, in any direction, in all inertial frames.

If one could send a message with a speed exceeding c, one could by choosing observers in different inertial frames in smart ways send messages to one another in such a way that the original message would end up at the sender before the message was sent in the first place. This would mean that betting on horse races, for example, would be very easy. If one could also send objects, like ourselves, with superluminal speeds one could travel in time. Travel backwards in time would have dramatic consequences; one could change the history, by, for example, preventing the murder of a president, even preventing one's own birth. No criminal would be safe; the police could go back and check what really happened. However, the mere fact that it would have strange effects can not be used as an evidence against it. There are examples of very strange effects that have been shown to be real. The twin paradox, for example in special relativity is resolved within the general relativity. But still remains the strange effect that if a twin were to go away with a space ship with near luminal velocity and later return he would be younger than his twin brother that stayed put.

From the postulate mentioned above one may derive the relation between the energy, E, and velocity, v, of an object:

$$E = \frac{mc^2}{\sqrt{1-(v/c)^2}}$$
(2.68)

From this relation follows that it would take an infinite energy to accelerate an object with finite mass, m, to the speed of light; it is thus impossible in practice to pass the speed barrier at c. It also follows that a particle with zero mass, like the photon, has to move with the speed of light; otherwise its energy would be zero.

The momentum of a particle is

$$p = \frac{mv}{\sqrt{1-(v/c)^2}} \quad (2.69)$$

and the velocity as function of momentum is

$$v = \frac{p}{\sqrt{m^2 + (p/c)^2}} \quad (2.70)$$

2.8.3 Tachyons

Now, the special theory of relativity only says that a particle can not be accelerated to the speed of light and beyond. There could still be particles having speed exceeding c. This idea was first proposed by professor Arnold Sommerfeld in Munich, Germany. The name Tachyons was coined by Gerald Feinberg. This name derives from the Greek *tachys*, meaning swift. These particles have an imaginary mass and the velocity is then governed by the equation:

$$v = \frac{p}{\sqrt{-m^2 + (p/c)^2}} \quad (2.71)$$

where m now is the imaginary part of the Tachyon mass. Thus the Tachyons form a second branch for the velocity as function of momentum (see Figure 2.28.) They have a strange behavior in that they slow down when gaining energy and the velocity approaches c from above. Their existence has not yet been verified.

2.8 Superluminal Speeds

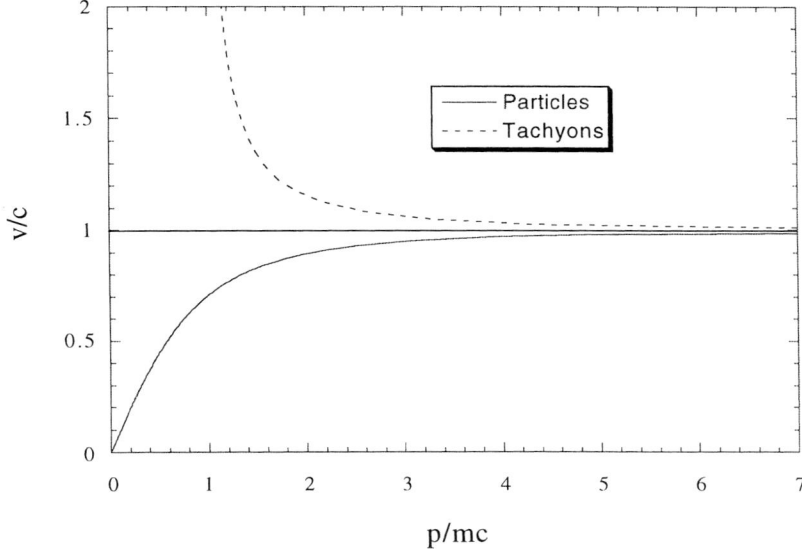

Figure 2.28: The speed as function of momentum of an ordinary particle (solid curve) and for a Tachyon (dashed curve).

2.8.4 Trivial examples where superluminal speeds seem to be possible but are not

We will here briefly list some examples and tell why they don't qualify as candidates.

a) *Rigid bodies*. If we give a truly rigid body a push at one end, the other end moves immediately. This means that a signal has moved from one end of the object to the other instantaneously, that is, with infinite speed. Where is the hatch? Well, there are no truly rigid bodies. A rigid body is an idealization that is useful in some situations but may in some, as in this example, lead to wrong results. If we look more carefully into this example we find that the other end will not move until the time it takes a sound wave to traverse the object. The speed of sound is much smaller than c. Other examples in the same category are the long swinging rod and the large circular rotating disc. One might think that a point at the periphery of a rotating disc would move with superluminal speed if one made the radius of the disc large enough. What would really happen is the following: Assume that one draws straight lines from the center of the disk outwards when the disc is still. When the disc rotates these lines will be bent into spiral form; the angular velocity of points on the disc decreases with distance from the center. What would happen with a rotating rod is that it would be bent into a spiral the same way as the straight lines on the disc. This is also what happens with large objects like galaxies. They are spirals.

b) *Torch*: If we instead of a long rod rotate a torch the light ray bends too. But if the ray were to sweep over the surface of a distant planet the light spot would move with superluminal speed. This case is different. In the rod case no point of the rod moves faster than c. In this case there is apart from the rotation also a movement outwards. This movement occurs with the speed of light. The light spot moves with superluminal speed but the light or the photons move with exactly c, or slightly slower than c if in some atmosphere. Photons hit the ground at different points in a systematic way that gives the impression that the light travels faster than c. We could produce an analogous effect in the following way: Assume that we have a long street with street lights and we turn the lights on successively with a slight time delay, smaller than the nearest neighbor distance divided by c. Then the position where the lights are turned on moves faster than c. Still nothing moves with superluminal speed in reality.

There are several examples similar to the torch: the speed of a shadow; the crossing point of two almost parallel lines that move relative each other; the notch where the blades of a pair of scissors with extremely long blades meet accelerates and may reach superluminal speed; the breaking point of a breaking wave that is almost parallel to a long straight beach. One can easily make up a long list of similar effects.

c) *The scalar potential*, when using Coulomb gauge, changes immediately if the position of charges are changed independent on at what distance the charges are. There is no retardation. Could this be used to send information with superluminal speeds? The answer is no! The potentials are only auxiliary functions. The real quantities are the fields. It turns out that when the charges are moved also the vector potential provides a contribution to the fields and the net result is that the fields are retarded; the changes of the fields occur after a time the distance divided by c, or later.

2.8.5 EPR paradox

Einstein, Podolsky and Rosen published a paper in 1935 where they theoretically treated a problem in experiments with entangled states. In a two-photon emission experiment a measurement on one of the photons means an immediate change in the state of the other – even if they are far away from each other. How this should be interpreted is still controversial. This example is closely connected to other cases like quantum mechanical jumps and the collapse of the wave function.

2.8.6 Phase velocity versus group velocity

A pure solution to MEs in a medium is a plane wave that moves throughout the system with the phase velocity. We can not transmit a signal or a message with a pure plane wave.

2.8 Superluminal Speeds

In order to transmit a signal we have to create a wave packet; a wave packet is formed as a superposition of many plane waves within a band of momenta centered around a given momentum. This wave packet will move with the group velocity at this given momentum.

The phase velocity of an electromagnetic wave in a medium may take on any value between zero and infinity. This possibility of superluminal speeds is not so interesting since it can not be utilized to send signals and it means no threat to causality. That the phase velocity may exceed c was discovered for the first time at the turn of the 20th century by radio engineers. Radio signals in the upper atmosphere traveled faster than c. The waves were moving through ionized gas – a plasma. In Section 2.4 we discussed the dielectric properties of a plasma; its dielectric behavior is very similar to a metallic system having a plasmon mode where the charged components move collectively. The light dispersion curve starts out above the plasmon energy with an initial infinite phase velocity.

Let us now return to the problem with unlimited group velocity we encountered in Section 2.1. Let us study a system that has a narrow absorption line at 4 eV. The dielectric function is given in Figure 2.29

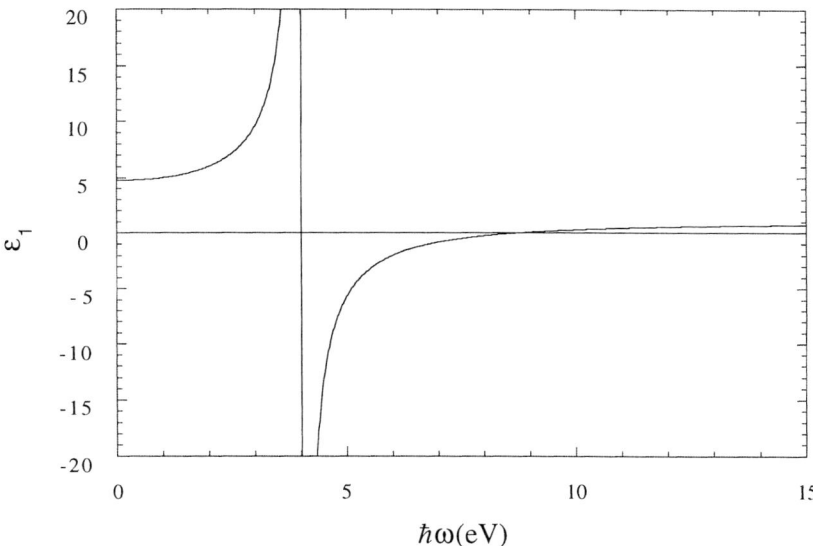

Figure 2.29: The real part of the dielectric function around an excitation level in absence of damping

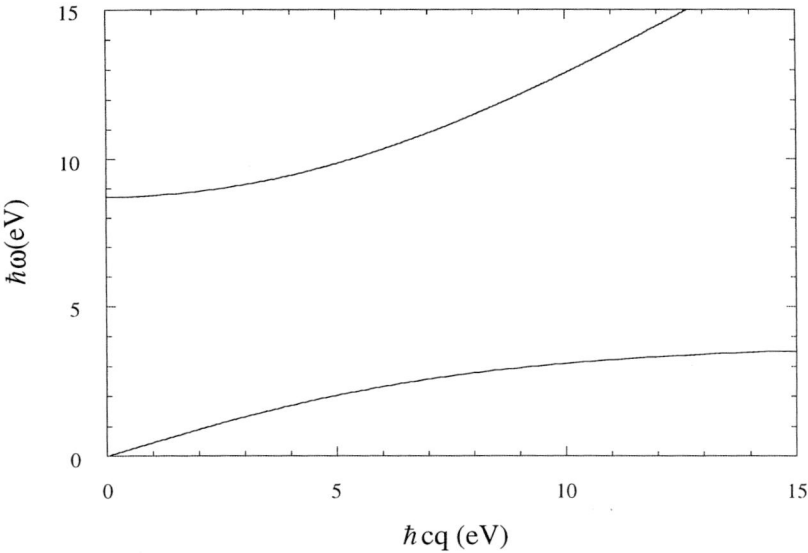

Figure 2.30: The dispersion of electromagnetic waves in a medium with dielectric function given in Figure 2.29. There are two branches separated by a gap in which electromagnetic waves can not travel.

We see there is a region of negative dielectric function above the absorption line. Light can not travel through the system in this region; there is an optical gap. The dispersion curve for electromagnetic waves in the system is given in Figure 2.30. There are two branches; one below the gap and one above. In the lower branch the phase velocity is always smaller than c. In the upper branch it is larger than c and even infinite for zero momentum. The group velocity, the slope of the curve, is everywhere smaller than c (slope smaller than unity in our scaled figure). Thus, in this system we have no problem with superluminal speeds.

Let us now see what happens if we have important damping in the system. The absorption line turns into a band as illustrated in Figure 2.2 in Section 2.1; the imaginary part of the dielectric function is finite in a band of frequencies; the real part is also changed. Remember that the real and imaginary parts are intimately connected. The light dispersion curve for such a system was given in Figure 2.3. We find the gap has closed and there are regions where the group velocity exceeds c. A wave packet in this region will travel with superluminal speeds; its amplitude will decrease continuously due to the absorption but we can not circumvent the fact that it moves with superluminal speeds. Has this ever been observed? The answer is yes! Steven Chu and Stephen Wong [5] at AT&T Bell Labs found in 1982 superluminal speeds for light pulses traveling through absorbing materials.

2.8.7 Surpassing the sonic speed barrier

We have said that according to Einstein's special relativity one may not accelerate a particle to the speed of light in vacuum and beyond but in principle particles could go faster than c. Maybe there are two separate worlds than can not mix; one with objects always going faster than c and our own world where objects always are slower than c.

Why is the velocity of light so special? Does the same apply to the speed of sound? Well, after the end of the second world war there was a struggle to exceed the sound velocity. One knew that objects could have super sonic speeds; the bullet of a rifle, for example moved faster than the sound velocity, but one had not been able to fly an aeroplane with supersonic speeds. Many believed that it was impossible. At first they seemed to be right. In several of the tests where one tried to pass the "magical" speed the planes disappeared and there were no sign of them. Maybe, they had entered the other world on the other side of the barrier? Pilots that came back but had failed to pass the critical speed told that it really felt like a barrier; they felt a resistance or a force preventing them to increase the speed. Eventually they succeeded. The first super sonic aeroplane was flown by Charles Yeager in 1947. It was an experimental rocket plane called Bell X-1 and managed Mach 1.02. The first jet plane, a Douglas "Sky rocket", that managed to overcome the barrier was flown by Gene May in 1949. It made Mach 1.03. Nowadays military aircrafts have maximum speeds in the range between 2 and 3 Mach. There is also a commercial aircraft, transporting people, that fly with super sonic speeds, the French Concorde. There has also been a corresponding Russian commercial aeroplane, which is no longer in use.

Thus, the sound barrier was possible to overcome. Now, what happens when this barrier is passed; why did the pilot feel the resistance? Why is there a bang when the barrier is passed? Well, when travelling with below sonic speeds the aeroplane excites individual air atoms or molecules. When the limit is exceeded it may also excite collective excitations in the air; sound waves are emitted. We may make analogies to effects that appear in condensed matter. We choose two examples; one is plasmon excitations in the metal aluminum; the other is optical phonon excitations in heavily doped CdS. We accelerate an electron in the system over the point where its energy above the Fermi level equals the energy of the collective excitation. When this happens plasmons and longitudinal optical phonons, respectively, are emitted. The result for Al is shown in Figure 2.31.

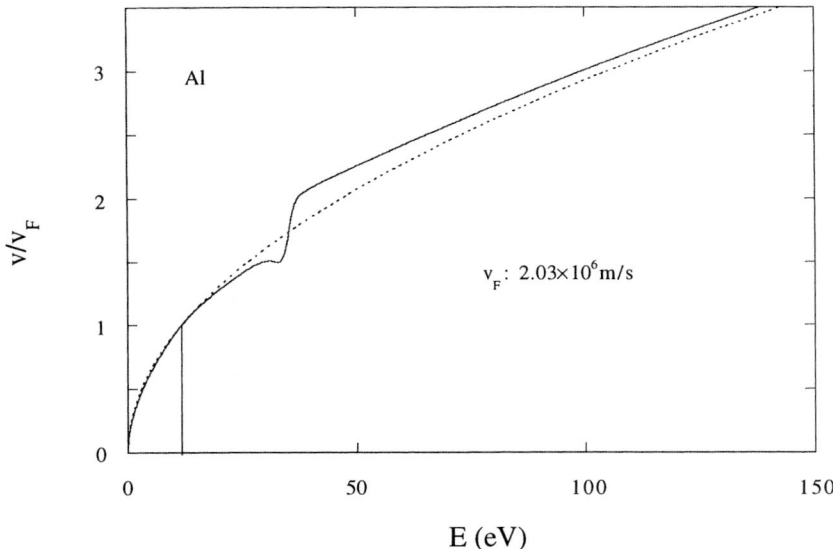

Figure 2.31: The velocity as function of energy for an electron in Al accelerated past the energy where plasmon emission sets in.

We have plotted the group velocity of the electron, in units of the Fermi velocity, as function of energy; the Fermi velocity is the velocity of an electron at the Fermi level. The dotted curve is the result one would get if there were only one electron in the conduction band. The solid curve is the full result including the interactions in the system. This is obtained using many-body theory. The vertical line indicates the position of the Fermi level.

We see that as the electron is accelerated to a certain point the velocity saturates and even dips down near the critical velocity, then it rapidly increases and becomes even larger than it would have been were the interactions absent. How can this be? We supply energy and still the velocity decreases when we come close to the barrier. Well, what happens is that there is a built up of electromagnetic fields surrounding the electron and the excess energy is stored in these fields. This built up continues until suddenly it is released in the form of plasmon emission and the electron gets even a kick forward. It is like an eruption. This explains what the pilots experienced. It explains the resistance felt, or force, that prevented the plane to increase its speed; it also explains why many planes were ripped apart. The energy build up around the airplane could not be handled by the plane; the planes did not have the same good quality of today's aircrafts; they were ripped apart. It also explains why there is a sound bang. There is a burst of emitted sound waves when the stored energy is released just as the barrier is crossed – the bang. When the speed of the plane exceeds the sound speed it continuously emits sound waves but at a slower rate.

In a polar semiconductor longitudinal optical phonons can be excited by a speeding electron. We present in Figure 2.32 the result in heavily n-type doped CdS. The right-most structure is the result for an electron, at an energy distance equal to the phonon energy above the Fermi level. We see that the behavior is similar to that of an electron in Al, but more

2.8 Superluminal Speeds

pronounced. The left-most structure represents a particle, a hole, at an energy distance equal to the phonon energy below the Fermi level. A hole moves towards the left in the figure, away from the Fermi level, when we supply energy (the energy scale has been chosen for electrons.) We see that its velocity goes down when the critical limit is approached in analogy with the electron.

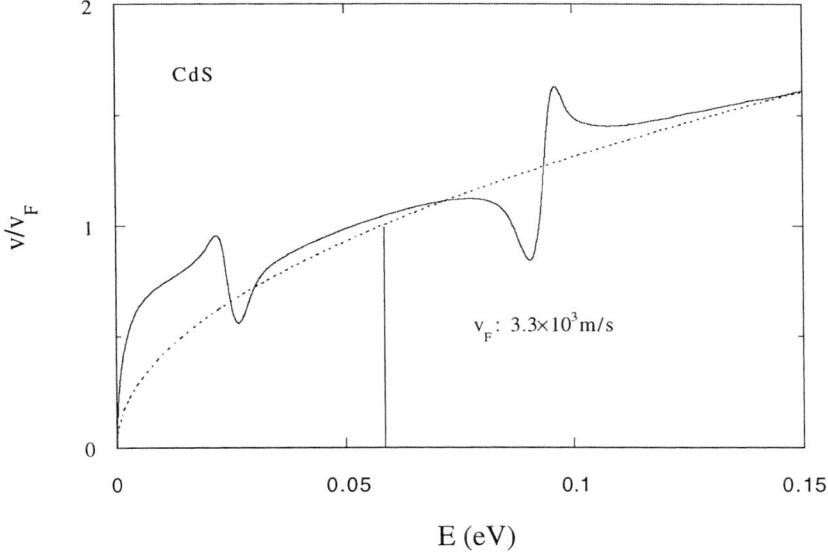

Figure 2.32: The velocity as function of energy for an electron in heavily doped CdS accelerated past the energy where longitudinal optical phonon emission sets in. The left-most structure is where a hole below the Fermi level excites phonons and the right-most structure where an electron above the Fermi level excites phonons.

Had we extended the figure to higher energies we would have seen yet another structure corresponding to emission of plasmons in the donor electron system.

2.8.8 Faster than the speed of light in a medium

In a medium there are regions where the speed (phase velocity) of an electromagnetic wave is smaller than the speed of light in vacuum. A fast particle may then have a speed that exceeds the speed of the wave. The situation is very similar to the two examples we gave above; here an electromagnetic wave is emitted. A relativistic particle, that is, a particle with a speed close to c, constantly emits electromagnetic fields; also in the visible range of the spectrum. This is called Cerenkov radiation and can be studied in the water of the reactor tanks at nuclear plants. There is a bluish glow that is clearly seen by the naked eye.

2.8.9 Superluminal speeds caused by changes in the vacuum

The mere presence of two thin conducting plates changes the density of photon modes in the region between the plates. This leads to the Casimir effect which will be discussed in Section 4.3. This has been shown by Scharnhorst [6] to furthermore lead to the possibility that light travels slightly faster than c. He found the following speed as function of separation, d, between the plates:

$$c_\perp = \left[1 + \frac{11}{2^6(45)^2} \frac{e^4}{(m_e d)^4}\right] c_0 \tag{2.72}$$

The speed parallel to the plates is unchanged as is the perpendicular speed in the regions outside the plates. To be noted are the following: the wave has been assumed not to "touch" the plates so the result has nothing to do with classical wave guide problems; furthermore, there is no dispersion so both the phase and group velocities are equal and equally enhanced. The effect is caused by the change of the density of states in the vacuum induced by the presence of the two plates. The expression means a very small correction to the speed, probably of no physical importance, but still of principal importance; for a separation of 1 μm the relative enhancement is of the order of 10^{-36}.

Another suggested possibility is so-called worm holes, short cuts through space-time, created by changes of the topology of space-time. We will not discuss this further.

2.8.10 Tunneling

One of the early triumphs of Quantum Mechanics was George Gamow's interpretation, in 1928, of the emission of α-particles in α-decay as quantum tunneling. The same idea was proposed also by Gurney and Condon in 1928.

In 1955 Eugene Wigner and his student L. Eisenbud at Princeton, studied how long it takes a wave packet representing the α-particle to tunnel through the barrier. They made a non-relativistic calculation. They found the very amazing result that if one increased the width of the barrier the tunnel- or transit-time first increased but then saturated. For very thick barriers the time did not increase when the thickness was increased. This means that the tunnel speed, that is, the width divided by the tunnel time, is unbounded. In other words one can achieve as high a velocity as one wants, by just increasing the width; at the same time the tunnel probability decreases. Their astounding result did not attract so much attention, probably due to the fact that they used the non-relativistic Schrödinger equation to treat a relativistic particle. However, the derivations were repeated later by Hartman [7] within the fully relativistic Dirac formalism, with the same result.

In the α-decay case the wave packet represents the wave function of a particle. It can

2.8 Superluminal Speeds

equally well represent an electromagnetic wave tunneling through an optical barrier. There have been several interesting experiments performed lately on electromagnetic waves tunneling through barriers with astounding results. These results have been possible due to technical improvements which make it possible to measure very short times.

In 1993 an American group [8] at the University of California in Berkeley, headed by Raymond Chiao, found that a photon travelled through their sample with the speed $1.7\ c$. The photon was in the visible range so they really demonstrated that light can travel faster than light in vacuum. They managed to measure the time with a resolution of one quarter of a femtosecond, which is an amazing resolution. Their sample was a filter consisting of many layers designed to totally reflect electromagnetic waves in the wavelength range of the photon. A wave that enters the filter is multiple reflected which, with the proper design, results in destructive interference and virtually no field penetrates the sample. Thus, the system has an optical gap but the origin of this gap is different from the one we have discussed before. One can view the experiment as tunneling through a barrier, but it is somewhat special. As Chiao has pointed out, interference takes some time to develop; when the wave packet representing the photon enters the system it takes some time to develop the destructive interference, which means that the front part of the wave packet will not experience any destructive interference. The front part is very little suppressed while the part that arrives later is much more suppressed. This means that the peak of the wave packet shifts forward which leads to an apparent increase in speed. The group has done more experiments on superluminal speeds in other types of system [9,10].

Similar experiments were performed in 1994 by Ferenc Krauss et al. at the Technical University of Vienna with varying barrier thickness. They found, just as theory had predicted, that the tunneling time saturated towards a maximum value.

Anedio Ranfagni et al. [11] at the National Institute for Research into Electromagnetic Waves, in Florence, Italy studied the speed of propagation of microwaves through forbidden zones inside square metal wave guides in 1991. They found speeds lower than c. In 1992 they repeated the experiments for thicker barriers and found superluminal speeds.

In march 1995 at a colloquium in Snowbird, Utah, Günter Nimtz [12] from the University of Cologne, made a very astounding announcement. He claimed that his research group had managed to send a signal through a gap of 11.4 cm with a speed of $4.7\ c$. When someone in the audience asked what kind of signal that was, his surprising answer was: Mozart's 40th Symphony. He had actually brought a recording to prove it. They had been working with sending 8.7 GHz (3.4 cm) microwaves through rectangular wave guides with barriers of varying widths. They found [13,14] that the transit time was about 81 ps and constant for gaps varying from 4.0 cm to 11.4 cm, in accordance with the theory of Wigner, Eisenbud and Hartman. The symphony had been transmitted as frequency modulated microwaves.

This last experiments prompts us to redefine what we really mean with sending a signal, information or message.

2.8.11 What do we mean by signals, information and message?

After the experiments by Nimtz et al. having Mozart's 40th Symphony making a 11.4 cm jump with 4.7 times the speed of light in vacuum, we all have to reconsider what we mean by a signal. Before the experiment at least most people would have considered music to qualify as a signal; but not after the experiment. It is now clear that in order for a signal to qualify in connection with violation of causality it must have a distinct front edge. The wave packets used in the experiments showing superluminal speeds do no qualify; they all have a slowly increasing front. The peak amplitude of the transmitted wave packet is not larger than the amplitude would have been in the same position in absence of the barrier. We can make the following analogy to explain why the experiments show superluminal speeds but still do not violate causality.

Assume that we have a car race with two cars; one short car and a ten times longer one. The drivers are sitting near the fronts. In this race the cars don't start in the usual way with their fronts aligned at the starting line. They start with their centers aligned. Somewhere during the race the long car drops it's 90 % back-most part after which it has the same size as the other car. Now, at the finishing line one notes the time again when the center of the cars pass the line. If the cars move with the same speed the longer car wins and it appears as if it had moved faster. This is very similar to what happens in the experiments. If instead the cars have a distinct front and one aligns the fronts at the starting line and notes the time when the fronts pass the finishing line one has a more fair race. This situation is what one would like to have in the experiments. Unfortunately a sharp edge at the front of the wave packet means a very broad distribution in momentum space which makes the experiment more difficult to design.

2.8.12 Conclusions

The recent experiments have shown that wave packets may move with speeds exceeding c; objects may have superluminal speeds; group velocities may exceed c. They have not shown that signals or information can be transmitted faster than c; causality has not been violated.

References

[1] N. W. Ashcroft and N. D. Mermin, *Solid State Physics*, Saunders College Publishing, 1976, Figure 15.12, p. 297, from H. Ehrenreich and H. R. Phillip, Phys. Rev. **B128T**, 1622 (1962).
[2] Bo E. Sernelius, Phys. Rev. B **43**, 7136 (1991).
[3] P. Debye, *Polar molecules*, Chemical Catalog Co., New York, 1929, Chapter V.
[4] M. R. Querry, D. M. Wieliczka, and D. J. Segelstein, *Handbook of optical constants of solids II*, (Academic Press 1991), p. 1059.
[5] S. Chu and S. Wong, Phys. Rev. Lett. **48**, 738 (1982).
[6] K. Scharnhorst, Physics Lett. **B236**, 354 (1990).
[7] T. E. Hartman, J. Appl. Phys. **33**, 3427 (1962).
[8] A. M. Steinberg, P. G. Kwiat, R. Y. Chiao, Phys. Rev. Lett. **71**, 708 (1993).
[9] R. Y. Chiao, Phys. Rev. **A48**, B34, (1993).
[10] R. Y. Chiao, Phys. Rev. Lett. **77**, 1254, (1996).
[11] A. Ranfagni et al., Applied Physics Lett. **587**, 774 (1991).
[12] G. Nimtz and W. Heitmann, Progress in Quantum Electronics **21**, 81 (1997).
[13] W. Heitmann and G. Nimtz, Phys. Lett. **A196**, 154 (1994).
[14] A. Enders and G. Nimtz, Phys. Rev. **E48**, 632 (1993).

3 Zero-point energy of modes

Why does the interaction between the electrons in a metallic system contribute to the binding of the system in spite of the fact that the interaction is repulsive?

The collective excitations of a system are massless bosons described by a Hamiltonian of the kind:

$$H = \sum_{\mathbf{q}} \hbar\omega_{\mathbf{q}}\left(n_{\mathbf{q}} + \tfrac{1}{2}\right) \tag{3.1}$$

where $\hbar\omega_{\mathbf{q}}$ and $n_{\mathbf{q}}$ represent the energy of the mode $\mathbf{q}$ and the operator for the occupation of the mode, respectively. The "1/2" terms represent the energy of the vacuum fluctuations or with another name the zero-point energy. The collective excitations can be phonons, photons, plasmons, magnons, oscillations in an LC circuit and so forth. The zero-point energy of modes has a central role in this book. We will in this chapter motivate why this is so. We will discuss the exchange- and correlation-energy in a bulk system. This energy is a central quantity in quantum mechanics and many-body theory and is the result of the correlated motion of a large number of interacting fermions. The motion is correlated both due to the fermion statistics and due to the repulsion between the particles. We will perform a demonstration where we stepwise go from a simple classical description to a more and more sophisticated treatment and finally end up in a picture where the energy is expressed in terms of classical electromagnetic fields, solutions to Maxwell's equations. These are the normal modes of the system. This last step may be looked upon as a paradigm shift. It has the important consequence that many problems can be discussed and solved by persons not mastering the very complicated theoretical framework making up the basis of many-body theory.

The energy density in an electric field is

$$w = \frac{1}{8\pi}|\mathbf{E}|^2 \tag{3.2}$$

Since the electron is surrounded by an electric field there is an energy, a self-energy, associated with the electron. Part, or all, of the particle rest energy consists of this energy. It turns out that if the electron is a true point particle this energy diverges. The divergence comes from the region very close to the electron. If we assume that the electron has a finite but small spatial extension the divergence goes away. Let us assume that the electron charge is homogeneously distributed on the surface of a small sphere of radius r_0 and that the rest energy is fully given by the energy in the field. Then we have

$$m_e c^2 = W = \int d^3 r w(r) = \int_{r_0}^{\infty} dr 4\pi r^2 \frac{1}{8\pi}\left(\frac{e}{r^2}\right)^2 = \frac{e^2}{2}\int_{r_0}^{\infty} dr \frac{1}{r^2} = \frac{e^2}{2r_0} \qquad (3.3)$$

and

$$r_0 = \frac{1}{2}\frac{e^2}{m_e c^2} \qquad (3.4)$$

If we instead had chosen the charge to be homogeneously distributed throughout the interior of the sphere the prefactor 1/2 had been 3/5. Thus, the result is relatively insensitive to the actual distribution of the charge. One has chosen to define the classical electron radius as the result without the prefactor ($r_0 = e^2/m_e c^2 \approx 2.8 \times 10^{-5}$ Å).

Let us now derive the electron self-energy in an alternative way. The interaction energy for charges distributed according to a charge distribution $\rho(\mathbf{r})$ is

$$W = \frac{1}{2}\iint d^3 \mathbf{r}_1 d^3 \mathbf{r}_2 \frac{\rho(\mathbf{r}_1)\rho(\mathbf{r}_2)}{|\mathbf{r}_1 - \mathbf{r}_2|} \qquad (3.5)$$

If the charge distribution represents the electron as a point particle, that is, if $\rho(\mathbf{r}) = -e\delta(\mathbf{r})$, we have

$$W = \frac{1}{2}v(0) = \frac{e^2}{2r}\bigg|_{r=0} \qquad (3.6)$$

which is a divergent result. If the electron charge is distributed on the surface of a sphere, we have

$$W = \frac{1}{2}v(r_0) = \frac{e^2}{2r_0} \qquad (3.7)$$

that is, the same as before in Equation (3.3).

Whichever way we distribute the charge within the radius r_0, as long as the distribution is spherically symmetric, the fields outside the sphere of radius r_0 are the same. When several electrons are present in the system the fields are added and result in forces between the electrons. The regions inside the small spherical particles have negligible effects on these forces. High-energy scattering experiments designed to probe the size of the electron reveal that it is at least not larger than this classical estimate. It is furthermore improbable that it has an internal structure. If it were to be made up by constituent particles analogous to quarks, for example, the size quantization would make the kinetic energy of these particles exceed the rest energy of the electron by several orders of magnitude.

3 Zero-Point Energy of Modes

Let us now study a system containing a large number of electrons, N, that are more or less free to move throughout the whole volume, Ω, of the system. This system is metallic. The simplest model for a metal is the so-called jellium model. The charges of the positive ions are in this model smeared out homogeneously throughout the whole volume; the system is charge neutral. The average electron density is constant throughout the system. On the average there is no electric field in the system. Does this mean that there is no energy in the system, apart from the kinetic energy of the electrons? The answer is no! The energy density is proportional to the square of the electric field, and the square does not vanish. At each instant of time, or for each distribution of the electrons throughout the system, there is an electric field. Furthermore, if there is no correlation in the motion, or distribution, of the electrons the energy is just the number of electrons times the self-energy that comes from the electron self-interaction just discussed above:

$$\overline{\left(\sum_i \mathbf{E}_i\right)^2} = \sum_i \overline{(\mathbf{E}_i)^2} + \sum_{\substack{i,j \\ i \neq j}} \overline{\mathbf{E}_i \cdot \mathbf{E}_j} = \sum_i \overline{(\mathbf{E}_i)^2} \tag{3.8}$$

Now, the motion of the electrons is correlated. The electrons are fermions. We know from *Pauli's exclusion principle* that two electrons can not have the same set of quantum numbers. The position of the electron is as good a quantum number as any. This means that two electrons with same spin can never be in the same position in space. Actually, the fermion property has a wider effect on the spatial electron distribution than provided by Pauli's exclusion principle; the probability to have two electrons with same spin close to each other is reduced. This means that the electrostatic repulsion energy is reduced and this results in a negative energy contribution. Thus the Fermi-Dirac statistics leads to a spatial correlation in the motion of the electrons; we call this correlation the statistical correlation. The energy that results from this effect is termed the *exchange energy*, E_x, for reasons discussed below. There is another effect that causes a correlated motion of the electrons. Since the electrons are of same charge they repel each other and any pair of electrons avoids to come close to each other. This repulsion is independent on if the two electrons have the same spin or not. This additional correlation also leads to a negative energy contribution. It is termed the *correlation energy*, E_c. If we are content by just including the statistical correlations the approximation is the *Hartree-Fock approximation* (*HF*). The result in *HF* can be calculated exactly. When going beyond *HF* one has to rely on approximations. By definition the energy contribution beyond *HF* is called the correlation energy. Thus, E_x is exact and E_c is approximate.

We will now calculate E_x on the first level of conceptual complexity. This energy is a quantum mechanical effect and can not be obtained in a purely classical derivation but the results will be described in a classical picture. We will at this level use *first quantization*. The single particle states are characterized by the momentum, **k**, and the spin, σ. The single particle wave functions are plane waves:

$$\varphi_{\mathbf{k}}(\mathbf{r}) = \frac{1}{\Omega^{1/2}} e^{i\mathbf{k}\cdot\mathbf{r}} \qquad (3.9)$$

There is equal probability to find the electron in state $(\mathbf{k},\sigma)$ anywhere in the system:

$$|\varphi_{\mathbf{k}}(\mathbf{r})|^2 = \frac{1}{\Omega} = \text{constant} \qquad (3.10)$$

and

$$\int d^3r |\varphi_{\mathbf{k}}(\mathbf{r})|^2 = 1 \qquad (3.11)$$

that is, the probability to find the electron anywhere in the system, given it is in this state, is unity. In our model there is no preferred place in the system for the electron to be. However, there is a correlation in the relative position of pairs of electrons and we want to find this. The single-particle wave function squared gives the probability distribution for each single electron. The N-particle wave function squared, the density matrix:

$$\rho_N(\mathbf{r}_1 \ldots \mathbf{r}_N) = |\psi(\mathbf{r}_1 \ldots \mathbf{r}_N)|^2 \qquad (3.12)$$

is the probability density for finding particle *1* at $\mathbf{r}_1$, particle *2* at $\mathbf{r}_2$ ... particle N at $\mathbf{r}_N$. It is normalized so that unity is obtained when integrating over all the coordinates:

$$1 = \int d^3r_1 \ldots d^3r_N \rho_N(\mathbf{r}_1 \ldots \mathbf{r}_N) \qquad (3.13)$$

We are interested in the correlation of the positions of a pair of particles. The probability density for finding particle *1* at $\mathbf{r}_1$ and particle *2* at $\mathbf{r}_2$ is obtained by integrating over the coordinates of all the other particles:

$$\rho_2(\mathbf{r}_1,\mathbf{r}_2) = \int d^3r_3 d^3r_4 \ldots d^3r_N \rho_N(\mathbf{r}_1 \ldots \mathbf{r}_N)$$
$$1 = \int d^3r_1 d^3r_2 \rho_2(\mathbf{r}_1,\mathbf{r}_2) \qquad (3.14)$$

For a homogeneous system this two-particle density matrix can only depend on the relative positions, $(\mathbf{r}_2-\mathbf{r}_1)$, of the two particles. This means that

$$\rho_2(\mathbf{r}_1,\mathbf{r}_2) = \rho_2(\mathbf{r}_2 - \mathbf{r}_1) = \rho_2(\mathbf{r})$$
$$1 = \Omega \int d^3r \rho_2(\mathbf{r}) \qquad (3.15)$$

Now, the *pair correlation function* is defined in the following way: If one of the particles in the system is at the origin the pair correlation function is the probability density for finding another particle at position $\mathbf{r}$. Now, this is not quite correct. The function is

3 Zero-Point Energy of Modes

scaled in such a way that it approaches unity for large distances. Thus, it is

$$g(\mathbf{r}) = \Omega^2 \rho_2(\mathbf{r}) \tag{3.16}$$

In the Hartree-Fock approximation the many-particle wave function is an N-dimensional Slater determinant and hence

$$\rho_N(\mathbf{r}_1...\mathbf{r}_N) = \frac{1}{N!} \begin{Vmatrix} \varphi_{\lambda_1}(\mathbf{r}_1) & \varphi_{\lambda_2}(\mathbf{r}_1) & \cdots & \varphi_{\lambda_N}(\mathbf{r}_1) \\ \varphi_{\lambda_1}(\mathbf{r}_2) & \varphi_{\lambda_2}(\mathbf{r}_2) & \cdots & \varphi_{\lambda_N}(\mathbf{r}_2) \\ \vdots & \vdots & \ddots & \\ \varphi_{\lambda_1}(\mathbf{r}_N) & \varphi_{\lambda_2}(\mathbf{r}_2) & & \varphi_{\lambda_N}(\mathbf{r}_N) \end{Vmatrix}^2 \tag{3.17}$$

where the single-particle orbitals $\varphi_\lambda(\mathbf{r})$ are assumed to be orthogonal for different states. The quantum number λ represents a complete set of quantum numbers, that is, includes in general the spin. In our case it represents the momentum and spin. There is one orbital for each occupied state. One can show that

$$\rho_2(\mathbf{r}_1,\mathbf{r}_2) = \frac{1}{N(N-1)} \sum_{\substack{\lambda_1,\lambda_2 \\ \lambda_1 \neq \lambda_2}} \frac{1}{2!} \begin{Vmatrix} \varphi_{\lambda_1}(\mathbf{r}_1) & \varphi_{\lambda_2}(\mathbf{r}_1) \\ \varphi_{\lambda_1}(\mathbf{r}_2) & \varphi_{\lambda_2}(\mathbf{r}_2) \end{Vmatrix}^2 \tag{3.18}$$

where the summation is over all occupied orbitals. The Slater determinant can be expanded to give the following, equivalent result:

$$\rho_2(\mathbf{r}_1,\mathbf{r}_2) = \frac{1}{2N(N-1)} \sum_{\substack{\lambda_1,\lambda_2 \\ \lambda_1 \neq \lambda_2}} \left| \varphi_{\lambda_1}(\mathbf{r}_1)\varphi_{\lambda_2}(\mathbf{r}_2) - \varphi_{\lambda_1}(\mathbf{r}_2)\varphi_{\lambda_2}(\mathbf{r}_1) \right|^2 \tag{3.19}$$

This leads to

$$g(\mathbf{r}_2 - \mathbf{r}_1) = \frac{1}{N^2} \sum_{\substack{\mathbf{k},\sigma \\ \mathbf{k}',\sigma'}} \left[1 - \delta_{\sigma,\sigma'} \cos[(\mathbf{k}-\mathbf{k}')\cdot(\mathbf{r}_1-\mathbf{r}_2)] \right] \tag{3.20}$$

and

$$g(R) = 1 - \frac{36}{R^6}\left[4 + R^2 - 4\cos(R) + R^2 \cos(R) - 4R\sin(R) \right] \tag{3.21}$$

where we have let $R = r2k_F$. This function is shown in Figure 3.1.

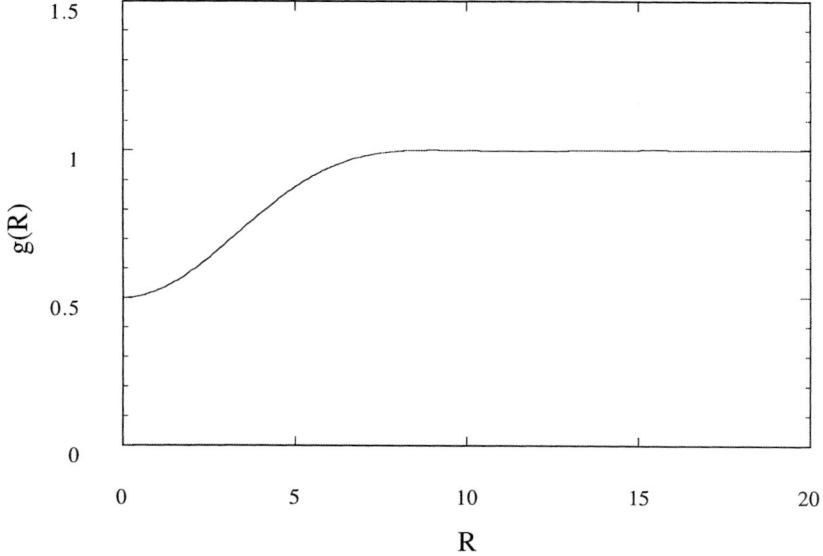

Figure 3.1: The pair correlation function in *HF* as function of separation.

As can be seen there is a depletion in the electron density near the electron at the origin. This *exchange hole* contains exactly one electron charge. The charge distribution as a function of distance, *P(R)*, is

$$P(R) = \frac{R^2}{6\pi}[1-g(R)] = \frac{6}{\pi R^4}\left[4 + R^2 - 4\cos(R) + R^2\cos(R) - 4R\sin(R)\right] ;$$

$$1 = \int d^3 rn[1-g(r)] = \int dR P(R)$$

(3.22)

and is shown in Figure 3.2.

3 Zero-Point Energy of Modes

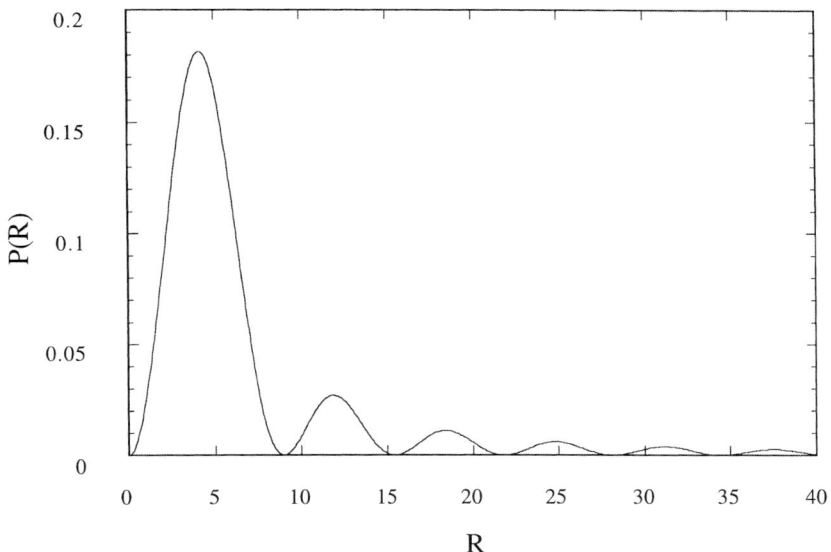

Figure 3.2: The charge distribution in the exchange hole, in *HF*, as function of separation.

As can be seen there is, apart from the main contribution near the electron, oscillating contributions for larger separations. The exchange hole we have considered is an average hole for the electrons within the Fermi sphere. The origin of the exchange energy is just the electrostatic interaction energy for this average electron and its exchange hole. This is

$$\overline{\Sigma}_x = -\int_0^\infty dR P(R) \frac{e^2}{R} 2k_F = -\frac{e^2 k_F}{\pi} \int_0^\infty dR P(R) \frac{2\pi}{R} = -\frac{3}{2} \frac{e^2 k_F}{\pi} \qquad (3.23)$$

This is the average self-energy, or rather change in the self-energy, for an electron due to the statistical correlated motion of the electrons. This energy contains contributions from all pairs of electrons. The exchange energy is just half of this:

$$E_x = -\frac{3}{4} \frac{e^2 k_F}{\pi} \qquad (3.24)$$

If it were not for the half all contributions would be counted twice. Let us pick out two electrons. This pair gives a certain contribution to the exchange energy. The same contribution is given to the self-energy of one of the electrons but also to the other electron. It is a matter of book keeping. In calculating the self-energy of an electron all contributions involving this electron is included.

Now one may pick out a particular electron state and determine the exchange hole for this state. The energy is then different for different states and one finds the relation

$$\Sigma_x(K) = -\frac{e^2 k_F}{\pi}\left[1 + \frac{1-4K^2}{4K}\ln\left|\frac{1+2K}{1-2K}\right|\right] \quad (3.25)$$

and in particular

$$\Sigma_x(0) = -2\frac{e^2 k_F}{\pi} \;;$$
$$\Sigma_x(k_F) = -\frac{e^2 k_F}{\pi} \quad (3.26)$$

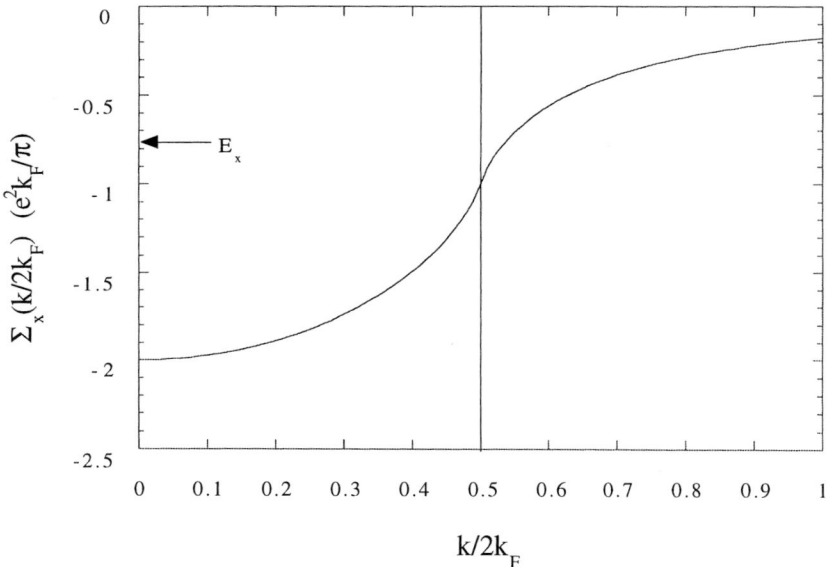

Figure 3.3: The electron self energy in *HF* as function of scaled electron momentum.

The vertical line in the figure separates occupied states to the left from unoccupied to the right. The exchange energy is indicated and one should in particular note that this energy is smaller in size than any of the self-energies for the occupied states.

The pair-correlation function for an electron in state $(\mathbf{k},\sigma)$ is

$$g(\mathbf{K},\mathbf{R}) = 1 - \cos(\mathbf{K}\cdot\mathbf{R})\frac{12}{R^3}\left[\sin\left(\frac{R}{2}\right) - \frac{R}{2}\cos\left(\frac{R}{2}\right)\right] \quad (3.27)$$

and in particular for a state at the bottom of the band:

$$g(0, R) = 1 - \frac{12}{R^3}\left[\sin\left(\frac{R}{2}\right) - \frac{R}{2}\cos\left(\frac{R}{2}\right)\right] \tag{3.28}$$

This function and the pair correlation function are shown in Figure 3.4.

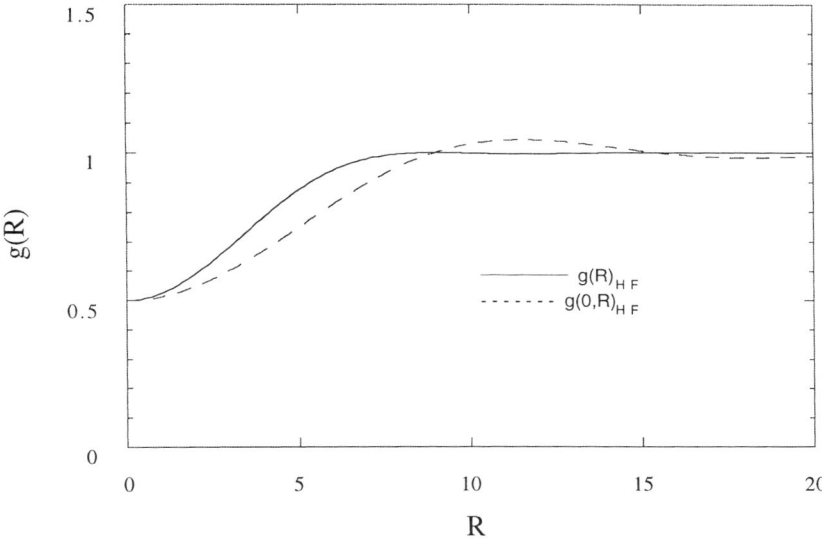

Figure 3.4: The pair correlation function in *HF* for the average electron and for one with zero momentum.

The result in Figure 3.5 is for different values of the projection of the momentum on **r**. The larger the momentum the faster the curves oscillate. The so-called exchange hole still contains exactly one electron charge, but this charge is on the average more distant from the electron which means that the self-energy shift is reduced in strength.

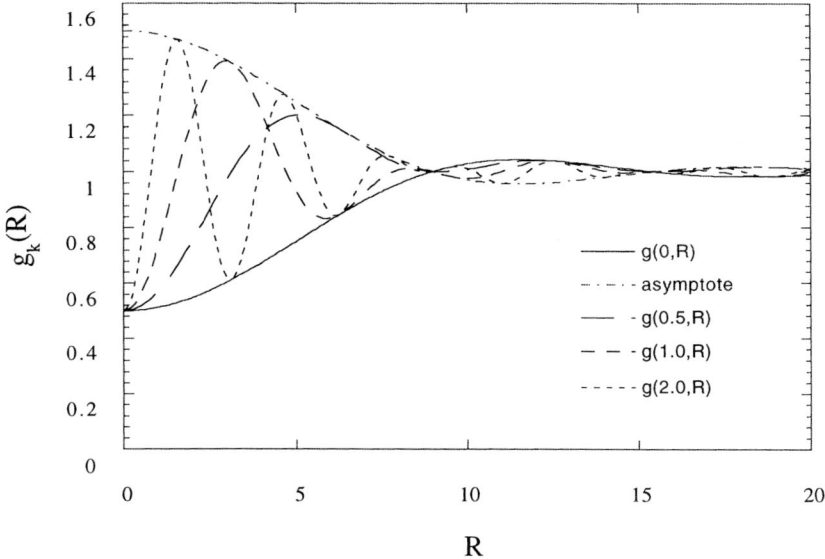

Figure 3.5: The pair correlation function in *HF* for electrons with various momenta.

We find that the result for different momentum oscillates between two asymptotes; one asymptote is the result for zero momentum and the other asymptote is related to this according to:

$$g_{asymptote} = 2 - g_{k=0} \tag{3.29}$$

The exchange hole contains exactly one unit of charge. When going beyond *HF* the exchange-correlation hole still contains one unit of charge but the hole is more closely concentrated near the electron. This means that the self-energy shift is larger in size. The pair correlation function is zero instead of 1/2 at zero separation because the probability is zero also for two electrons of opposite spin to be in the same position. We will not perform any calculations at this stage.

Let us now go to the next step and discuss the interactions in *second quantized* formalism. In this formalism the physics is described as moving particles around between single particle states. A particle is annihilated in one state and recreated in another. We start by returning to the self-interaction energy and reexpress it in momentum space:

$$\begin{aligned} W &= \frac{1}{2} \int\int d^3\mathbf{r}_1 d^3\mathbf{r}_2 \frac{\rho(\mathbf{r}_1)\rho(\mathbf{r}_2)}{|\mathbf{r}_1 - \mathbf{r}_2|} = \frac{1}{2e^2\Omega} \sum_\mathbf{q} \rho(-\mathbf{q})\rho(\mathbf{q}) v(\mathbf{q}) \\ &= \frac{1}{2\Omega} \sum_\mathbf{q} v(\mathbf{q}) = \frac{1}{2} \int \frac{d^3\mathbf{q}}{(2\pi)^3} v(\mathbf{q}) = \frac{e^2}{\pi} \int_0^\infty dq \end{aligned} \tag{3.30}$$

This result also diverges, of course. We have to make a cutoff at a large momentum to get a finite value. For $q_{max} = \pi/2r_0$ we get the same result as for the spherical shell. In the spherical shell approximation the actual weight in momentum space is not unity up to a sharp cutoff. It is $\sin(qr_0)/qr_0$. This shows that if the fields or potentials do not vary extremely fast in space, with variations within the size of the classical electron radius, the internal structure of the electron plays no role. We are only interested in changes to this self-interaction energy and we may keep the summation to infinity.

Let us look upon this expression for the energy

$$W = \frac{1}{2\Omega} \sum_{\mathbf{q}} v(\mathbf{q}) \tag{3.31}$$

in the following way:

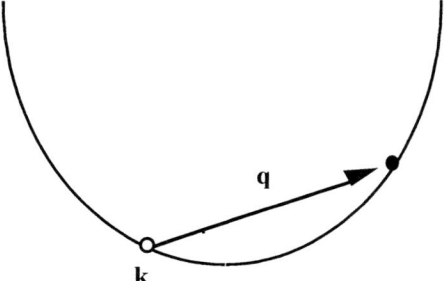

Figure 3.6: Processes leading to the self-interaction energy of the electron.

The electron in state $(\mathbf{k},\sigma)$ is scattered into all states $(\mathbf{k+q},\sigma)$ and each time it costs the energy $v(\mathbf{q})/2\Omega$. The result is the same for all states $(\mathbf{k},\sigma)$. When we bring all our N electrons together to form our system the electrons will fill up a sphere in momentum space, the Fermi sphere, with radius k_F.

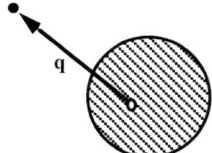

Figure 3.7: Allowed processes from within the Fermi sphere to states outside.

In this treatment the Pauli exclusion principle becomes really transparent. Our electron in state $(\mathbf{k},\sigma)$ is not allowed to scatter into $(\mathbf{k+q},\sigma)$ if this state is already taken by another electron. Thus we have

$$W_{\mathbf{k}} = \frac{1}{2\Omega}\sum_{\mathbf{q}} v(\mathbf{q})[1 - n(\mathbf{k}+\mathbf{q})] - \frac{1}{2\Omega}\sum_{\mathbf{q}} v(\mathbf{q})n(\mathbf{k}+\mathbf{q}) \tag{3.32}$$

The second term is because the presence of our electron in state $(\mathbf{k},\sigma)$ prevents the electron in state $(\mathbf{k}+\mathbf{q},\sigma)$ to scatter into this state. All energy changes are given to our electron. We see that the energy now is different for different states; it depends on the position in momentum space of $\mathbf{k}$ relative the Fermi sphere. Now, subtracting the energy due to the self-interaction we get the electron self-energy as

$$\begin{aligned}\Sigma_x(\mathbf{k}) &= \frac{1}{2\Omega}\sum_{\mathbf{q}} v(\mathbf{q})[1 - n(\mathbf{k}+\mathbf{q})] - \frac{1}{2\Omega}\sum_{\mathbf{q}} v(\mathbf{q})n(\mathbf{k}+\mathbf{q}) - \frac{1}{2\Omega}\sum_{\mathbf{q}} v(\mathbf{q}) \\ &= -\frac{1}{\Omega}\sum_{\mathbf{q}} v(\mathbf{q})n(\mathbf{k}+\mathbf{q}) = -\int \frac{d^3q}{(2\pi)^3} v(\mathbf{q})n(\mathbf{k}+\mathbf{q})\end{aligned} \tag{3.33}$$

Performing this integral gives us the result obtained earlier from the pair-correlation function:

$$\Sigma_x(K) = -\frac{e^2 k_F}{\pi}\left[1 + \frac{1 - 4K^2}{4K}\ln\left|\frac{1+2K}{1-2K}\right|\right] \; ; \; K = k/2k_F \tag{3.34}$$

The origin of the self-energy is that the presence of all other electrons prevents some of the scattering processes to occur, processes that were allowed when the electron was alone in the whole system. The energy is due to a subtraction of positive energy contributions. The total interaction energy is a sum over all processes where an electron within the Fermi sphere is scattered out of the sphere. The exchange energy is obtained when we subtract all processes where the electrons are allowed to scatter anywhere. The net result looks as if we sum all processes where electrons are scattered within the Fermi sphere. Since this is not allowed we tend to look upon it as if the electrons change places with each other; this is what gives this energy its name — exchange energy.

3 Zero-Point Energy of Modes

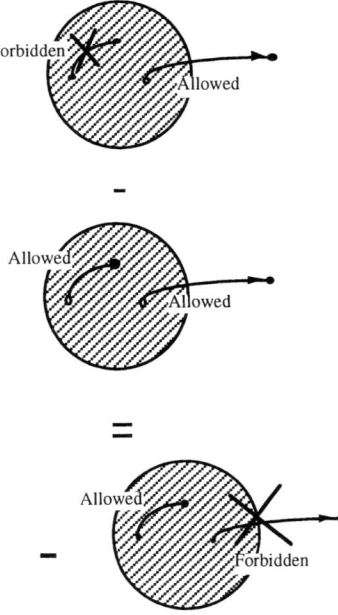

Figure 3.8: Processes leading to the self energy of the electron in *HF*.

When going beyond *HF*, the new effect is that the potential in each process is screened. Thus, the electrons are only allowed to be scattered out of the Fermi sphere and each of the positive energy contributions are now reduced due to the screening. Thus this causes a further reduction in the energy and the correlation energy is negative.

Schematically this is illustrated in Figure 3.9.

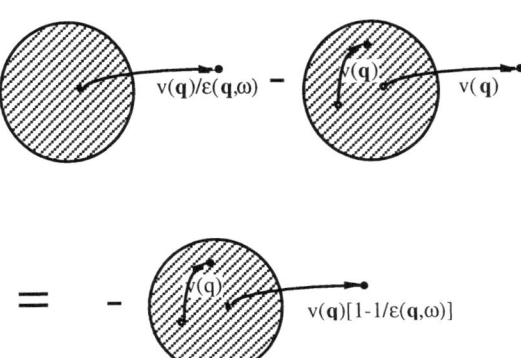

Figure 3.9: Processes leading to the self energy of the electron in treatments beyond *HF*.

The dielectric function is frequency dependent and we need to integrate over frequency as well as over momentum. When going beyond *HF* it turns out that also the kinetic energy, or better the expectation value of the kinetic energy operator, of the electrons changes when the interaction is "turned on". To see this we turn to the so-called *ground-state-energy-theorem*.

The Hamiltonian is written with a variable coupling constant, λ,

$$H(\lambda) = H_0 + \lambda V$$
$$H(1) = H \quad (3.35)$$
$$H(0) = H_0$$

Let,

$$H(\lambda)|\Psi_0(\lambda)\rangle = E(\lambda)|\Psi_0(\lambda)\rangle$$
$$\langle \Psi_0(\lambda)|\Psi_0(\lambda)\rangle = 1$$
$$\Downarrow \quad (3.36)$$
$$E(\lambda) = \langle \Psi_0(\lambda)|H(\lambda)|\Psi_0(\lambda)\rangle$$

where $|\Psi_0(1)\rangle$ and $|\Psi_0(0)\rangle$ are the ground-state state-vectors with and without interaction, respectively. In *HF* they are the same.

The energy in the interacting ground state is $E(1)$. It consists of two terms, one expectation value over the kinetic energy operator, H_0, and one over the interaction V, both with respect to the interacting ground state. The derivative of the final expression in Equation (3.36) with respect to the coupling constant reduces to

$$\frac{d}{d\lambda}E(\lambda) = \left\langle \frac{d\Psi_0(\lambda)}{d\lambda}\bigg|H(\lambda)\bigg|\Psi_0(\lambda)\right\rangle + \left\langle \Psi_0(\lambda)\bigg|H(\lambda)\bigg|\frac{d\Psi_0(\lambda)}{d\lambda}\right\rangle + \left\langle \Psi_0(\lambda)\bigg|\frac{dH(\lambda)}{d\lambda}\bigg|\Psi_0(\lambda)\right\rangle$$
$$= E(\lambda)\frac{d}{d\lambda}\langle \Psi_0(\lambda)|\Psi_0(\lambda)\rangle + \langle \Psi_0(\lambda)|V|\Psi_0(\lambda)\rangle$$
$$= \langle \Psi_0(\lambda)|V|\Psi_0(\lambda)\rangle. \quad (3.37)$$

Integrating this equation with respect to the coupling constant from 0 to 1 results in

$$E - E_0 = \int_0^1 \frac{d\lambda}{\lambda}\langle \Psi_0(\lambda)|\lambda V|\Psi_0(\lambda)\rangle \quad (3.38)$$

The shift in the ground-state energy is expressed solely in terms of the matrix element of the interaction λV. In *HF* the matrix element is proportional to λ which means that

3 Zero-Point Energy of Modes

$$E^{HF} - E_0 = \langle \Psi_0(0)|V|\Psi_0(0)\rangle \tag{3.39}$$

or

$$E^{HF} = \langle \Psi_0(0)|H_0|\Psi_0(0)\rangle + \langle \Psi_0(0)|V|\Psi_0(0)\rangle \tag{3.40}$$

Thus the kinetic energy is unchanged by the interaction. This is no longer true beyond HF, and we have to perform the integration over coupling constant. The result for the exchange and correlation energy per electron is

$$E_{xc} = +i\int_0^1 \frac{d\lambda}{\lambda} \frac{1}{2N} \sum_{\mathbf{q}}{}' \left\{ \int_{-\infty}^{\infty} \frac{d\omega}{2\pi} \hbar \left[\frac{1}{\varepsilon^{\lambda}(\mathbf{q},\omega)} - 1 \right] - \frac{N\lambda v_q}{i\Omega} \right\} \tag{3.41}$$

The index λ on the dielectric function means that all interactions v_q appearing in the dielectric function are to be multiplied by λ, the same way as is the last term of the summand. The frequency integral over the real part of the square bracket vanishes and the integrand is even in frequency. Thus we have

$$E_{xc} = \int_0^1 \frac{d\lambda}{\lambda} \frac{1}{2N} \sum_{\mathbf{q}}{}' \left\{ 2\int_0^{\infty} \frac{d\omega}{2\pi} \hbar \,\text{Im}\left[\frac{-1}{\varepsilon^{\lambda}(\mathbf{q},\omega)} \right] - \frac{N\lambda v_q}{\Omega} \right\} \tag{3.42}$$

In the *Random Phase Approximation* (RPA) the dielectric function is given by

$$\varepsilon(\mathbf{q},\omega) = 1 + \frac{v_q}{\hbar\Omega} \sum_{\mathbf{k},\sigma} n(\mathbf{k})[1 - n(\mathbf{k}+\mathbf{q})] \left[\frac{1}{\omega + \frac{1}{\hbar}(\varepsilon_{\mathbf{k}+\mathbf{q}} - \varepsilon_{\mathbf{k}}) - i\eta} - \frac{1}{\omega - \frac{1}{\hbar}(\varepsilon_{\mathbf{k}+\mathbf{q}} - \varepsilon_{\mathbf{k}}) + i\eta} \right] \tag{3.43}$$

where $n(\mathbf{k})$ is the Fermi-Dirac occupation number. We see that the imaginary part consists of a sum of δ-functions, infinitesimally spaced when the volume of the system goes to infinity. These form the single particle continuum. The real part passes through zero between each neighboring pairs of δ-functions. When the volume goes to infinity one can replace the summation by an integral. The imaginary part then turns into a smooth continuous function and the real part does no longer pass through zero. We note in passing that the dielectric function in Equation (3.43) is on *time-ordered* form; this is the form suitable for diagrammatic perturbation theory. The dielectric functions discussed in Chapter 2 were all on *retarded* form; functions describing actual response functions for real systems are on retarded form; they obey the causality rules — the response to a disturbance comes after the disturbance. These are the two most important forms. What distinguishes them is the signs in front of the infinitesimal imaginary terms in the denominators. The retarded form has the combination $\omega + i\eta$ in both terms. All forms are identical in the complex frequency plane except for on the real axis where the sign of their imaginary parts varies.

If we refrain from performing the integral over $\mathbf{k}$ in the calculation of the dielectric function, which smears out the imaginary part to form the continuum, we have for the total exchange-correlation energy

$$
\begin{aligned}
E_{xc}^{tot} &= \int_0^1 \frac{d\lambda}{\lambda} \frac{1}{2} {\sum_{\mathbf{q}}}' \left\{ 2\int_0^\infty \frac{d\omega}{2\pi} \hbar\pi \delta\left[\varepsilon^\lambda(\mathbf{q},\omega)\right] - \frac{N\lambda v_q}{\Omega} \right\} \\
&= \frac{1}{2} {\sum_{\mathbf{q}}}' \left\{ \int_0^\infty d\omega \hbar \int_0^1 \frac{d\lambda}{\lambda} \delta[1 + \lambda\alpha_0(\mathbf{q},\omega)] - \frac{Nv_q}{\Omega} \right\} \\
&= \frac{1}{2} {\sum_{\mathbf{q}}}' \left\{ \int_0^\infty d\omega \hbar \int_0^1 \frac{d\lambda}{\lambda|\alpha_0(\mathbf{q},\omega)|} \delta[\lambda + 1/\alpha_0(\mathbf{q},\omega)] - \frac{Nv_q}{\Omega} \right\} \\
&= \frac{1}{2} {\sum_{\mathbf{q}}}' \left\{ \int_0^\infty d\omega \hbar \, \theta[-\varepsilon(\mathbf{q},\omega)] - \frac{Nv_q}{\Omega} \right\} \\
&= {\sum_{\mathbf{q}}}' \sum_i \left[\frac{1}{2}\hbar\omega_1^i(\mathbf{q}) - \frac{1}{2}\hbar\omega_0^i(\mathbf{q}) \right] - \frac{N}{\Omega} {\sum_{\mathbf{q}}}' \frac{v_q}{2}
\end{aligned}
\tag{3.44}
$$

where $\omega_1(\mathbf{q})$ and $\omega_0(\mathbf{q})$ are solutions to the equation

$$\varepsilon^\lambda(\mathbf{q},\omega_\lambda(\mathbf{q})) = 0 \tag{3.45}$$

for the λ values 1 and 0, respectively. Thus we find that the exchange-correlation energy is the change in the zero-point energy of the normal modes of the system when the interaction is "turned on" minus the self-interaction energy of the electrons. Now, this self-interaction can be expressed in another way:

$$
\begin{aligned}
E_{self} &= N\int_0^1 \frac{d\lambda}{\lambda} \frac{1}{2} {\sum_{\mathbf{q}}}' 2\int_0^\infty \frac{d\omega}{2\pi} \hbar \, \mathrm{Im} \left[\frac{-1}{\varepsilon_{\text{single electron}}^\lambda(\mathbf{q},\omega)} \right] \\
&= N {\sum_{\mathbf{q}}}' \left[\frac{1}{2}\hbar\omega_1^{s.e.} - \frac{1}{2}\hbar\omega_0^{s.e.} \right]
\end{aligned}
\tag{3.46}
$$

where now the frequencies are solutions to the equation

3 Zero-Point Energy of Modes

$$\varepsilon^\lambda_{s.e.}(\mathbf{q}, \omega_\lambda) = 0 \tag{3.47}$$

This dielectric function is given by

$$\varepsilon^\lambda_{s.e.}(\mathbf{q}, \omega) = 1 + \lambda \underbrace{\frac{v_q}{\hbar\Omega}}_{a} \left\{ \left[\omega + \underbrace{\frac{1}{\hbar}(\varepsilon_{\mathbf{k}+\mathbf{q}} - \varepsilon_{\mathbf{k}})}_{\Delta_\mathbf{k}} - i\eta \right]^{-1} - \left[\omega - \underbrace{\frac{1}{\hbar}(\varepsilon_{\mathbf{k}+\mathbf{q}} - \varepsilon_{\mathbf{k}})}_{\Delta_\mathbf{k}} + i\eta \right]^{-1} \right\} \tag{3.48}$$

if the electron is in state k. The zeros of this function are

$$\omega_\lambda = \pm \Delta_\mathbf{k} \sqrt{1 + \lambda \frac{2a}{\Delta_\mathbf{k}}} \approx \pm \Delta_\mathbf{k}\left(1 + \lambda \frac{a}{\Delta_\mathbf{k}}\right) = \pm(\Delta_\mathbf{k} + \lambda a) \tag{3.49}$$

where we have used the fact that a is very small. It is inversely proportional to the volume of the system which we let go to infinity. Thus we have

$$\frac{1}{2}\hbar\omega_1^{s.e.} - \frac{1}{2}\hbar\omega_0^{s.e.} = a = \frac{1}{2}\frac{v_q}{\Omega} \tag{3.50}$$

and

$$E_{self} = \frac{N}{\Omega} {\sum_\mathbf{q}}' \frac{v_q}{2} \tag{3.51}$$

We may write

$$E^{tot}_{xc} = {\sum_\mathbf{q}}' \sum_i \left[\frac{1}{2}\hbar\omega_1^i(\mathbf{q}) - \frac{1}{2}\hbar\omega_0^i(\mathbf{q})\right] - N{\sum_\mathbf{q}}' \left[\frac{1}{2}\hbar\omega_1^{s.e.} - \frac{1}{2}\hbar\omega_0^{s.e.}\right] \tag{3.52}$$

Thus we have found that the interaction energy is just the change in the zero-point energy of the normal modes of the system caused by the interaction. Thus if we know the dielectric properties of the system we can determine the normal modes and the interaction energy without involving the complicated many-body theory and even without the use of quantum mechanics. We may get away with using the theory of classical electromagnetic waves, the Maxwell's equations. If we have a continuum of excitation modes like in our free-electron system this is not a practical method to use. There is a normal mode squeezed in between each nearest pair of single-particle excitation modes, and these come infinitely close to each other. Thus it is in practice impossible to use this approach in this situation. However, often one may, to a good approximation, replace the continuum with one or a few discrete modes and the approach can be applied. We should note that in the interacting system there are no longer any pure single-particle excitations. In each excitation, characterized by a change in momentum and frequency, all electrons are involved. When an electron is promoted from one state to another, all other electrons adjust to this change. Thus all excitations are really collective and in principle not different in this respect from the plasmon excitation. The difference is that all but the plasmon are still bound to the original region of the single-particle continuum.

We now use a simple example to illustrate what we have found:

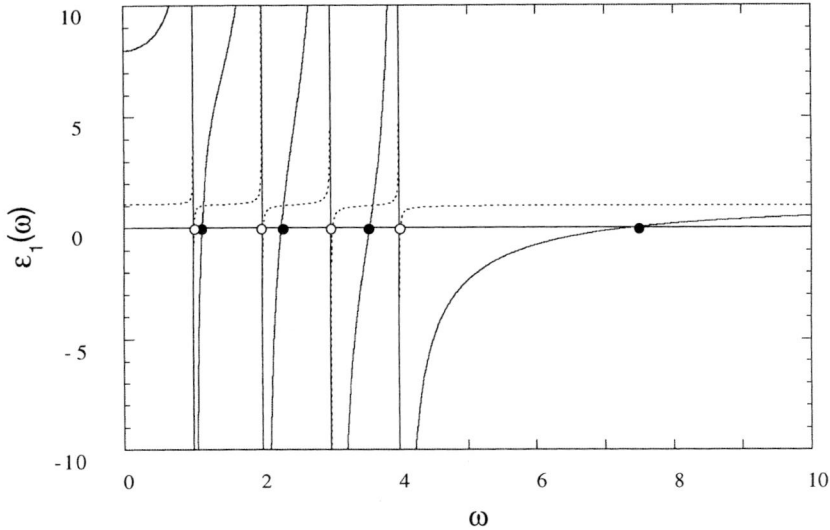

Figure 3.10: The dielectric function with λ equal to unity (solid curve) and equal to 0.01 (dashed curve). The interaction shifts the modes from the open circles to the filled ones.

Figure 3.10 illustrates the following dielectric function:

$$\varepsilon_1^\lambda(\omega) = 1 - \lambda \left[\frac{1}{\omega - 1} + \frac{2}{\omega - 2} + \frac{3}{\omega - 3} + \frac{2}{\omega - 4} + \frac{1}{-\omega - 1} + \frac{2}{-\omega - 2} + \frac{3}{-\omega - 3} + \frac{2}{-\omega - 4} \right] \tag{3.53}$$

which gets its contributions from the four excitation energies 1, 2, 3 and 4 with weights 1, 2, 3 and 2 respectively. The three left-most modes are supposed to represent the continuum and the forth the split off plasmon mode. The normal modes are shifted from the white to the black positions indicated in the figure by circles. The solid curve is for $\lambda = 1$ and the dotted one is for $\lambda = 0.01$.

It would be very nice if one could find a simple recipe for replacing the modes with a single mode and still get the same interaction energy:

$$\varepsilon_1^\lambda(\omega) = 1 - \lambda \sum_i a_i \frac{\omega_i}{\omega_0}\left[\frac{1}{\omega-\omega_0} - \frac{1}{\omega+\omega_0}\right] = 1 - \lambda\left[2\sum_i a_i \omega_i \Big/ \left(\omega^2 - \omega_0^2\right)\right] \quad (3.54)$$

In doing so one has to consider the following: There is a sum rule, the *f-sum rule*, that demands that

$$\sum_i 2a_i\omega_i = \text{constant} = \omega_{pl}^2 \quad (3.55)$$

One can furthermore show that

$$\sum_i \left\{\left[\omega_1^i(\mathbf{q})\right]^2 - \left[\omega_0^i(\mathbf{q})\right]^2\right\} = \omega_{pl}^2 \quad (3.56)$$

that is, the total change in the square of the frequency of all modes due to the interaction is constant; it is constant in **q** and also in the number of modes. We leave this open for anyone to try.

To summarize this chapter, we have demonstrated that the interaction energy in a metal, the exchange and correlation energy, can be expressed as the change in zero-point energy of the longitudinal electromagnetic modes in the system when the interaction is turned on.

If we expand or compress this system the zero-point energies of these modes change and this change is an important factor that determines the stability of the system. For metals it is the modes associated with intraband transitions of the free carriers, the plasmons, that are most important. We found in Chapter 2 that there are many other types of bulk mode in different materials. For each type of bulk mode there is a corresponding surface mode. The zero-point energies of these surface modes contribute to the surface energy and is important for the stability of the surface among other things. We will come back to this many times throughout the book. In Chapter 1 we found that there are also transverse modes apart from the longitudinal ones treated in this chapter. Their zero-point energies also contribute to the energy of the system and to its stability. It turns out that they play a much smaller role. We will demonstrate in Chapters 4 and 6 that these modes, or rather corresponding modes outside the material and between objects, are responsible for the Casimir force. The longitudinal bulk modes and their surface counterparts are responsible for the van der Waals force. This force is much stronger for small separations between the objects but drops suddenly in strength at a certain distance, characteristic for the material the objects are made from. At this separation the Casimir force gains strength and replaces the van der Waals force.

4 Modes at flat interfaces

Can the vacuum have any affects on our lives?

Of all the different interfaces we treat in this book the flat interface is the most important type of interface. If the object we study is not extremely small we may also treat a curved surface, for example that of a sphere, as flat. This effect is similar to what we experience in our every-day life: that the surface of earth appears to be flat and not spherical. This is why, we treat flat surfaces and interfaces in a separate chapter. We will first study a single interface. The results from this treatment can be used to calculate the surface energy for a material in vacuum or in another medium. Then we will treat systems with two interfaces. The results from this treatment can be used to calculate the force between two objects, the strength obtained when gluing or soldering, the cohesion energy and the adhesion energy.

We will be concerned with three types of material: vacuum; metal; semiconductor. The semiconductor can be a polar semiconductor or ionic insulator. As we have discussed earlier there is no important difference between the various materials. The longitudinal and transverse excitations always come in pairs. What is special with metals is that the transverse excitation energy is zero. We will find surface modes in the regions between all pairs of transverse and longitudinal excitation energies. Thus, in general there will be many regions of this type and hence many modes at an interface between two different materials. Usually only a few of the regions are important for the energy and forces. We will here only consider the contribution from one of the regions.

Before we study the modes we repeat some general results from basic electromagnetic theory. Study the interface between two dielectrics, *1* and *2*, in Figure 4.1.

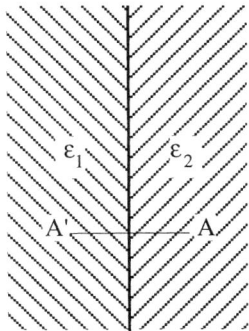

Figure 4.1: The interface between two dielectrics as discussed in the text.

Assume that we place a charge q in A. The fields in dielectric 2 is then the same as they would be if we filled the whole space with dielectric 2 and had the charge q in A and an extra charge q' in A' (A' is positioned in the *1*-side at the same distance from the interface as A is). In dielectric *1* the fields are the same as they would be if we filled the whole space with dielectric *1* and instead of the charge q had a charge q'' in position A. Now, the charges q' and q'' are

$$q' = -\left(\frac{\varepsilon_1 - \varepsilon_2}{\varepsilon_1 + \varepsilon_2}\right) q$$
$$q'' = \left(\frac{2\varepsilon_1}{\varepsilon_1 + \varepsilon_2}\right) q \qquad (4.1)$$

We are here interested in the special case when the charge q is placed at the interface. From the *1*-side the fields are the same as if q'' was placed at the interface and from the *2*-side the fields are the same as if the charge $q+q'$ were placed at the interface, that is,

$$q'' = \left(\frac{2\varepsilon_1}{\varepsilon_1 + \varepsilon_2}\right) q$$
$$q + q' = \left[1 - \left(\frac{\varepsilon_1 - \varepsilon_2}{\varepsilon_1 + \varepsilon_2}\right)\right] q = \left(\frac{2\varepsilon_2}{\varepsilon_1 + \varepsilon_2}\right) q \qquad (4.2)$$

We find that from both sides the electric fields and the scalar potentials are the same as they would be if the charge q were placed in a dielectric with dielectric constant $\tilde{\varepsilon} = (\varepsilon_1 + \varepsilon_2)/2$ or the charge $q/\tilde{\varepsilon}$ were placed in vacuum. The screening of charges at the interface is by the mean value of the dielectric functions on the two sides of the interface.

Let us now study the fields from a Fourier component of charge oscillation at the interface, travelling in the *x*-direction ($\mathbf{k} = k\hat{x}$):

$$\rho_{\mathbf{k},\omega}(\mathbf{r},t) = \frac{1}{\tilde{\varepsilon}} q \delta(z) e^{i(kx - \omega t)} \qquad (4.3)$$

Note that the Fourier decomposition is made in the plane of the interface, the *xy*-plane; we keep the spatial dependence in the *z*-direction.

What we derive below is correct as long as the wavelength of the oscillation is short compared to the thicknesses of the dielectric layers. The potential, in Coulomb gauge, is in general given by

$$\phi(\mathbf{r},t) = \int d^3 r' \frac{\rho(\mathbf{r}',t)}{|\mathbf{r} - \mathbf{r}'|} \qquad (4.4)$$

and in in our example we have the Fourier component

$$\phi_{\mathbf{k},\omega}(\mathbf{r},t) = \frac{1}{\tilde{\varepsilon}}\frac{2\pi q}{k}e^{-k|z|}e^{i(kx-\omega t)} \qquad (4.5)$$

The Fourier component is complex valued. It is important to remember that for an actual time dependent charge distribution, and resulting time dependent potential, there are always also Fourier components for negative momentum and frequency. Adding up these and the ones we have given above leads to real-valued results. We have

$$\rho_{-\mathbf{k},-\omega}(\mathbf{r},t) = \rho^*_{\mathbf{k},\omega}(\mathbf{r},t)$$
$$\phi_{-\mathbf{k},-\omega}(\mathbf{r},t) = \phi^*_{\mathbf{k},\omega}(\mathbf{r},t) \qquad (4.6)$$

where the "*" means complex conjugation.

From the first of Equations (1.39) we get the electric field: $\mathbf{E} = -\nabla\phi - \frac{1}{c}\partial \mathbf{A}/\partial t$. If now $\omega/c \ll k$ we may neglect the second term. When doing this we neglect retardation effects; this would always be permissible if the speed of light were infinite. Neglecting retardation effects, we have

$$\begin{aligned}E_x(\mathbf{r},t) &= -i\frac{1}{\tilde{\varepsilon}}2\pi q e^{-k|z|}e^{i(kx-\omega t)}\\ E_y(\mathbf{r},t) &= 0\\ E_z(\mathbf{r},t) &= \frac{1}{\tilde{\varepsilon}}2\pi q e^{-k|z|}e^{i(kx-\omega t)} \;; z \geq 0\\ &= -\frac{1}{\tilde{\varepsilon}}2\pi q e^{-k|z|}e^{i(kx-\omega t)} \;; z \leq 0\end{aligned} \qquad (4.7)$$

The collective excitation modes for the system are obtained if we have an induced electric field that survives although the external perturbation vanishes. In the present case the dispersion of the surface plasmons is obtained as the solution to the equation $\tilde{\varepsilon}(k,\omega) = 0$. We should also notice that the potential and electric field decay exponentially away from the interface. This is illustrated in Figures 4.2-4.4. The interface is defined by the horizontal and vertical lines in the plot cube. The potential is independent of the y-coordinate and dies out exponentially in the z-direction.

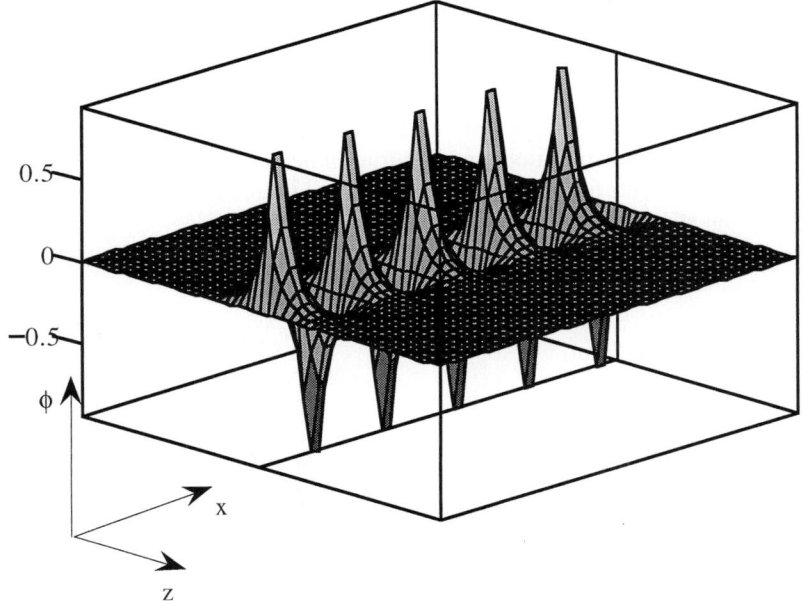

Figure 4.2: Schematic illustration of the potential from a longitudinal surface mode moving in the x-direction

The electric field in the x-direction will behave as the potential but will be half a wave length out of phase: when the potential has a maximum or minimum the field has a node and vice versa.

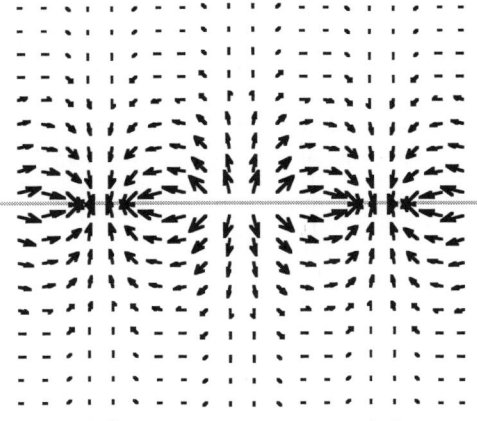

Figure 4.3: The electric field distribution in a cross sectional plane through the interface. The plane is perpendicular to the y-axis.

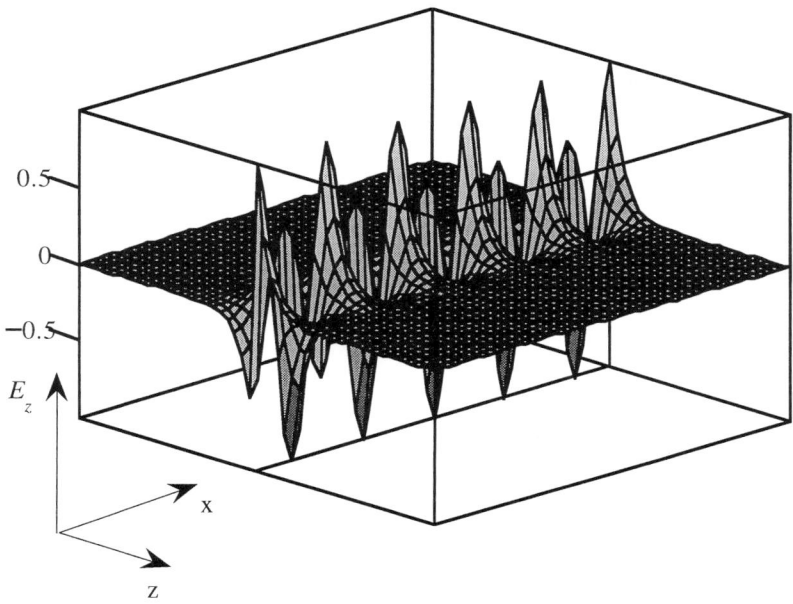

Figure 4.4: Same as Figure 4.2 but for the electric-field component in the z-direction.

The rest of this chapter is arranged in the following way: In Section 4.1 and subsections we derive the surface modes at a single interface and show explicit results for different combinations of materials on the two sides of the interface; we study vacuum, semiconductors and metals. The dielectric function for vacuum is unity; for the semiconductor we use the result from Equation (2.30) neglecting the interband contributions; for metals we use the simple Drude result from Equation (2.14). In both the semiconductor and metal cases we neglect damping. We will demonstrate later in Section 6.8 that damping can be important for the force between metal objects at finite temperature, at large separations where retardation effects are important. In Section 4.2 and subsections we extend the treatment to the slab geometry, where we have two parallel interfaces. In Section 4.3 we discuss the Casimir effect; in Section 4.4 metal surfaces; in Section 4.5 the interaction between two two-dimensional metallic sheets.

4.1 Modes at a single interface

Now we make a more strict derivation. The Maxwell's equations read

$$\nabla \cdot \mathbf{D} = 4\pi\rho$$
$$\nabla \cdot \mathbf{B} = 0$$
$$\nabla \times \mathbf{E} = -\frac{1}{c}\frac{\partial \mathbf{B}}{\partial t} \tag{4.8}$$
$$\nabla \times \mathbf{H} = \frac{4\pi}{c}\mathbf{J} + \frac{1}{c}\frac{\partial \mathbf{D}}{\partial t}$$

We also have the constitutive equations:

$$\mathbf{D} = \varepsilon\mathbf{E}$$
$$\mathbf{B} = \mu\mathbf{H} \tag{4.9}$$

We assume that there are no external charges or currents and that the system is nonmagnetic, thus

$$\nabla \cdot \varepsilon\mathbf{E} = 0$$
$$\nabla \cdot \mathbf{H} = 0$$
$$\nabla \times \mathbf{E} = -\frac{1}{c}\frac{\partial \mathbf{H}}{\partial t} \tag{4.10}$$
$$\nabla \times \mathbf{H} = +\frac{\varepsilon}{c}\frac{\partial \mathbf{E}}{\partial t}$$

We use our experience from the previous section and expect the following polarization charge distribution to be produced at the interface:

$$\rho(\mathbf{r},t) = q\delta(z)e^{i(kx-\omega t)} \tag{4.11}$$

and make the following ansatz for the solution for the fields:

$$\mathbf{E} = \begin{cases} \mathbf{E}^1 e^{\gamma_1 z} e^{i(kx-\omega t)} \; ; \; z \leq 0 \\ \mathbf{E}^2 e^{-\gamma_2 z} e^{i(kx-\omega t)} \; ; \; z \geq 0 \end{cases}$$
$$\mathbf{H} = \begin{cases} \mathbf{H}^1 e^{\gamma_1 z} e^{i(kx-\omega t)} \; ; \; z \leq 0 \\ \mathbf{H}^2 e^{-\gamma_2 z} e^{i(kx-\omega t)} \; ; \; z \geq 0 \end{cases} \tag{4.12}$$

4.1 Modes at a Single Interface

We insert these relations in the MEs and use the matching conditions at the interface:

$$\begin{aligned} \mathbf{D}_\perp^1 &= \mathbf{D}_\perp^2 \\ \mathbf{B}_\perp^1 &= \mathbf{B}_\perp^2 \\ \mathbf{E}_{//}^1 &= \mathbf{E}_{//}^2 \\ \mathbf{H}_{//}^1 &= \mathbf{H}_{//}^2 \end{aligned} \qquad (4.13)$$

This gives

$$M1, M2: \begin{cases} \mathbf{E}_x^1(ik) + \mathbf{E}_z^1(\gamma_1) = 0 \\ \mathbf{E}_x^2(ik) + \mathbf{E}_z^2(-\gamma_2) = 0 \\ \mathbf{H}_x^1(ik) + \mathbf{H}_z^1(\gamma_1) = 0 \\ \mathbf{H}_x^2(ik) + \mathbf{H}_z^2(-\gamma_2) = 0 \end{cases}$$

$$M3: \begin{cases} -\mathbf{E}_y^1(\gamma_1) = \dfrac{i\omega}{c}\mathbf{H}_x^1 \\ -\mathbf{E}_y^2(-\gamma_2) = \dfrac{i\omega}{c}\mathbf{H}_x^2 \\ \mathbf{E}_x^1(\gamma_1) - \mathbf{E}_z^1(ik) = \dfrac{i\omega}{c}\mathbf{H}_y^1 \\ \mathbf{E}_x^2(-\gamma_2) - \mathbf{E}_z^2(ik) = \dfrac{i\omega}{c}\mathbf{H}_y^2 \\ \mathbf{E}_y^1(ik) = \dfrac{i\omega}{c}\mathbf{H}_z^1 \\ \mathbf{E}_y^2(ik) = \dfrac{i\omega}{c}\mathbf{H}_z^2 \end{cases}$$

$$M4: \begin{cases} -\mathbf{H}_y^1(\gamma_1) = -\dfrac{i\omega}{c}\varepsilon_1\mathbf{E}_x^1 \\ -\mathbf{H}_y^2(-\gamma_2) = -\dfrac{i\omega}{c}\varepsilon_2\mathbf{E}_x^2 \\ \mathbf{H}_x^1(\gamma_1) - \mathbf{H}_z^1(ik) = -\dfrac{i\omega}{c}\varepsilon_1\mathbf{E}_y^1 \\ \mathbf{H}_x^2(-\gamma_2) - \mathbf{H}_z^2(ik) = -\dfrac{i\omega}{c}\varepsilon_2\mathbf{E}_y^2 \\ \mathbf{H}_y^1(ik) = -\dfrac{i\omega}{c}\varepsilon_1\mathbf{E}_z^1 \\ \mathbf{H}_y^2(ik) = -\dfrac{i\omega}{c}\varepsilon_2\mathbf{E}_z^2 \end{cases}$$

$$M.C.: \begin{cases} \varepsilon_1\mathbf{E}_z^1 = \varepsilon_2\mathbf{E}_z^2 \\ \mathbf{E}_x^1 = \mathbf{E}_x^2 \\ \mathbf{E}_y^1 = \mathbf{E}_y^2 \\ \mathbf{H}_z^1 = \mathbf{H}_z^2 \\ \mathbf{H}_x^1 = \mathbf{H}_x^2 \\ \mathbf{H}_y^1 = \mathbf{H}_y^2 \end{cases} \qquad (4.14)$$

We find that $\mathbf{E}_x$, $\mathbf{E}_z$, and $\mathbf{H}_y$ couple to each other and $\mathbf{H}_x$, $\mathbf{H}_z$, and $\mathbf{E}_y$ couple to each other. We are interested in the first set of quantities. The second set does not produce any modes for non-magnetic materials. Let us start with the *1*-side. We have

$$\mathbf{E}_x^1(ik) + \mathbf{E}_z^1(\gamma_1) = 0$$

$$\mathbf{E}_x^1(\gamma_1) - \mathbf{E}_z^1(ik) = \frac{i\omega}{c}\mathbf{H}_y^1$$

$$-\mathbf{H}_y^1(\gamma_1) = -\frac{i\omega}{c}\varepsilon_1\mathbf{E}_x^1 \quad (4.15)$$

$$\mathbf{H}_y^1(ik) = -\frac{i\omega}{c}\varepsilon_1\mathbf{E}_z^1$$

This leads to

$$\mathbf{E}_x^1 = \frac{i\gamma_1}{k}C_1$$

$$\mathbf{E}_z^1 = C_1$$

$$\mathbf{H}_y^1 = -\frac{\varepsilon_1\omega}{ck}C_1 \quad (4.16)$$

$$(\gamma_1)^2 = k^2 - \varepsilon_1\left(\frac{\omega}{c}\right)^2$$

where C_1 is an arbitrary constant. Similarly, we get for the 2-side

$$\mathbf{E}_x^2 = -\frac{i\gamma_2}{k}C_2$$

$$\mathbf{E}_z^2 = C_2$$

$$\mathbf{H}_y^2 = -\frac{\varepsilon_2\omega}{ck}C_2 \quad (4.17)$$

$$(\gamma_2)^2 = k^2 - \varepsilon_2\left(\frac{\omega}{c}\right)^2$$

The matching conditions give

$$C_2 = -\frac{\gamma_1}{\gamma_2}C_1$$

$$\gamma_1\varepsilon_2 = -\gamma_2\varepsilon_1 \quad (4.18)$$

The last of these relations determines the modes. Here we can draw some general conclusions. If the γ functions are real valued, that is, the fields decay exponentially away from the interface (which is what we have assumed) the dielectric functions on the two sides of the interface *must have different sign*; one must be positive and one negative. The side with the negative dielectric function is called the *active side* and the other the *passive side*. Since in general all materials have several regions with negative dielectric function both sides are active in some frequency regions but never in the same region.

4.1 Modes at a Single Interface

Light has a finite velocity. This means that electromagnetic signals take a finite time to reach from one point in space to another. The effects from this are called retardation effects. These effects are often (but not always) of minor importance for the energies and forces we are concerned with here. If retardation effects are neglected the equations governing the normal modes are simplified. Thus in most cases we may use these simpler equations. To get the non-retarded results we just let the speed of light go to infinity. In the present problem we get that the γ functions both equal k and the equation determining the modes turns into $\varepsilon_1(\omega) + \varepsilon_2(\omega) = 0$, that is, the same result as in the simplified treatment of the previous section.

4.1.1 Metal-vacuum interface

We are interested in a metal surface in vacuum. In this case we have for small momenta

$$\varepsilon_1 \approx 1 - \frac{\omega_{pl}^2}{\omega^2} \, ; \, \varepsilon_2 = 1 \tag{4.19}$$

which gives for the surface plasmon energy, the energy of the surface mode obtained neglecting retardation effects,

$$\omega_s = \frac{\omega_{pl}}{\sqrt{2}} \tag{4.20}$$

The equation determining the modes when retardation effects are taken into account has the solution

$$\begin{aligned}\omega^2 &= \tfrac{1}{2}\left[\left(\omega_{pl}^2 + 2c^2k^2\right) - \sqrt{\omega_{pl}^4 + 4c^4k^4}\right] \\ &= \left[\left(\omega_s^2 + c^2k^2\right) - \sqrt{\omega_s^4 + c^4k^4}\right]\end{aligned} \tag{4.21}$$

This result gives the surface plasmon dispersion. It is an important result if we want to calculate surface energies of metals. The mode approaches the non-retarded result for large momentum. The deviation from this result only occurs for very small momentum, near the light dispersion curve. We will find later that the retardation effects are important for the force between objects at large separation only. The surface plasmon dispersion is shown in Figure 4.5.

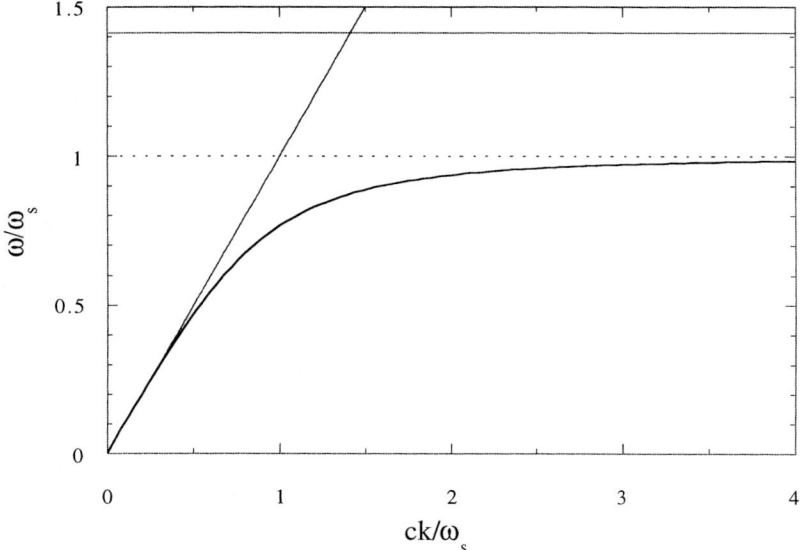

Figure 4.5: The surface mode for a metal-vacuum interface. The solid horizontal line is the plasmon energy and the dotted is the surface plasmon energy. The diagonal straight line is the light dispersion curve in vacuum and the curved solid curve is the surface plasmon dispersion.

We see that the modes are close to the non-retarded result for large momentum and are pushed below the light dispersion curve for small momentum. They are only present in the frequency range where the dielectric function of the metal is negative and they are confined to the region to the right of the light dispersion curve for the medium outside the metal surface.

The modes are by some named *surface polaritons* for general materials, or *surface plasmon polaritons* in this special case, since they contain a photon component; some reserve the name polariton to the solution near the light dispersion curve and call them surface plasmons for larger momentum; some call them surface plasmons in the whole momentum range.

It could be of interest to study what fields are involved in these modes. We have

4.1 Modes at a Single Interface

$$\mathbf{E}_x(k,z;\omega) = \begin{cases} iC\dfrac{\sqrt{k^2-\varepsilon(\omega/c)^2}}{k} e^{\sqrt{k^2-\varepsilon(\omega/c)^2}\,z} e^{i(kx-\omega t)} & ;\ z \leq 0 \\ iC\dfrac{\sqrt{k^2-\varepsilon(\omega/c)^2}}{k} e^{-\sqrt{k^2-(\omega/c)^2}\,z} e^{i(kx-\omega t)} & ;\ z \geq 0 \end{cases}$$

$$\mathbf{E}_z(k,z;\omega) = \begin{cases} C e^{\sqrt{k^2-\varepsilon(\omega/c)^2}\,z} e^{i(kx-\omega t)} & ;\ z \leq 0 \\ C\varepsilon\, e^{-\sqrt{k^2-(\omega/c)^2}\,z} e^{i(kx-\omega t)} & ;\ z \geq 0 \end{cases} \qquad (4.22)$$

$$\mathbf{H}_y(k,z;\omega) = \begin{cases} -C\dfrac{\varepsilon\omega}{ck} e^{\sqrt{k^2-\varepsilon(\omega/c)^2}\,z} e^{i(kx-\omega t)} & ;\ z \leq 0 \\ -C\dfrac{\varepsilon\omega}{ck} e^{-\sqrt{k^2-(\omega/c)^2}\,z} e^{i(kx-\omega t)} & ;\ z \geq 0 \end{cases}$$

We will make use of the plasmon dispersion to simplify the results. We have

$$\sqrt{k^2-\varepsilon(\omega/c)^2} = k\sqrt{\sqrt{1+\xi^4}+\xi^2}$$

$$\sqrt{k^2-(\omega/c)^2} = k\sqrt{\sqrt{1+\xi^4}-\xi^2}$$

$$\varepsilon = \frac{\sqrt{1+\xi^4}+\xi^2-1}{\sqrt{1+\xi^4}-\xi^2-1} = -\left(\sqrt{1+\xi^4}+\xi^2\right)\left[=-\frac{\sqrt{k^2-\varepsilon(\omega/c)^2}}{\sqrt{k^2-(\omega/c)^2}}\right] \qquad (4.23)$$

$$\frac{\varepsilon\omega}{ck} = \varepsilon\sqrt{-\sqrt{1+\xi^4}+\xi^2+1} = -\frac{\sqrt{1+\xi^4}+\xi^2-1}{\sqrt{-\sqrt{1+\xi^4}+\xi^2+1}}$$

where $\xi = \dfrac{\omega_s}{ck}$

Now, it might be interesting to see how the degree of longitudinality and transversality varies for the plasmon field along the dispersion and also how the energy flux varies. We see above that when the surface plasmon dispersion gets close to the dispersion curve for light the longitudinal electric field component goes to zero and the fields become purely transverse. To get the flux we study the time averaged energy densities:

$$u = \frac{1}{16\pi}\left(\mathbf{D}\cdot\mathbf{E}^* + \mathbf{H}\cdot\mathbf{B}^*\right) \qquad (4.24)$$

as functions of momentum; we give the two-dimensional density, we have integrated in the z-direction.

We find

$$u_x = \frac{1}{16\pi} \mathbf{D}_x \cdot \mathbf{E}_x^*$$
$$= \frac{1}{16\pi} C^2 \gamma_1^2 \frac{1}{k^2} \left[\frac{\varepsilon}{2\gamma_1} + \frac{1}{2\gamma_0} \right] = \frac{1}{16\pi} C^2 \gamma_1^2 \frac{1}{k^2} \left[-\frac{1}{2\gamma_0} + \frac{1}{2\gamma_0} \right] = 0 \ ; \quad (4.25)$$

$$u_z = \frac{1}{16\pi} \mathbf{D}_z \cdot \mathbf{E}_z^*$$
$$= \frac{1}{16\pi} C^2 \left[\frac{\varepsilon}{2\gamma_1} + \frac{\varepsilon^2}{2\gamma_0} \right] = \frac{1}{16\pi} C^2 \frac{1}{2\gamma_0} \left[\varepsilon^2 - 1 \right] = \frac{1}{16\pi} C^2 \frac{\xi^2}{k} \left(\sqrt{1+\xi^4} + \xi^2 \right)^{\frac{3}{2}} ;$$
$$(4.26)$$

$$u_y = \frac{1}{16\pi} \mathbf{H}_y \cdot \mathbf{B}_y^*$$
$$= \frac{1}{16\pi} C^2 \left(\frac{\varepsilon\omega}{ck} \right)^2 \left[\frac{1}{2\gamma_1} + \frac{1}{2\gamma_0} \right] = \frac{1}{16\pi} C^2 \left(\frac{\omega}{ck} \right)^2 \varepsilon \left[\frac{\varepsilon}{2\gamma_1} + \frac{\varepsilon}{2\gamma_0} \right]$$
$$= \frac{1}{16\pi} C^2 \frac{\varepsilon}{2\gamma_0} \left(\frac{\omega}{ck} \right)^2 (\varepsilon - 1) = \frac{1}{16\pi} C^2 \frac{1}{2\gamma_0} (\varepsilon+1)(\varepsilon-1) = \frac{1}{16\pi} C^2 \frac{1}{2\gamma_0} \left(\varepsilon^2 - 1 \right)$$
$$= u_z = \frac{1}{16\pi} C^2 \frac{\xi^2}{k} \left(\sqrt{1+\xi^4} + \xi^2 \right)^{\frac{3}{2}}$$
$$(4.27)$$

The complex valued Poynting vector is

$$\mathbf{S} = \frac{c}{8\pi} \mathbf{E} \times \mathbf{H}^* \qquad (4.28)$$

and its real part is the time-averaged energy flux. Thus,

$$\mathrm{Re}\,\mathbf{S}_z = \frac{c}{8\pi} \mathrm{Re}\,\mathbf{E}_x \mathbf{H}_y^* = \mathrm{Re}\,i\frac{c}{8\pi} C^2 \frac{\varepsilon}{2k}\left(\frac{\varepsilon\omega}{ck}\right) = 0 \qquad (4.29)$$

and

4.1 Modes at a Single Interface

$$\text{Re}\, S_x = -\frac{c}{8\pi}\text{Re}\, \mathbf{E}_z\mathbf{H}_y^* = \frac{c}{8\pi}C^2\left(\frac{\omega}{ck}\right)\left[\frac{\varepsilon}{2\gamma_1}+\frac{\varepsilon^2}{2\gamma_0}\right] = \frac{c}{8\pi}C^2\left(\frac{\omega}{ck}\right)\frac{1}{2\gamma_0}\left(\varepsilon^2-1\right)$$

$$= \frac{c}{16\pi}C^2\frac{\sqrt{-\sqrt{1+\xi^4}+\xi^2+1}\left(2\xi^2\right)\left(\sqrt{1+\xi^4}+\xi^2\right)}{k\sqrt{\sqrt{1+\xi^4}-\xi^2}}$$

$$= \frac{\omega}{k}\left(u_z+u_y\right) = v_{phase}\left(u_z+u_y\right)$$

(4.30)

We should note that there are electric field components in the x- and z-directions but none in the y-direction. In other words, the excitation has p-character. We have now derived what is needed for a calculation of how the flux varies along the dispersion curve of the surface plasmon polariton.

4.1.2 Semiconductor-vacuum interface

The expressions in Section 4.1 are valid for all types of material. In the present case we have

$$\varepsilon_1 = \varepsilon_\infty \frac{\omega^2-\omega_L^2}{\omega^2-\omega_T^2} \,;\, \varepsilon_2 = 1 \qquad (4.31)$$

The frequencies ω_L and ω_T are the long-wavelength–limiting frequencies of the longitudinal optical and transverse optical vibration modes of the crystal.

The important relations for the modes are the following:

$$\gamma_1\varepsilon_2 = -\gamma_2\varepsilon_1 \,;$$
$$(\gamma_1)^2 = k^2 - \varepsilon_1\left(\frac{\omega}{c}\right)^2 \qquad (4.32)$$
$$(\gamma_2)^2 = k^2 - \varepsilon_2\left(\frac{\omega}{c}\right)^2$$

Let us introduce the variables:

$$x = \left(\frac{\omega}{ck}\right)^2$$

$$a = \left(\frac{\omega_T}{ck}\right)^2 \tag{4.33}$$

$$b = \left(\frac{\omega_L}{ck}\right)^2$$

With these inserted we have

$$(\gamma_1)^2 = (\gamma_2)^2 \varepsilon_1^2$$

$$1 - \varepsilon_1 x = [1 - x]\varepsilon_1^2 \tag{4.34}$$

$$1 - \varepsilon_\infty \frac{x-b}{x-a} x = (\varepsilon_\infty)^2 [1-x]\left(\frac{x-b}{x-a}\right)^2$$

and

$$x^3\left((\varepsilon_\infty)^2 - \varepsilon_\infty\right) + x^2\left(1 + a\varepsilon_\infty + b\varepsilon_\infty - (\varepsilon_\infty)^2 - 2b(\varepsilon_\infty)^2\right)$$
$$+ x\left(-2a - ab\varepsilon_\infty + b^2(\varepsilon_\infty)^2 + 2b(\varepsilon_\infty)^2\right) + \left(a^2 - b^2(\varepsilon_\infty)^2\right) = 0 \tag{4.35}$$

After rearrangements we have

$$x^3 + x^2 \frac{\left(1 + a\varepsilon_\infty + b\varepsilon_\infty - (\varepsilon_\infty)^2 - 2b(\varepsilon_\infty)^2\right)}{\left((\varepsilon_\infty)^2 - \varepsilon_\infty\right)}$$
$$+ x \frac{\left(-2a - ab\varepsilon_\infty + b^2(\varepsilon_\infty)^2 + 2b(\varepsilon_\infty)^2\right)}{\left((\varepsilon_\infty)^2 - \varepsilon_\infty\right)} + \frac{\left(a^2 - b^2(\varepsilon_\infty)^2\right)}{\left((\varepsilon_\infty)^2 - \varepsilon_\infty\right)} = 0 \tag{4.36}$$

This is of the form:

$$x^3 + x^2 A + xB + C = 0 \tag{4.37}$$

where

4.1 Modes at a Single Interface

$$A = \frac{\left(1 + a\varepsilon_\infty + b\varepsilon_\infty - (\varepsilon_\infty)^2 - 2b(\varepsilon_\infty)^2\right)}{\left((\varepsilon_\infty)^2 - \varepsilon_\infty\right)}$$

$$B = \frac{\left(-2a - ab\varepsilon_\infty + b^2(\varepsilon_\infty)^2 + 2b(\varepsilon_\infty)^2\right)}{\left((\varepsilon_\infty)^2 - \varepsilon_\infty\right)} \tag{4.38}$$

$$C = \frac{\left(a^2 - b^2(\varepsilon_\infty)^2\right)}{\left((\varepsilon_\infty)^2 - \varepsilon_\infty\right)}$$

Let

$$\begin{aligned} q &= \tfrac{1}{3}B - \tfrac{1}{9}A^2 \\ r &= \tfrac{1}{6}(BA - 3C) - \tfrac{1}{27}A^3 \end{aligned} \tag{4.39}$$

and

$$\begin{aligned} s_1 &= \left[r + \left(q^3 + r^2\right)^{\frac{1}{2}}\right]^{\frac{1}{3}} \\ s_2 &= \left[r - \left(q^3 + r^2\right)^{\frac{1}{2}}\right]^{\frac{1}{3}} \end{aligned} \tag{4.40}$$

In terms of these functions the three solutions to Equation (4.36) are

$$\begin{aligned} z_1 &= (s_1 + s_2) - \frac{A}{3} \\ z_2 &= -\tfrac{1}{2}(s_1 + s_2) - \frac{A}{3} + \frac{i\sqrt{3}}{2}(s_1 - s_2) \\ z_3 &= -\tfrac{1}{2}(s_1 + s_2) - \frac{A}{3} - \frac{i\sqrt{3}}{2}(s_1 - s_2) \end{aligned} \tag{4.41}$$

The solution for CdS is shown in Figure 4.6. Just one of the solutions is real. The quantities γ_1, γ_2, and $-\varepsilon_1$ must all be positive. They are so between the ω_L- and ω_T- dispersion curves, but only in the region to the right of the $\omega = ck$ curve.

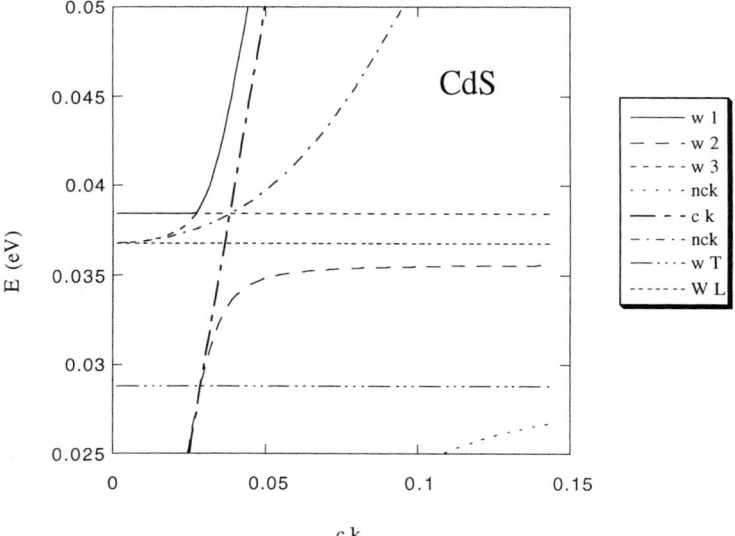

Figure 4.6: Solutions to equations generating the surface modes for CdS together with the light dispersion curves inside and outside the semiconductor.

The solution (dashed curve) approaches the surface phonon energy for large wave numbers. This is obtained from the solution to the equation: $\varepsilon_1 + 1 = 0$, and is given by

$$\omega_{sph} = \sqrt{\frac{\omega_T^2 + \varepsilon_\infty \omega_L^2}{1 + \varepsilon_\infty}} = \omega_T \sqrt{\frac{1 + \varepsilon_0}{1 + \varepsilon_\infty}} \tag{4.42}$$

From the numerical results we find that one solution is a constant. If we find that solution we may divide it out of the equation and thereby obtain a second order equation instead of a third order. Since the solution is a constant the dielectric function is also constant for this solution. From

$$1 - \varepsilon_1 x = [1 - x]\varepsilon_1^2 \tag{4.43}$$

we immediately find that $\varepsilon_1 \equiv 1$ solves the equation. This corresponds to

$$x = \frac{\varepsilon_\infty b - a}{\varepsilon_\infty - 1} \tag{4.44}$$

We may divide Equation (4.44) by $(\varepsilon_1 - 1)$. We find

4.1 Modes at a Single Interface

$$\varepsilon_1 = \frac{1}{x-1} \tag{4.45}$$

or

$$\varepsilon_\infty \left(\frac{x-b}{x-a} \right) = \frac{1}{x-1} \tag{4.46}$$

with the solutions

$$x = \frac{1}{2}\left[\left(b+1+\frac{1}{\varepsilon_\infty}\right) \pm \sqrt{\left(b+1+\frac{1}{\varepsilon_\infty}\right)^2 - 4\left(b+\frac{a}{\varepsilon_\infty}\right)} \right] \tag{4.47}$$

or [1]

$$\omega^2(q) = \frac{1}{2}\left[\left(\omega_q^2 + \omega_L^2\right) \pm \sqrt{\left(\omega_q^2 + \omega_L^2\right)^2 - 4\omega_q^2\omega_{sph}^2} \right] \tag{4.48}$$

where

$$\omega_q^2 = \frac{\varepsilon_\infty + 1}{\varepsilon_\infty}(ck)^2$$

$$\omega_L^2 = \frac{\varepsilon_0}{\varepsilon_\infty}\omega_T^2 \tag{4.49}$$

$$\omega_{sph}^2 = \frac{\varepsilon_0 + 1}{\varepsilon_\infty + 1}\omega_T^2$$

The result is shown in Figure 4.7

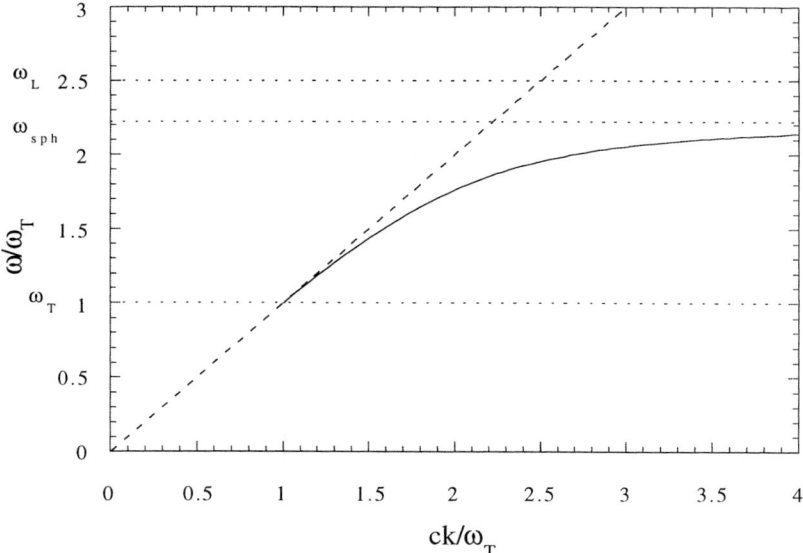

Figure 4.7: Surface phonon mode in a polar semiconductor.

We find that the surface phonon mode only exists in the frequency range between the transverse and longitudinal phonon modes. It approaches the non-retarded result for high momentum and is pushed down below the light dispersion curve for small momentum. The modes are called surface phonons or surface phonon polaritons.

4.2 Modes in slab geometry

Slab geometry is shown in Figure 4.8. There are two parallel interfaces separating material *1*, in the center, from material *2*, on both sides. The slab geometry can be used in calculating the energy of a thin film; the energy of a gap inside a material; the force between two half spaces of the same material, in vacuum or in a medium; the energy of cohesion; the strength of gluing, or soldering, two pieces of the same material together.

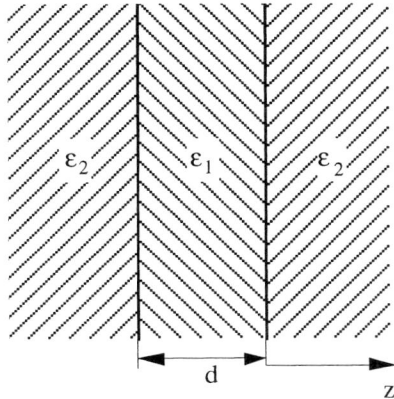

Figure 4.8:   Slab geometry.

In this geometry we make the following ansatz for the fields in the three regions:

$$\mathbf{E} = \begin{cases} \left(\mathbf{E}^{BR} e^{\gamma_1 z} + \mathbf{E}^{BL} e^{-\gamma_1 (z+d)}\right) e^{i(kx-\omega t)} \; ; \; -d \leq z \leq 0 \\ \mathbf{E}^{R} e^{-\gamma_2 z} e^{i(kx-\omega t)} \; ; \; z \geq 0 \\ \mathbf{E}^{L} e^{+\gamma_2 (z+d)} e^{i(kx-\omega t)} \; ; \; z \leq -d \end{cases} \qquad (4.50)$$

$$\mathbf{H} = \begin{cases} \left(\mathbf{H}^{BR} e^{\gamma_1 z} + \mathbf{H}^{BL} e^{-\gamma_1 (z+d)}\right) e^{i(kx-\omega t)} \; ; \; -d \leq z \leq 0 \\ \mathbf{H}^{R} e^{-\gamma_2 z} e^{i(kx-\omega t)} \; ; \; z \geq 0 \\ \mathbf{H}^{L} e^{+\gamma_2 (z+d)} e^{i(kx-\omega t)} \; ; \; z \leq -d \end{cases} \qquad (4.51)$$

We insert these relations in the MEs and use the matching conditions at the interface:

$$\begin{aligned}
\mathbf{D}_\perp^{BR} + \mathbf{D}_\perp^{BL} e^{-\gamma_1 d} &= \mathbf{D}_\perp^{R} \\
\mathbf{D}_\perp^{BL} + \mathbf{D}_\perp^{BR} e^{-\gamma_1 d} &= \mathbf{D}_\perp^{L} \\
\mathbf{B}_\perp^{BR} + \mathbf{B}_\perp^{BL} e^{-\gamma_1 d} &= \mathbf{B}_\perp^{R} \\
\mathbf{B}_\perp^{BL} + \mathbf{B}_\perp^{BR} e^{-\gamma_1 d} &= \mathbf{B}_\perp^{L} \\
\mathbf{E}_{//}^{BR} + \mathbf{E}_{//}^{BL} e^{-\gamma_1 d} &= \mathbf{E}_{//}^{R} \\
\mathbf{E}_{//}^{BL} + \mathbf{E}_{//}^{BR} e^{-\gamma_1 d} &= \mathbf{E}_{//}^{L} \\
\mathbf{H}_{//}^{BR} + \mathbf{H}_{//}^{BL} e^{-\gamma_1 d} &= \mathbf{H}_{//}^{R} \\
\mathbf{H}_{//}^{BL} + \mathbf{H}_{//}^{BR} e^{-\gamma_1 d} &= \mathbf{H}_{//}^{L}
\end{aligned} \qquad (4.52)$$

This gives from the first two MEs in Equation (4.10):

$$M1, M2 : \begin{cases} \mathbf{E}_x^{BL}(ik) - \mathbf{E}_z^{BL}(\gamma_1) = 0 \\ \mathbf{E}_x^{BR}(ik) + \mathbf{E}_z^{BR}(\gamma_1) = 0 \\ \mathbf{E}_x^{R}(ik) + \mathbf{E}_z^{R}(-\gamma_2) = 0 \\ \mathbf{E}_x^{L}(ik) + \mathbf{E}_z^{L}(\gamma_2) = 0 \\ \mathbf{H}_x^{BL}(ik) - \mathbf{H}_z^{BL}(\gamma_1) = 0 \\ \mathbf{H}_x^{BR}(ik) + \mathbf{H}_z^{BR}(\gamma_1) = 0 \\ \mathbf{H}_x^{R}(ik) + \mathbf{H}_z^{R}(-\gamma_2) = 0 \\ \mathbf{H}_x^{L}(ik) + \mathbf{H}_z^{L}(\gamma_2) = 0 \end{cases} \qquad (4.53)$$

4.2 Modes in Slab Geometry

from the third:

$$M3: \begin{cases} -\mathbf{E}_y^{BR}(\gamma_1) = \dfrac{i\omega}{c}\mathbf{H}_x^{BR} \\ -\mathbf{E}_y^{BL}(-\gamma_1) = \dfrac{i\omega}{c}\mathbf{H}_x^{BL} \\ -\mathbf{E}_y^{R}(-\gamma_2) = \dfrac{i\omega}{c}\mathbf{H}_x^{R} \\ -\mathbf{E}_y^{L}(\gamma_2) = \dfrac{i\omega}{c}\mathbf{H}_x^{L} \\ \mathbf{E}_x^{BR}(\gamma_1) - \mathbf{E}_z^{BR}(ik) = \dfrac{i\omega}{c}\mathbf{H}_y^{BR} \\ \mathbf{E}_x^{BL}(-\gamma_1) - \mathbf{E}_z^{BL}(ik) = \dfrac{i\omega}{c}\mathbf{H}_y^{BL} \\ \mathbf{E}_x^{R}(-\gamma_2) - \mathbf{E}_z^{R}(ik) = \dfrac{i\omega}{c}\mathbf{H}_y^{R} \\ \mathbf{E}_x^{L}(\gamma_2) - \mathbf{E}_z^{L}(ik) = \dfrac{i\omega}{c}\mathbf{H}_y^{L} \\ \mathbf{E}_y^{BR}(ik) = \dfrac{i\omega}{c}\mathbf{H}_z^{BR} \\ \mathbf{E}_y^{BL}(ik) = \dfrac{i\omega}{c}\mathbf{H}_z^{BL} \\ \mathbf{E}_y^{R}(ik) = \dfrac{i\omega}{c}\mathbf{H}_z^{R} \\ \mathbf{E}_y^{L}(ik) = \dfrac{i\omega}{c}\mathbf{H}_z^{L} \end{cases} \quad (4.54)$$

from the fourth:

$$M4: \begin{cases} -\mathbf{H}_y^{BR}(\gamma_1) = -\frac{i\omega}{c}\varepsilon_1 \mathbf{E}_x^{BR} \\ -\mathbf{H}_y^{BL}(-\gamma_1) = -\frac{i\omega}{c}\varepsilon_1 \mathbf{E}_x^{BL} \\ -\mathbf{H}_y^R(-\gamma_2) = -\frac{i\omega}{c}\varepsilon_2 \mathbf{E}_x^R \\ -\mathbf{H}_y^L(\gamma_2) = -\frac{i\omega}{c}\varepsilon_2 \mathbf{E}_x^L \\ \mathbf{H}_x^{BR}(\gamma_1) - \mathbf{H}_z^{BR}(ik) = -\frac{i\omega}{c}\varepsilon_1 \mathbf{E}_y^{BR} \\ \mathbf{H}_x^{BL}(-\gamma_1) - \mathbf{H}_z^{bL}(ik) = -\frac{i\omega}{c}\varepsilon_1 \mathbf{E}_y^{BL} \\ \mathbf{H}_x^R(-\gamma_2) - \mathbf{H}_z^R(ik) = -\frac{i\omega}{c}\varepsilon_2 \mathbf{E}_y^R \\ \mathbf{H}_x^L(\gamma_2) - \mathbf{H}_z^L(ik) = -\frac{i\omega}{c}\varepsilon_2 \mathbf{E}_y^L \\ \mathbf{H}_y^{BR}(ik) = -\frac{i\omega}{c}\varepsilon_1 \mathbf{E}_z^{BR} \\ \mathbf{H}_y^{BL}(ik) = -\frac{i\omega}{c}\varepsilon_1 \mathbf{E}_z^{BL} \\ \mathbf{H}_y^R(ik) = -\frac{i\omega}{c}\varepsilon_2 \mathbf{E}_z^R \\ \mathbf{H}_y^L(ik) = -\frac{i\omega}{c}\varepsilon_2 \mathbf{E}_z^L \end{cases} \qquad (4.55)$$

4.2 Modes in Slab Geometry

and from the boundary conditions:

$$\text{B.C.}: \begin{cases} \varepsilon_1\left(\mathbf{E}_z^{BR} + e^{-\gamma_1 d}\mathbf{E}_z^{BL}\right) = \varepsilon_2 \mathbf{E}_z^R \\ \varepsilon_1\left(\mathbf{E}_z^{BL} + e^{-\gamma_1 d}\mathbf{E}_z^{BR}\right) = \varepsilon_2 \mathbf{E}_z^L \\ \left(\mathbf{E}_x^{BR} + e^{-\gamma_1 d}\mathbf{E}_x^{BL}\right) = \mathbf{E}_x^R \\ \left(\mathbf{E}_x^{BL} + e^{-\gamma_1 d}\mathbf{E}_x^{BR}\right) = \mathbf{E}_x^L \\ \left(\mathbf{E}_y^{BR} + e^{-\gamma_1 d}\mathbf{E}_y^{BL}\right) = \mathbf{E}_y^R \\ \left(\mathbf{E}_y^{BL} + e^{-\gamma_1 d}\mathbf{E}_y^{BR}\right) = \mathbf{E}_y^L \\ \left(\mathbf{H}_z^{BR} + e^{-\gamma_1 d}\mathbf{H}_z^{BL}\right) = \mathbf{H}_z^R \\ \left(\mathbf{H}_z^{BL} + e^{-\gamma_1 d}\mathbf{H}_z^{BR}\right) = \mathbf{H}_z^L \\ \left(\mathbf{H}_x^{BR} + e^{-\gamma_1 d}\mathbf{H}_x^{BL}\right) = \mathbf{H}_x^R \\ \left(\mathbf{H}_x^{BL} + e^{-\gamma_1 d}\mathbf{H}_x^{BR}\right) = \mathbf{H}_x^L \\ \left(\mathbf{H}_y^{BR} + e^{-\gamma_1 d}\mathbf{H}_y^{BL}\right) = \mathbf{H}_y^R \\ \left(\mathbf{H}_y^{BL} + e^{-\gamma_1 d}\mathbf{H}_y^{BR}\right) = \mathbf{H}_y^L \end{cases} \quad (4.56)$$

As before the equations decouple into two sets of equations. The ones we are mostly interested in are the so-called **TM** modes and are obtained from the equations that couple $\mathbf{E}_x$, $\mathbf{E}_z$, and $\mathbf{H}_y$ to each other. The equations that couple the three other vector fields have the so-called **TE** modes as solutions. These last modes do not survive in the non-retarded treatment if the materials are non-magnetic; thus they do not contribute to the van der Waals interaction. They are important for the Casimir force, though. They are of a different type, not exponentially decaying in region *1* but of oscillating, standing-wave type. The two mode types are also known as wave-guide modes. We need to address them both. The **TM** modes derive from the longitudinal bulk modes in the material on the active side of the interfaces and so-called *p-polarized* transverse bulk modes. The TE modes derive from the so-called *s-polarized* transverse bulk modes. The terms *p-* and *s-* polarized waves refer to the direction of the **E** vector. The first type has the **E** vector in the plane of incidence and the second in the direction perpendicular to this plane.

We start with the equations leading to **TM** modes:

$$(\gamma_1)^2 = k^2 - \varepsilon_1\left(\frac{\omega}{c}\right)^2$$

$$(\gamma_2)^2 = k^2 - \varepsilon_2\left(\frac{\omega}{c}\right)^2$$

$$\mathbf{E}_x^{BL} = \mathbf{E}_z^{BL} \frac{\gamma_1}{ik}$$

$$\mathbf{E}_x^{BR} = -\mathbf{E}_z^{BR} \frac{\gamma_1}{ik}$$

$$\mathbf{E}_x^{R} = \mathbf{E}_z^{R} \frac{\gamma_2}{ik}$$

$$\mathbf{E}_x^{L} = -\mathbf{E}_z^{L} \frac{\gamma_2}{ik} \tag{4.57}$$

$$\mathbf{H}_y^{BR} = -\mathbf{E}_z^{BR} \frac{\varepsilon_1\left(\frac{\omega}{c}\right)}{k}$$

$$\mathbf{H}_y^{BL} = -\mathbf{E}_z^{BL} \frac{\varepsilon_1\left(\frac{\omega}{c}\right)}{k}$$

$$\mathbf{H}_y^{R} = -\mathbf{E}_z^{R} \frac{\varepsilon_2\left(\frac{\omega}{c}\right)}{k}$$

$$\mathbf{H}_y^{L} = -\mathbf{E}_z^{L} \frac{\varepsilon_2\left(\frac{\omega}{c}\right)}{k}$$

The boundary conditions give

$$\mathbf{E}_z^{BR} + e^{-\gamma_1 d}\mathbf{E}_z^{BL} = \mathbf{E}_z^{R} \frac{\varepsilon_2}{\varepsilon_1}$$

$$\mathbf{E}_z^{BL} + e^{-\gamma_1 d}\mathbf{E}_z^{BR} = \mathbf{E}_z^{L} \frac{\varepsilon_2}{\varepsilon_1}$$

$$-\mathbf{E}_z^{BR} + e^{-\gamma_1 d}\mathbf{E}_z^{BL} = \mathbf{E}_z^{R} \frac{\gamma_2}{\gamma_1}$$

$$-\mathbf{E}_z^{BL} + e^{-\gamma_1 d}\mathbf{E}_z^{BR} = \mathbf{E}_z^{L} \frac{\gamma_2}{\gamma_1} \tag{4.58}$$

$$\mathbf{E}_z^{BR} + e^{-\gamma_1 d}\mathbf{E}_z^{BL} = \mathbf{E}_z^{R} \frac{\varepsilon_2}{\varepsilon_1}$$

$$\mathbf{E}_z^{BL} + e^{-\gamma_1 d}\mathbf{E}_z^{BR} = \mathbf{E}_z^{L} \frac{\varepsilon_2}{\varepsilon_1}$$

Two of them are redundant. We are left with

4.2 Modes in Slab Geometry

$$\begin{aligned}
\mathbf{E}_z^{BR} + e^{-\gamma_1 d}\mathbf{E}_z^{BL} &= \mathbf{E}_z^{R}\frac{\varepsilon_2}{\varepsilon_1} \\
\mathbf{E}_z^{BL} + e^{-\gamma_1 d}\mathbf{E}_z^{BR} &= \mathbf{E}_z^{L}\frac{\varepsilon_2}{\varepsilon_1} \\
-\mathbf{E}_z^{BR} + e^{-\gamma_1 d}\mathbf{E}_z^{BL} &= \mathbf{E}_z^{R}\frac{\gamma_2}{\gamma_1} \\
-\mathbf{E}_z^{BL} + e^{-\gamma_1 d}\mathbf{E}_z^{BR} &= \mathbf{E}_z^{L}\frac{\gamma_2}{\gamma_1}
\end{aligned} \tag{4.59}$$

This leads to

$$\begin{aligned}
\mathbf{E}_z^{R} &= C \\
2\mathbf{E}_z^{BR} &= C\left(\frac{\varepsilon_2}{\varepsilon_1} - \frac{\gamma_2}{\gamma_1}\right) \\
2e^{-\gamma_1 d}\mathbf{E}_z^{BR} &= \mathbf{E}_z^{L}\frac{\varepsilon_2}{\varepsilon_1} + \mathbf{E}_z^{L}\frac{\gamma_2}{\gamma_1} \\
-\mathbf{E}_z^{BR} + e^{-\gamma_1 d}\mathbf{E}_z^{BL} &= C\frac{\gamma_2}{\gamma_1} \\
-\mathbf{E}_z^{BL} + e^{-\gamma_1 d}\mathbf{E}_z^{BR} &= \mathbf{E}_z^{L}\frac{\gamma_2}{\gamma_1}
\end{aligned} \tag{4.60}$$

and

$$\begin{aligned}
\mathbf{E}_z^{BR} &= C\frac{1}{2}\left(\frac{\varepsilon_2}{\varepsilon_1} - \frac{\gamma_2}{\gamma_1}\right) \\
\mathbf{E}_z^{L} &= e^{-\gamma_1 d}C\frac{\left(\dfrac{\varepsilon_2}{\varepsilon_1} - \dfrac{\gamma_2}{\gamma_1}\right)}{\left(\dfrac{\varepsilon_2}{\varepsilon_1} + \dfrac{\gamma_2}{\gamma_1}\right)} \\
\mathbf{E}_z^{BL} &= e^{+\gamma_1 d}C\left[\frac{\gamma_2}{\gamma_1} + \frac{1}{2}\left(\frac{\varepsilon_2}{\varepsilon_1} - \frac{\gamma_2}{\gamma_1}\right)\right] \\
\left(\frac{\varepsilon_2}{\varepsilon_1} + \frac{\gamma_2}{\gamma_1}\right)^2 - e^{-2\gamma_1 d}&\left(\frac{\varepsilon_2}{\varepsilon_1} - \frac{\gamma_2}{\gamma_1}\right)^2 = 0
\end{aligned} \tag{4.61}$$

where the last equation determines the dispersion of the modes. When retardation is neglected this equation is reduced into

$$\frac{\varepsilon_1(\omega) - \varepsilon_2(\omega)}{\varepsilon_1(\omega) + \varepsilon_2(\omega)} = \pm e^{kd} \tag{4.62}$$

The **TE** modes are derived in an analogous way. The modes are determined from the equation:

$$(\gamma_2 + \gamma_1)^2 - e^{-2\gamma_1 d}(\gamma_2 - \gamma_1)^2 = 0 \tag{4.63}$$

where

$$\begin{aligned}(\gamma_1)^2 &= k^2 - \varepsilon_1(\omega/c)^2 \\ (\gamma_2)^2 &= k^2 - \varepsilon_2(\omega/c)^2\end{aligned} \tag{4.64}$$

4.2.1 Metal slab in vacuum

For a metal slab in vacuum we have

$$\begin{aligned}&\left(\frac{x}{x-y} + \frac{\sqrt{1-x}}{\sqrt{1-(x-y)}}\right)^2 - e^{-2\sqrt{1-(x-y)}kd}\left(\frac{x}{x-y} - \frac{\sqrt{1-x}}{\sqrt{1-(x-y)}}\right)^2 = 0 \\ &\varepsilon_1 = 1 - \frac{\omega_{pl}^2}{\omega^2} = 1 - \frac{y}{x} \\ &\varepsilon_2 = 1 \\ &x = \left(\frac{\omega}{ck}\right)^2 \\ &y = \left(\frac{\omega_{pl}}{ck}\right)^2\end{aligned} \tag{4.65}$$

Let us for the moment neglect retardation effects. Then we have

$$\begin{aligned}&\left(\frac{x}{x-y} + 1\right)^2 - e^{-2kd}\left(\frac{x}{x-y} - 1\right)^2 = 0 \\ &\left(\frac{x}{x-y} + 1\right) \pm e^{-kd}\left(\frac{x}{x-y} - 1\right) = 0 \\ &x = \tfrac{1}{2}y\left(1 \pm e^{-kd}\right) \\ &\omega = \omega_s \sqrt{1 \pm e^{-kd}}\end{aligned} \tag{4.66}$$

and

4.2 Modes in Slab Geometry

$$\mathbf{E}_z^{BR} = C\frac{1}{2}\left(\frac{x}{x-y}-1\right) = C\frac{1}{2}\left(\frac{y}{x-y}\right) = C\frac{1}{2}\left(\frac{2\omega_s^2}{\omega^2-2\omega_s^2}\right)$$

$$\mathbf{E}_z^{BL} = e^{+kd}C\left[1+\frac{1}{2}\left(\frac{x}{x-y}-1\right)\right] = e^{+kd}C\left[\frac{1}{2}\left(\frac{x}{x-y}+1\right)\right]$$

$$= e^{+kd}C\left[\frac{1}{2}\left(\frac{\omega^2}{\omega^2-2\omega_s^2}+1\right)\right] \quad (4.67)$$

$$\mathbf{E}_z^R = C$$

$$\mathbf{E}_z^L = e^{-kd}C\frac{\left(\frac{x}{x-y}-1\right)}{\left(\frac{x}{x-y}+1\right)} = e^{-kd}C\frac{\left(\frac{\omega^2}{\omega^2-2\omega_s^2}-1\right)}{\left(\frac{\omega^2}{\omega^2-2\omega_s^2}+1\right)}$$

Let us now insert the two solutions found in Equation (4.66) into Equations (4.67) to see which is odd and which is even. The result is

$$\mathbf{E}_z^{BR} = C\frac{1}{2}\left(\frac{2\omega_s^2}{\omega^2-2\omega_s^2}\right) = C\frac{1}{2}\left(\frac{1\pm e^{-kd}}{-1\pm e^{-kd}}-1\right) = C\left(\frac{1}{-1\pm e^{-kd}}\right)$$

$$\mathbf{E}_z^{BL} = e^{+kd}C\left[\frac{1}{2}\left(\frac{\omega^2}{\omega^2-2\omega_s^2}+1\right)\right] = \pm C\left[\left(\frac{1}{-1\pm e^{-kd}}\right)\right]$$

$$\mathbf{E}_z^R = C \quad (4.68)$$

$$\mathbf{E}_z^L = e^{-kd}C\frac{\left(\frac{\omega^2}{\omega^2-2\omega_s^2}-1\right)}{\left(\frac{\omega^2}{\omega^2-2\omega_s^2}+1\right)} = e^{-kd}C\frac{\left(\frac{1\pm e^{-kd}}{-1\pm e^{-kd}}-1\right)}{\left(\frac{1\pm e^{-kd}}{-1\pm e^{-kd}}+1\right)} = \pm C$$

Now, let d be very small

$$(\omega_+)^2 = (\omega_s)^2\left(1+e^{-kd}\right) \approx (\omega_{pl})^2$$

$$(\omega_-)^2 = (\omega_s)^2\left(1-e^{-kd}\right) \approx (\omega_s)^2 kd = \frac{2\pi(nd)e^2}{m}k \quad (4.69)$$

The low-frequency mode (-) approaches the 2D-plasmon and is the anti-symmetric solution, while the high-frequency mode (+) has the energy of a plasmon and is the symmetric solution. Note that the symmetry considerations are made for the fields. The 2D-plasmon is the plasmon in a strictly two-dimensional metallic system. For a system like that the plasmon dispersion has a square-root-of-momentum dependence. We will discuss this type of system in Section 4.5. If we study the distribution of polarization charge the low-frequency mode is symmetric and the high-frequency mode is anti-symmetric. For d very large both solutions approach the surface plasmon dispersion.

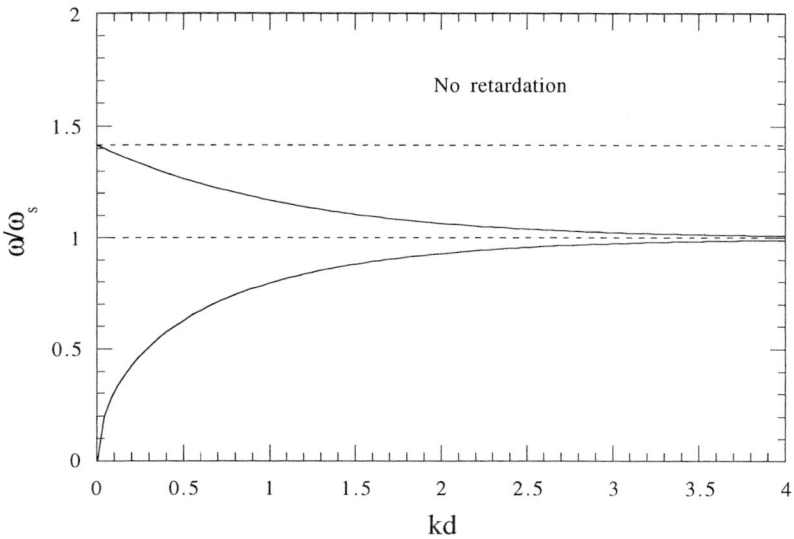

Figure 4.9: The two surface modes for a metal slab in vacuum: retardation neglected.

The results in Figure 4.9 are valid for a general slab thickness. If we let the thickness go to infinity the modes decouple and there are two independent modes, one on each face of the slab. The dispersion is for both constant and equal to the surface plasmon energy. When we go in the other extreme and have a strictly 2D metallic system, will we have two plasmon modes then? The answer is no! The reason is that in this case the 3D carrier density goes to infinity when d goes to zero which means that the upper branch is pushed up to infinity and plays no role. It would be interesting to see if this argument still holds when retardation effects are included.

The results with retardation is shown in the Figures 4.10 and 4.11 for two different thicknesses. The retardation effects push the curves below the light dispersion curve. We see that if we compress the slab, making it infinitely thin, one mode turns into the mode for the strictly 2D metallic system and the other turns into a p-polarized light wave in the vacuum surrounding the slab.

That we should obtain the strictly two-dimensional result when compressing the slab towards zero thickness is not obvious. If we do that in a real system quantum mechanical effects appear; effects that we have not taken into consideration here. The kinetic energy of the carriers in the direction perpendicular to the slab becomes quantized. This size quantization is taken into account in our treatment of quantum wells in Section 4.5.

4.2 Modes in Slab Geometry

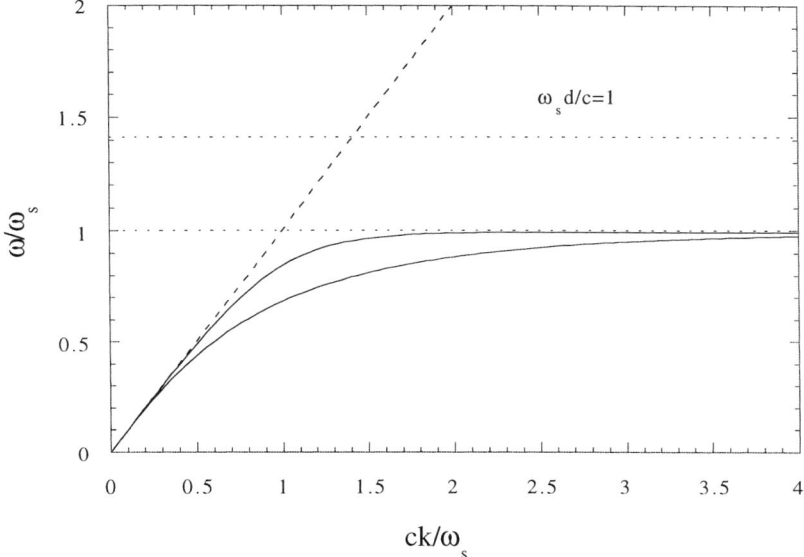

Figure 4.10: The two surface modes for a metal slab in vacuum: retardation included.

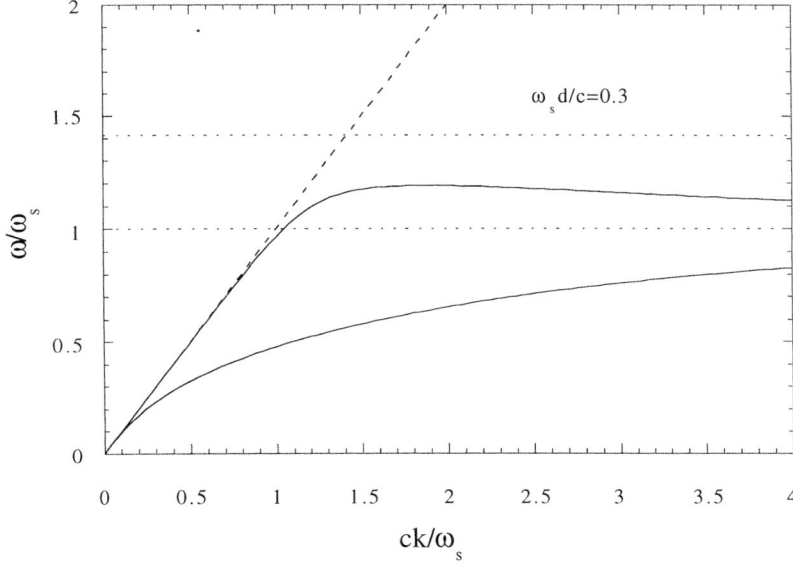

Figure 4.11: The same as Figure 4.10 but for a thinner slab.

4.2.2 Semiconductor slab in vacuum

For a semiconductor slab in vacuum we have

$$\left(\frac{x-a}{\varepsilon_\infty(x-b)} + \frac{\sqrt{1-x}}{\sqrt{1-\varepsilon_\infty\frac{x-b}{x-a}x}}\right)^2 - e^{-2\sqrt{1-\varepsilon_\infty\frac{x-b}{x-a}}xkd}\left(\frac{x-a}{\varepsilon_\infty(x-b)} - \frac{\sqrt{1-x}}{\sqrt{1-\varepsilon_\infty\frac{x-b}{x-a}x}}\right)^2 = 0$$

$$\varepsilon_1 = \varepsilon_\infty\frac{x-b}{x-a} \;;\; \varepsilon_2 = 1 \;;\; x = \left(\frac{\omega}{ck}\right)^2 \;;\; a = \left(\frac{\omega_T}{ck}\right)^2 \;;\; b = \left(\frac{\omega_L}{ck}\right)^2$$

(4.70)

Let us for the moment neglect retardation effects. Then we have

$$\left(\frac{x-a}{\varepsilon_\infty(x-b)} + 1\right)^2 - e^{-2kd}\left(\frac{x-a}{\varepsilon_\infty(x-b)} - 1\right)^2 = 0$$

$$\left(\frac{x-a}{\varepsilon_\infty(x-b)} + 1\right) = \pm e^{-kd}\left(\frac{x-a}{\varepsilon_\infty(x-b)} - 1\right)$$

$$x = \frac{a + b\varepsilon_\infty \pm e^{-kd}(b\varepsilon_\infty - a)}{1 + \varepsilon_\infty \pm e^{-kd}(\varepsilon_\infty - 1)}$$

$$\omega_\pm^2 = \frac{\omega_T^2 + \omega_L^2\varepsilon_\infty \pm e^{-kd}(\omega_L^2\varepsilon_\infty - \omega_T^2)}{1 + \varepsilon_\infty \pm e^{-kd}(\varepsilon_\infty - 1)}$$

(4.71)

and

$$\mathbf{E}_z^{BR} = C\frac{1}{2}\left(\frac{x-a}{\varepsilon_\infty(x-b)} - 1\right)$$

$$\mathbf{E}_z^{BL} = e^{+kd}C\frac{1}{2}\left[\frac{x-a}{\varepsilon_\infty(x-b)} + 1\right]$$

$$\mathbf{E}_z^L = e^{-kd}C\frac{\left(\frac{x-a}{\varepsilon_\infty(x-b)} - 1\right)}{\left(\frac{x-a}{\varepsilon_\infty(x-b)} + 1\right)}$$

$$\mathbf{E}_z^R = C$$

(4.72)

Let us now insert the two solutions found in Equation (4.71) into Equations (4.72) to see which is odd and which is even. We have

4.2 Modes in Slab Geometry

$$x = \frac{a + b\varepsilon_\infty \pm e^{-kd}(b\varepsilon_\infty - a)}{1 + \varepsilon_\infty \pm e^{-kd}(\varepsilon_\infty - 1)}$$

$$x - a = \frac{(b-a)\varepsilon_\infty\left(1 \pm e^{-kd}\right)}{1 + \varepsilon_\infty \pm e^{-kd}(\varepsilon_\infty - 1)} \quad (4.73)$$

$$x - b = \frac{(a-b)\left(1 \mp e^{-kd}\right)}{1 + \varepsilon_\infty \pm e^{-kd}(\varepsilon_\infty - 1)}$$

$$\frac{x-a}{\varepsilon_\infty(x-b)} = -\left(1 \pm e^{-kd}\right)\Big/\left(1 \mp e^{-kd}\right)$$

and this results in

$$\mathbf{E}_z^{BR} = C\tfrac{1}{2}\left[-\left(1 \pm e^{-kd}\right)\Big/\left(1 \mp e^{-kd}\right) - 1\right] = -C\Big/\left(1 \mp e^{-kd}\right)$$

$$\mathbf{E}_z^{BL} = \mp C\Big/\left(1 \mp e^{-kd}\right) \quad (4.74)$$

$$\mathbf{E}_z^L = \pm C$$

$$\mathbf{E}_z^R = C$$

In the limit kd small we have

$$\omega_\pm = \begin{cases} \omega_L \\ \omega_T\sqrt{1 + kd(\varepsilon_0 - \varepsilon_\infty)} \end{cases} \quad (4.75)$$

and in the high kd limit both modes approach the surface phonon energy:

$$\omega_\pm = \sqrt{\frac{\omega_T^2 + \omega_L^2 \varepsilon_\infty}{1 + \varepsilon_\infty}} = \omega_{sph} \quad (4.76)$$

The low-frequency mode (-) approaches the 2D-phonon and is the anti-symmetric solution, while the high-frequency mode (+) has the energy of a longitudinal phonon and is the symmetric solution. Note that the symmetry considerations are made for the fields. If we study the distribution of polarization charge the low-frequency mode is symmetric and the high-frequency mode is anti-symmetric. For d very big both solutions approach the surface phonon dispersion. In Figure 4.12 we show the results neglecting retardation and in Figure 4.13 the results including retardation.

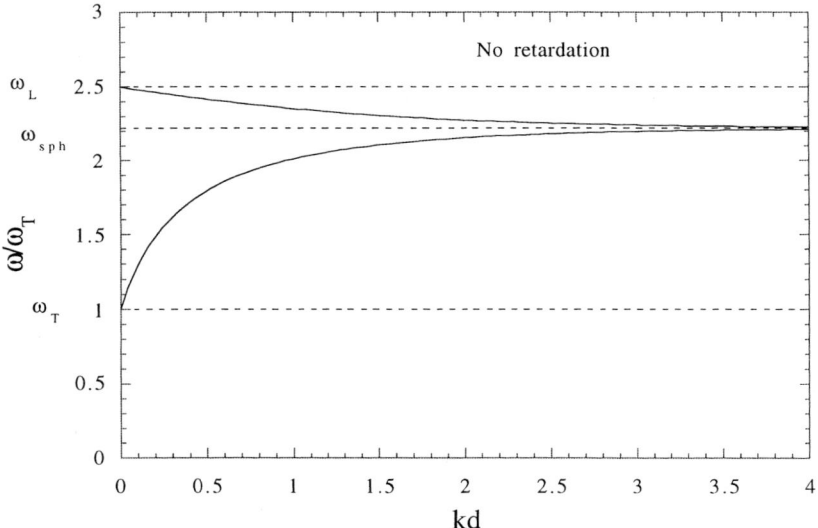

Figure 4.12: The two surface modes for a semiconductor slab in vacuum: retardation neglected.

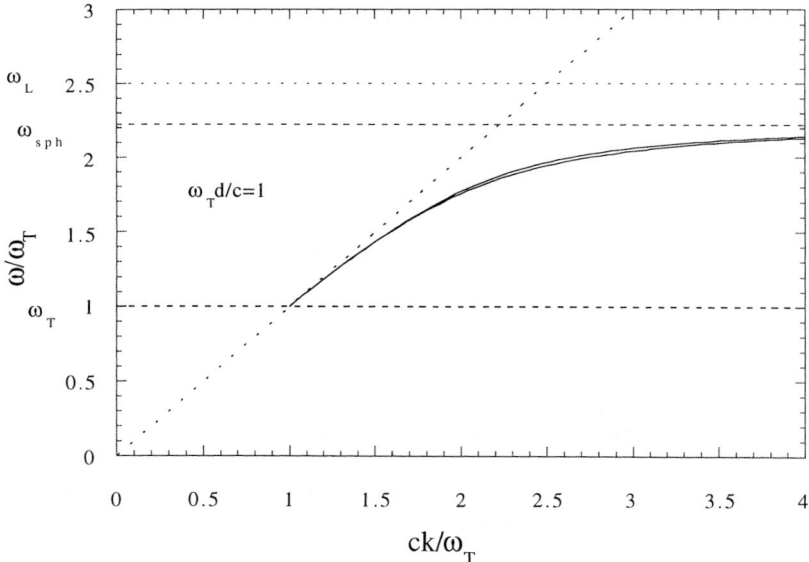

Figure 4.13: The two surface modes for a semiconductor slab in vacuum: retardation included.

4.2 Modes in Slab Geometry

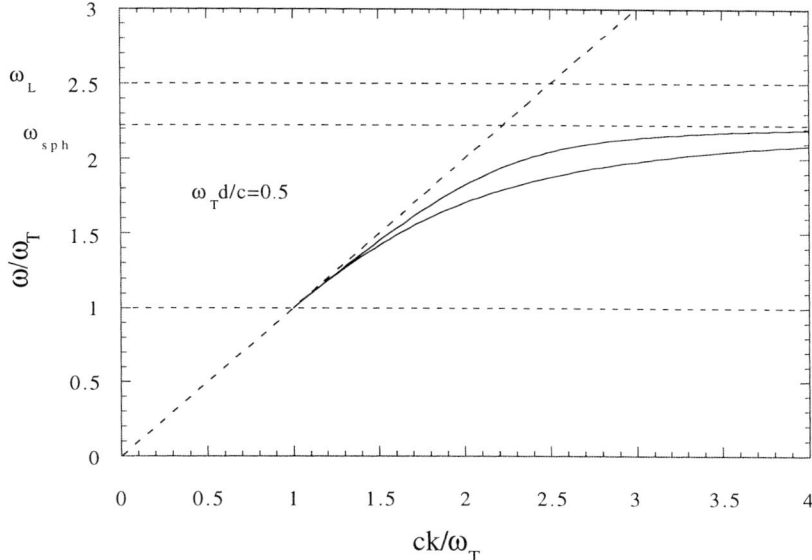

Figure 4.14: The same as Figure 4.13 but for a thinner slab.

In the examples we have used the values 3 and 2.5 for ε_∞ and ω_L/ω_T, respectively.

4.2.3 Vacuum gap in a metal

We now reverse the metal slab geometry and study a vacuum gap in a metal. Then we have

$$\left(\frac{x-y}{x} + \frac{\sqrt{1-(x-y)}}{\sqrt{1-x}}\right)^2 - e^{-2\sqrt{1-x}kd}\left(\frac{x-y}{x} - \frac{\sqrt{1-(x-y)}}{\sqrt{1-x}}\right)^2 = 0$$

$$\varepsilon_1 = 1$$

$$\varepsilon_2 = 1 - \frac{\omega_{pl}^2}{\omega^2} = 1 - \frac{y}{x} \qquad (4.77)$$

$$x = \left(\frac{\omega}{ck}\right)^2$$

$$y = \left(\frac{\omega_{pl}}{ck}\right)^2$$

Let us for the moment neglect retardation effects. Then we find

$$\left(\frac{x-y}{x} + 1\right)^2 - e^{-2kd}\left(\frac{x-y}{x} - 1\right)^2 = 0$$

$$\left(\frac{x-y}{x} + 1\right) \pm e^{-kd}\left(\frac{x-y}{x} - 1\right) = 0$$

$$x = \tfrac{1}{2}y\left(1 \mp e^{-kd}\right) \qquad (4.78)$$

$$\omega = \omega_s\sqrt{1 \mp e^{-kd}}$$

We find that when retardation is neglected the modes have the same dispersion as for a metal slab. With retardation effects included they are not. This is illustrated in Figures 4.15-4.18.

4.2 Modes in Slab Geometry

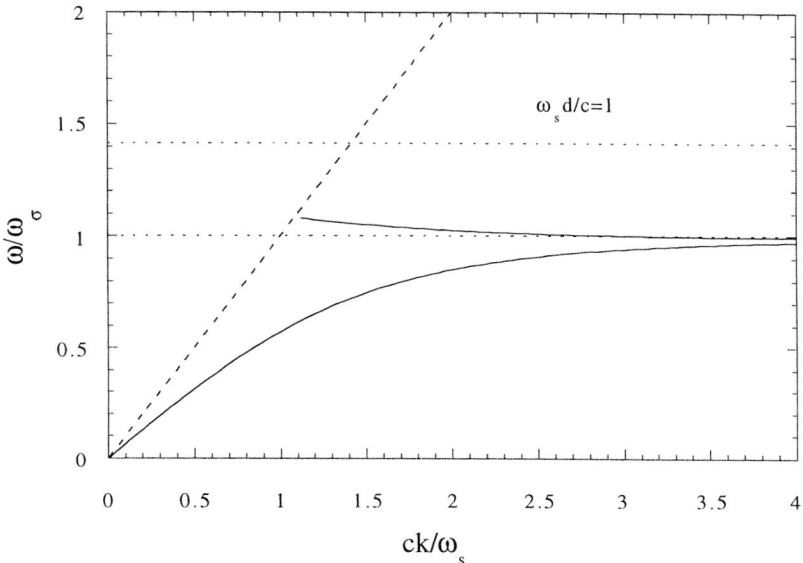

Figure 4.15: The two surface modes for a vacuum gap in a metal: retardation included.

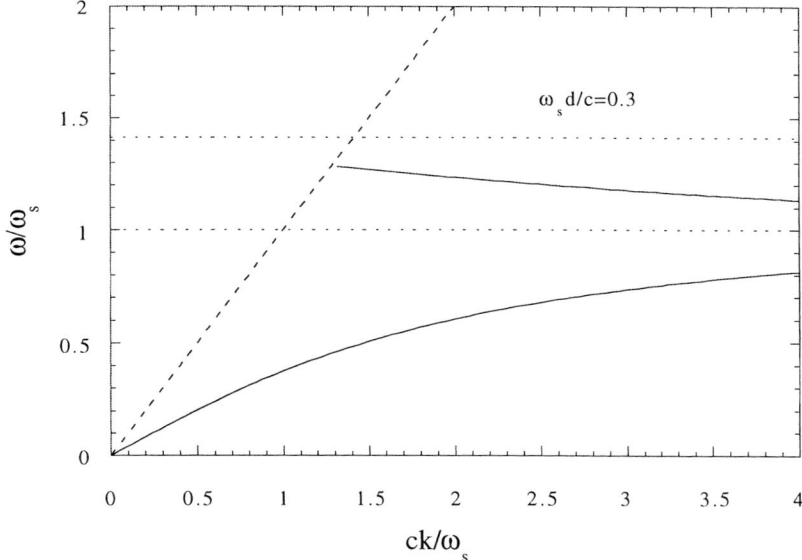

Figure 4.16: The same as Figure 4.15 but for a thinner gap.

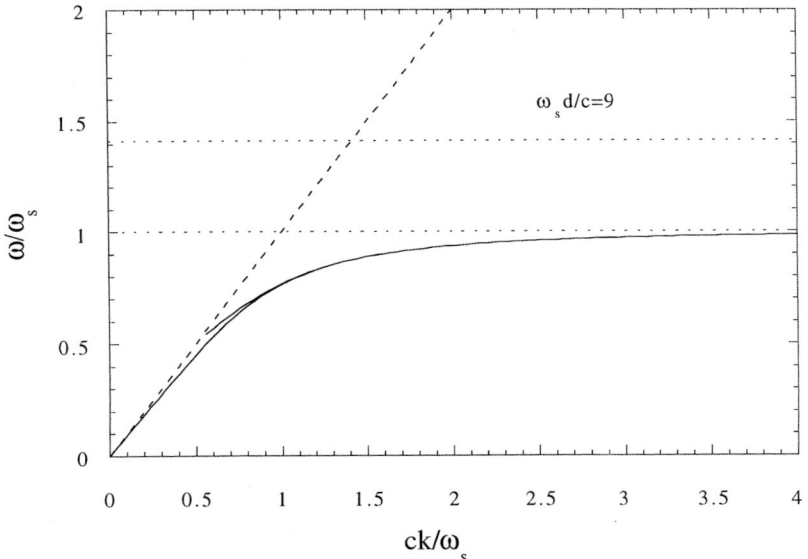

Figure 4.17: The same as Figures 4.15 and 4.16 but for a thicker gap.

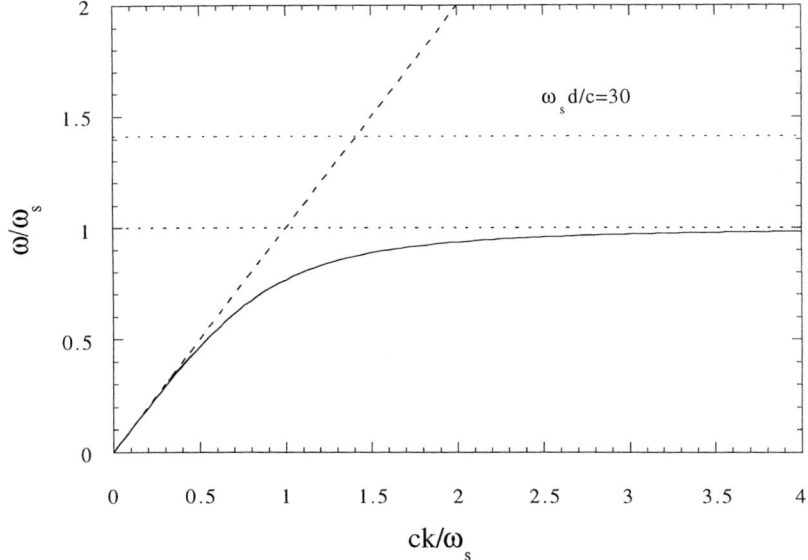

Figure 4.18: The same as Figures 4.15 - 4.17 but for a very thick gap.

We have found that in the gap geometry the upper mode is no longer in general pushed below the light dispersion curve. However, for thick gaps the retardation effects become

4.2 Modes in Slab Geometry

important also for this mode. This is bound to be the case since the modes on the two surfaces decouple when the distance increases. The two modes will also become degenerate. We should further note that we actually should have no solution above the light-dispersion curve with our assumptions. We may have other solutions that decay inside the metal but are oscillating outside. These modes are actually the ones that cause the Casimir force. The upper mode ends up at the light-dispersion curve at the value:

$$\frac{ck}{\omega_s} = \sqrt{\frac{2}{1+\frac{1}{\sqrt{2}}\frac{d\omega_s}{c}}} \qquad (4.79)$$

These values are shown in Figure 4.19.

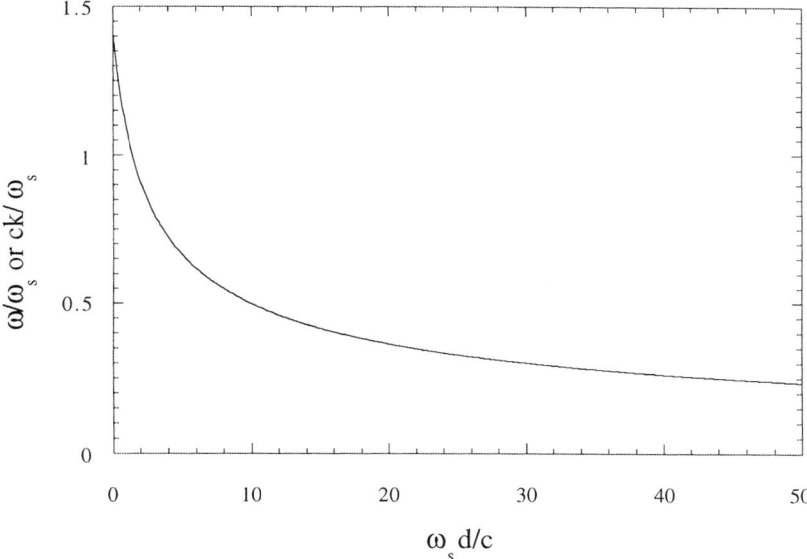

Figure 4.19: The normalized frequency or equivalently normalized momentum where the upper mode ends up at the light dispersion curve.

4.2.4 Vacuum gap in a semiconductor

For the configuration with a vacuum gap in a semiconductor we have

$$\left(\frac{\varepsilon_\infty(x-b)}{x-a} + \frac{\sqrt{1-\varepsilon_\infty\frac{x-b}{x-a}x}}{\sqrt{1-x}}\right)^2 - e^{-2\sqrt{1-x}kd}\left(\frac{\varepsilon_\infty(x-b)}{x-a} - \frac{\sqrt{1-\varepsilon_\infty\frac{x-b}{x-a}x}}{\sqrt{1-x}}\right)^2 = 0$$

$$\varepsilon_1 = 1$$
$$\varepsilon_2 = \varepsilon_\infty \frac{x-b}{x-a}$$
$$x = \left(\frac{\omega}{ck}\right)^2$$
$$a = \left(\frac{\omega_T}{ck}\right)^2$$
$$b = \left(\frac{\omega_L}{ck}\right)^2$$

(4.80)

Let us for the moment neglect retardation effects. Then we get

$$\left(\frac{\varepsilon_\infty(x-b)}{x-a} + 1\right)^2 - e^{-2kd}\left(\frac{\varepsilon_\infty(x-b)}{x-a} - 1\right)^2 = 0$$

$$\left(\frac{\varepsilon_\infty(x-b)}{x-a} + 1\right) = \pm e^{-kd}\left(\frac{\varepsilon_\infty(x-b)}{x-a} - 1\right)$$

$$x = \frac{a + b\varepsilon_\infty \mp e^{-kd}(b\varepsilon_\infty - a)}{1 + \varepsilon_\infty \mp e^{-kd}(\varepsilon_\infty - 1)}$$

$$\omega_\pm^2 = \frac{\omega_T^2 + \omega_L^2\varepsilon_\infty \mp e^{-kd}(\omega_L^2\varepsilon_\infty - \omega_T^2)}{1 + \varepsilon_\infty \mp e^{-kd}(\varepsilon_\infty - 1)}$$

(4.81)

We find that when retardation is neglected the modes have the same dispersion as for a semiconductor slab. With retardation effects included they are not. This is illustrated in Figures 4.20 and 4.21.

4.2 Modes in Slab Geometry

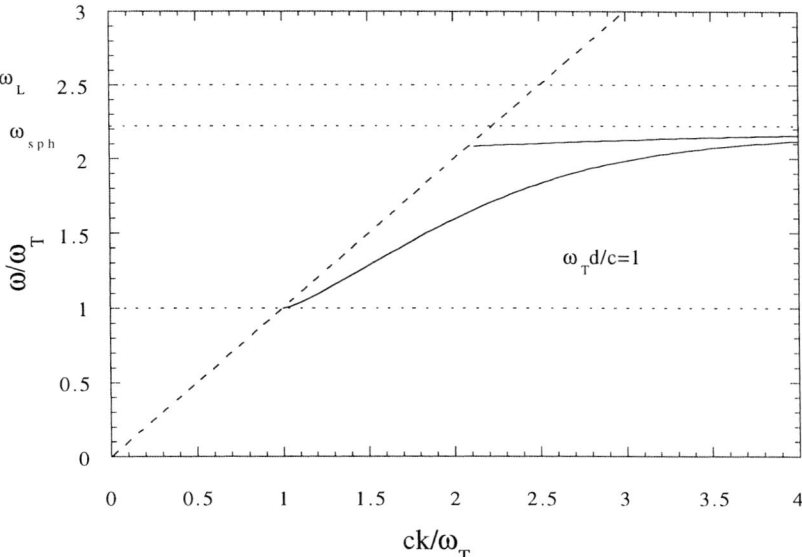

Figure 4.20: The two surface modes for a vacuum gap in a semiconductor: retardation included.

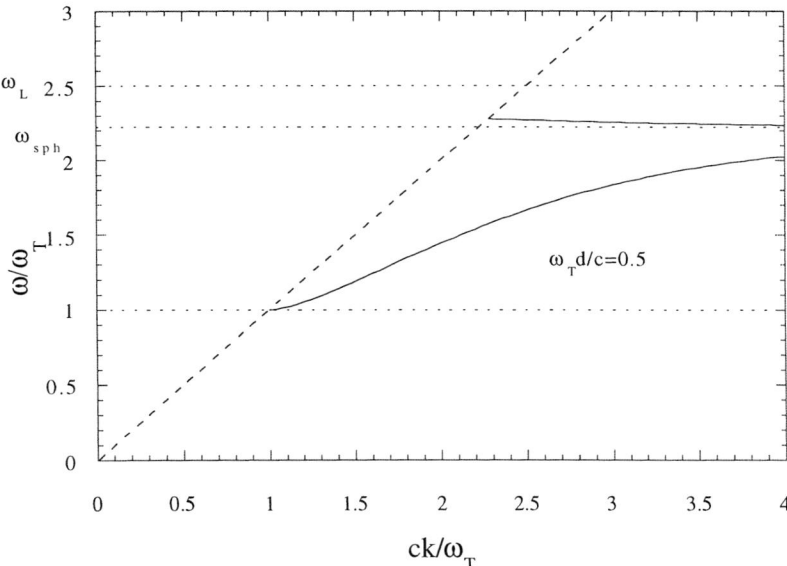

Figure 4.21: The same as in Figure 4.20 but for a thinner gap.

We find that just as for a thin gap between two metals the upper mode is not affected very much by retardation.

4.3 The Casimir effect

It all started in the Netherlands in the 1940s. J. T. G. Overbeek performed experiments on suspensions of quartz powder. He and E. J. W. Verwey had developed a theory for the stability of colloids. The theory relied on the long range interaction between the particles. The experiments suggested that the long range forces did not fall off as r^{-7}, which is the standard van der Waals interaction, but faster. H. B. G. Casimir and D. Polder were given this problem and they found that retardation effects, effects from the finite speed of light, made the force vary as r^{-8} instead of as r^{-7} (interaction energy as r^{-7} instead of as r^{-6}). They found the results:

$$E(r) = -\frac{23\hbar c \alpha_1(0)\alpha_2(0)}{4\pi r^7} \tag{4.82}$$

and

$$F(r) = -\frac{161\hbar c \alpha_1(0)\alpha_2(0)}{4\pi r^8} \tag{4.83}$$

for the energy and force, respectively. Casimir was intrigued by the simplicity in the actual expression for the force. The only material parameters that enter the expressions are the static polarizabilities of the atoms. The expressions contain the speed of light and $\hbar$. Why is this? He thought a lot about these things for many years. He then studied a simpler system, namely the force between two perfectly conducting plates. By assuming that the force was due to the change in the total zero point energy of all electromagnetic modes as a function of separation between the plates he managed to get the simple result:

$$F(d) = -\frac{\pi^2 \hbar c}{240 d^4} \tag{4.84}$$

The modes he used were the transverse modes. The result did neither contain the electron effective mass, nor the electron charge nor the density of electrons. Instead it contained the speed of light. To date there have been only three attempts to verify the Casimir force experimentally. The first was performed by Sparnaay [2] in 1958. The result was not conclusive since the uncertainty was 100 % of the value; one could only say that the presence of the force could not be ruled out. The next was performed in 1997 by Lamoreaux [3] using a torsion pendulum. The uncertainty was here reduced to 5 % and the presence of the Casimir force was clearly demonstrated. In 1998 Mohideen and Roy [4] measured the force with an atomic force microscope and found deviations between experimental and theoretical values for the force of 1 %.

We will now go through a derivation of the force between two metallic plates.

4.3.1 Casimir effect at zero temperature

In the Casimir effect other modes than the ones contributing to the van der Waals interaction are responsible. In the van der Waals case the modes are localized to the surfaces of the objects and decay exponentially away from the surfaces. In the Casimir case freely propagating electromagnetic modes or photons are responsible. The boundary conditions at the surfaces of the objects have effects on these modes. It is the change in the zero-point energy of these modes when we move the objects with respect to each other that causes the force between the objects. We have a problem with quantum field theory. The whole universe is full of these kind of modes. This means that they contain an infinite energy. However if we only consider the change in this energy we are safe. This change is finite.

We will now treat the Casimir energy between two metallic plates. In Section 6.6 we use another, simpler, method to derive the Casimir energy. Here, we treat the plates as perfectly reflecting. However, all real material can only reflect electromagnetic waves of long wave-lengths or small frequencies. Thus the reflecting power decreases with frequency and for high enough frequency the waves are unaffected by the plates. We actually need this behavior to get finite results for the force.

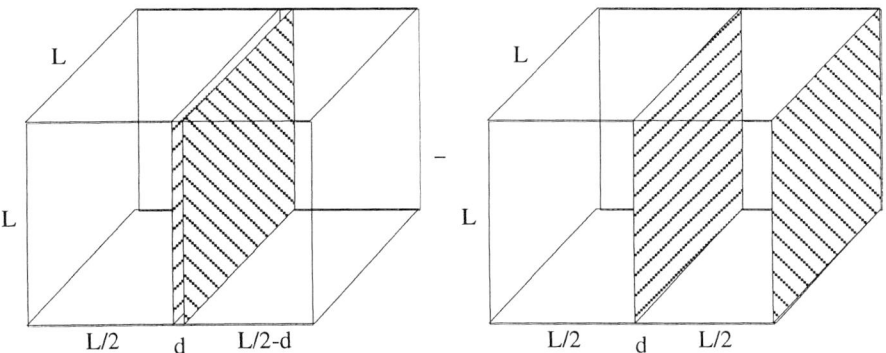

Figure 4.22: The configuration used in deriving the force between two plates.

Let the two plates be placed in a large cube with reflecting walls, according to the figure. We want to calculate the energy of the left system with the energy of the right one as our reference energy. We have

$$E_C(d) = \lim_{L \to \infty} \left\{ [E(L/2) + E(d) + E(L/2 - d)] - [E(L/2) + E(L/2)] \right\}$$
$$= \lim_{L \to \infty} [E(d) + E(L/2 - d) - E(L/2)]$$
(4.85)

The modes in the cavity of volume $L^2 d$ between the plates will be characterized by the eigenfrequencies

$$\omega = c\sqrt{k_{//}^2 + k_z^2} \ ; \ k_z^2 = (n\pi/d)^2 \ ; \ k_{//}^2 = (n_x\pi/L)^2 + (n_y\pi/L)^2 \tag{4.86}$$

The wave numbers parallel to the plates can be considered continuous while the ones perpendicular are treated as discrete. The eigenfrequencies of the **TE** and **TM** modes are the same. There are two standing waves for each wavenumber except for $n = 0$ where there is only one.

We simulate the effect of the reduced reflectance efficiency with frequency by introducing a weighting function $f(k/k_c)$ which is unity for $k \ll k_c$ and zero for $k \gg k_c$ where k_c is some effective cut-off wave-number. The simplest choice is

$$f(k/k_c) = \exp(-k/k_c) \tag{4.87}$$

We will find below that the leading term in the result is independent of the choice of function. Now, we have

$$E_C(d, k_c) = L^2 \hbar c \int \frac{d^2 k_{//}}{(2\pi)^2} \left\{ \frac{1}{2} |k_{//}| f(k_{//}/k_c) + \sum_{n=1}^{\infty} \sqrt{k_{//}^2 + (n\pi/d)^2} \, f\!\left(\sqrt{k_{//}^2 + (n\pi/d)^2}/k_c\right) \right.$$
$$\left. + [(L/2 - d) - (L/2)] \int \frac{dk_z}{2\pi} k f(k/k_c) \right\} \tag{4.88}$$

We perform the angular integration and introduce the dimensionless variable

$$s = d^2 k_{//}^2 / \pi^2 + n^2 \tag{4.89}$$

and obtain

$$E_C(d, k_c) = L^2 \hbar c \frac{\pi^2}{4d^3} \left\{ \int_{n^2=0}^{\infty} ds \frac{1}{2} \sqrt{s} f(\pi \sqrt{s}/dk_c) + \sum_{n=1}^{\infty} \int_{n^2}^{\infty} ds \sqrt{s} f(\pi \sqrt{s}/dk_c) \right.$$
$$\left. - \int_0^{\infty} dn \int_{n^2}^{\infty} ds \sqrt{s} f(\pi \sqrt{s}/dk_c) \right\} \tag{4.90}$$

Let us define the function

$$F(n) = \int_{n^2}^{\infty} ds \sqrt{s} f(\pi \sqrt{s}/dk_c) \tag{4.91}$$

Then we may write

$$E_C(d, k_c) = \frac{L^2 \hbar c \pi^2}{4d^3} \left[\frac{1}{2} F(0) + \sum_{n=1}^{\infty} F(n) - \int_0^{\infty} dn F(n) \right] \tag{4.92}$$

4.3 The Casimir Effect

We can handle the last two terms with help of the Euler-Maclaurin summation formula[5]:

$$\sum_{n=1}^{\infty} F(n) - \int_0^{\infty} dx F(x) = -\frac{1}{2} F(0) - \frac{1}{12} F'(0) - \frac{1}{720} F'''(0) \ldots \tag{4.93}$$

In our case we have $F'(0) = 0$, $F'''(0) = -4$, and all higher derivatives are zero if we assume that all derivatives of the cutoff function vanish for zero argument. Thus, we have

$$E_C(d) = -\frac{L^2 \hbar c \pi^2}{720 d^3} \tag{4.94}$$

and from this follows

$$F_C(d) = -\frac{d}{dd} E_C(d) = -\frac{L^2 \hbar c \pi^2}{240 d^4} \tag{4.95}$$

An alternative way to take care of the numerical problems is the following: we return to the equation:

$$E_C(d, k_c) = L^2 \hbar c \frac{\pi^2}{4d^3} \left\{ \int_{n^2=0}^{\infty} ds \frac{1}{2} \sqrt{s} f\left(\pi\sqrt{s}/dk_c\right) + \sum_{n=1}^{\infty} \int_{n^2}^{\infty} ds \sqrt{s} f\left(\pi\sqrt{s}/dk_c\right) \right. \\ \left. - \int_0^{\infty} dn \int_{n^2}^{\infty} ds \sqrt{s} f\left(\pi\sqrt{s}/dk_c\right) \right\} \tag{4.96}$$

and remove the cut-off function and change variables $s \to (sd/\pi)^2$. Then we have

$$E_C(d) = L^2 \hbar c \frac{1}{2\pi} \left\{ \int_{\frac{n\pi}{d}=0}^{\infty} ds \frac{1}{2} s^2 + \sum_{n=1}^{\infty} \int_{\frac{n\pi}{d}}^{\infty} dss^2 - \int_0^{\infty} dn \int_{\frac{n\pi}{d}}^{\infty} dss^2 \right\} \\ = L^2 \frac{\hbar c}{2\pi} \left\{ \sum_{n=0}^{\infty}{}' \int_{\frac{n\pi}{d}}^{\infty} dss^2 - \int_0^{\infty} dn \int_{\frac{n\pi}{d}}^{\infty} dss^2 \right\} \tag{4.97}$$

where the prime indicates that only half of the $n = 0$ term should be included. We now replace the integrals over s by:

$$\int_{\frac{n\pi}{d}}^{\infty} dss^2 = \lim_{\alpha \to 0} \frac{\partial^2}{\partial \alpha^2} \int_{\frac{n\pi}{d}}^{\infty} ds e^{-\alpha s} \tag{4.98}$$

This leads to

$$E_C(d) = -L^2 \frac{\hbar c}{2\pi} \lim_{\alpha \to 0} \frac{\partial^2}{\partial \alpha^2} \frac{1}{\alpha} \left\{ \sum_{n=0}^{\infty}{}' e^{-\frac{\alpha n \pi}{d}} - \int_0^{\infty} dn\, e^{-\frac{\alpha n \pi}{d}} \right\}$$

$$= -L^2 \frac{\hbar c}{2\pi} \lim_{\alpha \to 0} \frac{\partial^2}{\partial \alpha^2} \frac{1}{\alpha} \left\{ \frac{1}{2} \coth\left(\frac{\alpha \pi}{2d}\right) - \frac{d}{\alpha \pi} \right\}$$

$$= -L^2 \frac{\hbar c}{4\pi} \lim_{\alpha \to 0} \frac{\partial^2}{\partial \alpha^2} \frac{1}{\alpha} \left\{ \coth\left(\frac{\alpha \pi}{2d}\right) - \frac{2d}{\alpha \pi} \right\} \qquad (4.99)$$

$$= -L^2 \frac{\hbar c}{4\pi} \lim_{\alpha \to 0} \frac{\partial^2}{\partial \alpha^2} \frac{1}{\alpha} \left\{ \frac{2d}{\alpha \pi} + \frac{1}{3}\frac{\alpha \pi}{2d} - \frac{1}{45}\left(\frac{\alpha \pi}{2d}\right)^3 + O(\alpha^4) - \frac{2d}{\alpha \pi} \right\}$$

$$= -L^2 \frac{\hbar c}{4\pi} \lim_{\alpha \to 0} \frac{\partial^2}{\partial \alpha^2} \frac{1}{\alpha} \left\{ \frac{1}{3}\frac{\alpha \pi}{2d} - \frac{1}{45}\left(\frac{\alpha \pi}{2d}\right)^3 + O(\alpha^4) \right\} = -L^2 \frac{\hbar c}{4\pi} \frac{2\pi^3}{45 \cdot 8 d^3}$$

$$= -L^2 \frac{\hbar c \pi^2}{720 d^3}$$

4.3.2 Casimir effect at finite temperature

For finite temperature, the force is obtained as

$$F_C = -\frac{d}{dd} \mathfrak{F} \qquad (4.100)$$

where $\mathfrak{F}$ is the *Helmholtz free energy*:

$$\mathfrak{F} = -\frac{1}{\beta} \ln \mathfrak{Z} \qquad (4.101)$$

The *partition function*, $\mathfrak{Z}$, is

$$\mathfrak{Z} = tr\left(e^{-\beta H}\right) = \prod_{\mathbf{k}} \sum_{n=0}^{\infty} e^{-\beta \hbar \omega_{\mathbf{k}}\left(n+\frac{1}{2}\right)}$$

$$= \prod_{\mathbf{k}} e^{-\beta \frac{1}{2}\hbar \omega_{\mathbf{k}}} \sum_{n=0}^{\infty} e^{-\beta \hbar \omega_{\mathbf{k}} n} = \prod_{\mathbf{k}} e^{-\beta \frac{1}{2}\hbar \omega_{\mathbf{k}}} \frac{1}{1 - e^{-\beta \hbar \omega_{\mathbf{k}}}} \qquad (4.102)$$

4.3 The Casimir Effect

Thus we have

$$\mathfrak{F} = -\frac{1}{\beta}\sum_{\mathbf{k}}\left[-\beta\frac{1}{2}\hbar\omega_{\mathbf{k}} - \ln\left(1 - e^{-\beta\hbar\omega_{\mathbf{k}}}\right)\right] = \sum_{\mathbf{k}}\left[\frac{1}{2}\hbar\omega_{\mathbf{k}} + \frac{1}{\beta}\ln\left(1 - e^{-\beta\hbar\omega_{\mathbf{k}}}\right)\right] \quad (4.103)$$

The first part of this expression is the same as the zero temperature result and the temperature effects are all in the second term. This term is simpler to treat than the first. For two plates, the second part is given by

$$\mathfrak{F}_{C,2}(d,T) = 2L^2 \frac{1}{\beta}\int\frac{d^2k_{//}}{(2\pi)^2}\left\{\sum_{n=0}^{\infty}{}' \ln\left[1 - \exp\left(-\hbar\beta c\sqrt{k_{//}^2 + (n\pi/d)^2}\right)\right]\right.$$
$$\left. + \left[(L/2 - d) - (L/2)\right]\int\frac{dk_z}{2\pi}\ln\left[1 - \exp\left(-\hbar\beta c\sqrt{k_{//}^2 + k_z^2}\right)\right]\right\}$$
(4.104)

where the prime on the summation sign indicates that the $n = 0$ term should be multiplied by a factor of one half. In terms of the function:

$$b(d,T,n) = \frac{1}{2}\frac{1}{\beta}\int_{n^2}^{\infty} ds\,\ln\left[1 - \exp\left(-\pi\hbar\beta c\sqrt{s}/d\right)\right] \quad (4.105)$$

we can write

$$\mathfrak{F}_{C,2}(d,T) = \frac{L^2\pi}{d^2}\left[\frac{1}{2}b(d,T,0) + \sum_{n=1}^{\infty}b(d,T,n) - \int_0^{\infty}dn\,b(d,T,n)\right]$$
$$= \frac{L^2\pi}{d^2}\left[-\frac{1}{2\beta}\left(\frac{d}{\pi\hbar\beta c}\right)^2\zeta(3) + \sum_{n=1}^{\infty}b(d,T,n) + \frac{2}{\beta}\left(\frac{d}{\pi\hbar\beta c}\right)^3\zeta(4)\right]$$
(4.106)

In the low-temperature limit, that is, for $d/\hbar\beta c \ll 1$ we have

$$\sum_{n=1}^{\infty} b(d,T,n) = \frac{1}{2}\frac{1}{\beta}\sum_{n=1}^{\infty}\frac{1}{n^2}\int_{0}^{\infty} ds \ln\left[1-\exp\left(-\pi\hbar\beta c\sqrt{s}/d\right)\right]$$

$$= -\frac{1}{2}\frac{1}{\beta}\sum_{n=1}^{\infty}\frac{1}{n^2}\int_{0}^{\infty} ds \sum_{m=1}^{\infty}\frac{1}{m}\left[\exp\left(-m\pi\hbar\beta c\sqrt{s}/d\right)\right]$$

$$= -\sum_{m=1}^{\infty}\frac{1}{m\beta}\left(\frac{d}{m\pi\hbar\beta c}\right)^2 \sum_{n=1}^{\infty}(nm\pi\hbar\beta c/d+1)\exp(-nm\pi\hbar\beta c/d)$$

$$\approx -\frac{1}{\beta}\left(\frac{d}{\pi\hbar\beta c}\right)^2 (\pi\hbar\beta c/d+1)\exp(-\pi\hbar\beta c/d)$$

(4.107)

The total result is for low temperatures

$$\mathfrak{F}_C(d,T) = -\frac{L^2\pi^2\hbar c}{720 d^3}\Bigg\{1+360\zeta(3)\left(\frac{d}{\pi\hbar\beta c}\right)^3 - \left(\frac{2d}{\hbar\beta c}\right)^4$$

$$+ 720\left(\frac{d}{\pi\hbar\beta c}\right)^3\left(1+\frac{\pi\hbar\beta c}{d}\right)e^{-\pi\hbar\beta c/d}\Bigg\}$$

(4.108)

This gives rise to the following force per unit area:

$$F_C(d,T) = -\frac{L^2\pi^2\hbar c}{240 d^4}\Bigg\{1+\frac{48}{9}\left(\frac{d}{\hbar\beta c}\right)^4 - 240\left(\frac{d}{\pi\hbar\beta c}\right)e^{-\pi\hbar\beta c/d}\Bigg\}$$

(4.109)

In the high temperature limit, that is for $d/\hbar\beta c \gg 1$ we find:

$$\mathfrak{F}_C(d,T) \approx -\frac{L^2 1.2}{8\pi\beta d^2}; \quad F_C(d,T) \approx -\frac{L^2 2.4}{8\pi\beta d^3}$$

(4.110)

We have in this section described how the Casimir interaction can be calculated between two plates made up from perfect metals; we assumed that the plates were perfectly reflecting. We will in Section 6.6 derive the same results using a more realistic description of the dielectric properties of the metal plates, taking the finite conductivity into account. Then we do not need the artificial cutoff parameter we had to introduce above. In Section 6.8 we will briefly discuss some very recent results showing that inclusion of dissipation has very important effects on the temperature dependence of the Casimir effect.

4.4 Metal surfaces

We stated already in Chapter 3 that the collective excitations of a system are massless Bosons described by a Hamiltonian of the kind:

$$H = \sum_{\mathbf{q}} \hbar \omega_{\mathbf{q}} \left(n_{\mathbf{q}} + \tfrac{1}{2} \right) \tag{4.111}$$

where the "1/2" terms represent the energy of the vacuum fluctuations. The collective excitations can be phonons, photons, plasmons, magnons, oscillations in an *LC* circuit and so forth. In the majority of physical experiments it is only the occupation numbers and changes in these that characterize the state of the system and the excitations. Thus, the presence of the vacuum fluctuations does not show up. One is often taught that one can neglect this contribution since it only represents a constant energy term. However, there are other types of experiments or situations where they do show up! One example is when two systems with collective excitations are brought together and are allowed to interact. Then the energy of the collective modes might change. This change manifests itself in an attractive or repulsive force. We demonstrated in Section 4.3 that the change of the zero-point energy of the wave-guide modes between two metal plates, when they are moved relative each other, are responsible for the Casimir force.

We discussed plasmons in Chapter 3. If we start from isolated metal atoms and create the metal by adding the atoms one at a time we automatically create the plasmon modes as well and the energy of the vacuum fluctuations has to be provided. This means that the energy of the vacuum fluctuations is a part of the cohesive energy of metals. This is exactly what one finds, a part of the correlation energy is the energy of the vacuum fluctuations. If we treat the single particle continuum as discrete all the correlation energy is the energy of the vacuum fluctuations.

If we break up the metal into two pieces there will be new modes localized to the two surfaces. These modes contain energy and their zero-point energy make up an important part of the surface energy of the metal. This we will discuss in the following section.

4.4.1 Surface energy of metals

We will as an example study the surface energy, σ, of a metal, that is, the energy per unit area of a newly created surface. To be more specific it is half the energy needed to split the solid in two along a plane and to separate the two halves to infinite distance. The "1/2" comes from the fact that we create two new surfaces. In splitting the solid we also create surface plasmons; or rather bulk-plasmon modes are pealed off and form surface plasmons. This change in collective modes costs energy. This energy constitutes, as we shall see, an important part of the surface energy.

We follow a derivation by Schmit and Lucas [6] which attracted much attention when it was published. A very similar calculation was performed independently by Craig [7]. The derivation is valid for both polar semiconductors or insulators and metals. Let the system have longitudinal and transverse oscillations of frequency ω_L and ω_T, respectively. For simple metals ω_L corresponds to the bulk plasmon frequency and the corresponding ω_T is 0. Let **k** be the two-dimensional wave vector parallel to the surface. For each **k** there exists an infinite number of bulk modes corresponding to the third component k_z of the wave vector. When two surfaces are close together there are two surface modes, ω_+ and ω_-, split off from the "$k_z = 0$" components of the two bulk modes. These surface modes are obtained as solutions to the equation

$$\frac{\varepsilon(\omega)-1}{\varepsilon(\omega)+1} = \pm e^{kd} \qquad (4.112)$$

which was derived in Section 4.2, Equation (4.62). The **k**-dependence of $\varepsilon(\omega)$ is neglected. The dispersion curves of the surface modes are illustrated in Figure 4.23 for both a small separation (a) and a large (b).

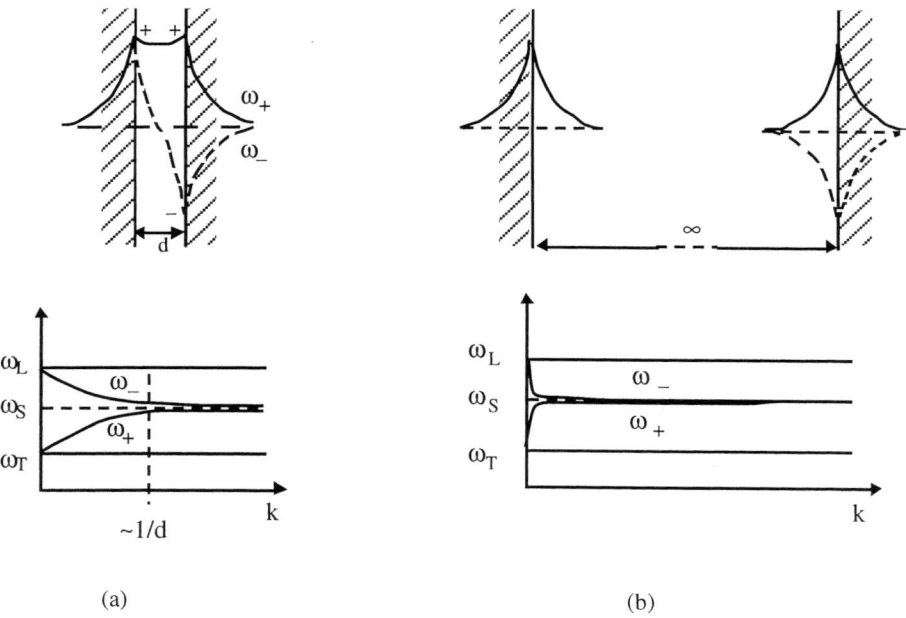

(a) \hspace{4cm} (b)

Figure 4.23: Surface modes and their dispersion for small (a) and large separations (b).

The surface energy is calculated for infinite separation of the surfaces in which case the dispersions of the two surface modes are degenerate and horizontal with the constant frequency ω_s; in the metal case there is also a momentum cutoff k_c when the modes enter

4.4 Metal Surfaces

the single particle continuum. According to the calculation of Schmit and Lucas the surface energy consists of the change in the total energy for the vacuum fluctuations of the surface modes, that is,

$$\sigma_{SL} = \frac{1}{2}\left[\frac{1}{A}\sum_{\mathbf{k}}\frac{1}{2}\hbar(\omega_s - \omega_L + \omega_s - \omega_T)\right] \tag{4.113}$$

where the first "1/2" comes from the fact that the energy is divided on two surfaces. For a metal this results in:

$$\sigma_{SL} = \frac{\sqrt{2}-1}{16\pi}\hbar\omega_{pl}k_c^2 \tag{4.114}$$

since

$$\omega_L = \omega_{pl}\ ;$$
$$\omega_T = 0\ ; \tag{4.115}$$
$$\omega_s = \frac{\omega_{pl}}{\sqrt{2}}$$

They chose

$$k_c = \frac{\omega_{pl}}{\sqrt{2}v_F} \tag{4.116}$$

which led to

$$\sigma_{SL} = Cr_s^{-5/2} \tag{4.117}$$

where C is a constant.

This result is plotted in Figure 4.24 together with experimental results. As can be seen this very crude calculation gives extremely good agreement with experiments. However, we should not take this agreement too seriously because we know that many important contributions have been left out. The second curve in the figure is the result from a theoretical calculation on the jellium model by Lang and Kohn [8]. In the jellium model (see Chapter 3) for this problem the surface energy has three parts. When the surface is created the electron distribution is no longer abruptly cut off at the surface. Electrons are spilling out. This lowers the kinetic energy and a dipole layer is created which means a positive electrostatic energy contribution. The third contribution comes from a change in the exchange and correlation energy. This contribution was by Lang and Kohn calculated in the local density approximation. The contribution from the surface modes was not explicitly included. The calculation by Lang and Kohn was much more elaborate than the one by

Schmit and Lucas and still was less successful in comparison with experiments. In a more realistic calculation one should also include the effects from the ion potentials. It is furthermore known that the relaxation of the outermost atomic layers is important for the surface energy.

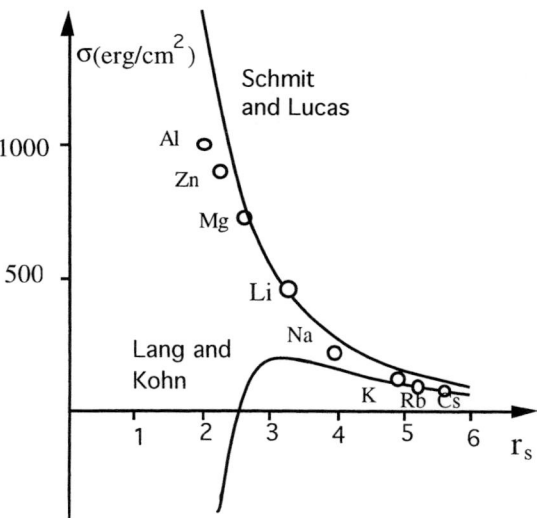

Figure 4.24: Surface energy of metals from theories of Schmit and Lucas and Lang and Kohn compared to the experimental values

Although we know that the success of Schmit and Lucas to a part must be coincidental, we have learned that the vacuum fluctuations have an important part in the surface energy. The force between conducting planes at large separation due to the vacuum fluctuations has been known since long and it is called the Casimir effect after the important work by Casimir [9]. Similar approaches can be used in many situations. It can for example be used to explain the van der Waals force. The reader is referred to an article by D. Kleppner [10] for a brief review. We will as closing remark for this section give one quotation from this article: *Casimir's thinking also underlies the most plausible explanation yet for Creation – namely, generation of the universe from vacuum fluctuations.*

Let us now study a special metal, mercury, a metal that is in liquid form at room temperature; this means that it is easier to measure its surface energy or surface tension.

4.4.2 Optical properties of mercury

The metal mercury (Hg) is a very special metal in that it is liquid. The normal method to measure the surface energy or surface tension of a metal is to melt it, measure the tension as a function of temperature and extrapolate down to room temperature. In the case of mercury we do not need any extrapolation and hence avoid one cause of uncertainty. Mercury does not wet glass. The capillary force is opposite to that of water. Mercury used to be a standard case to study in schools to demonstrate capillary forces, but due to its toxic properties it is no longer used. Let us study the optical properties of mercury. These have been compiled in the literature.

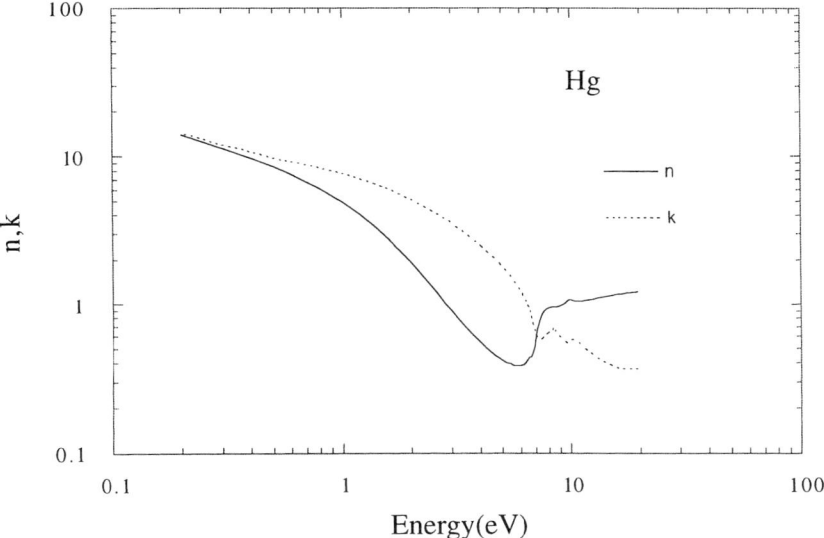

Figure 4.25: The optical constants of mercury.

The optical constants for mercury [11] as found in the Handbook of Optical Constants of Solids II are presented in Figure 4.25. From these it is easy to get the dielectric function. It is displayed in Figures 4.26 and 4.27.

The behavior is not quite the same as for an electron gas. There is still some absorption for high energies. This makes the plasmon energy shift towards lower energies. The theoretical value for the plasmon energy is given through the relation:

$$\hbar\omega_{pl} = \hbar\sqrt{\frac{4\pi n e^2}{\kappa m}} \tag{4.118}$$

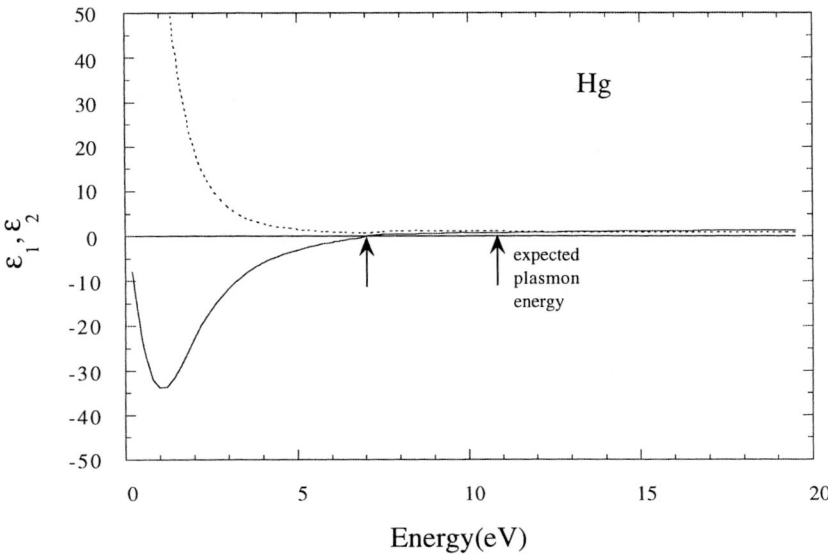

Figure 4.26: The real (solid curve) and imaginary (dotted curve) part of the dielectric function of mercury. The left (right) arrow indicates the experimental (theoretical) value of the plasmon energy.

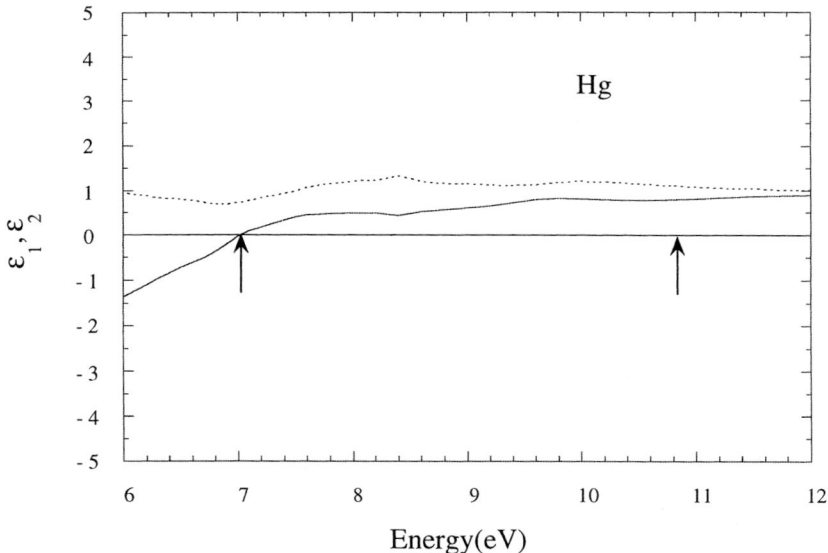

Figure 4.27: The real (solid curve) and imaginary (dotted curve) part of the dielectric function of mercury: expanded view.

4.4 Metal Surfaces

Mercury has two valence electrons per atom and the density of atoms is 4.26×10^{22}cm^{-3}. If there is no background dielectric screening (see the discussion in connection with Equation (2.29)) the theoretical plasmon energy is 10.84 eV. The experimental value is 7.0 eV.

Just by putting the background screening to 2.4 we reproduce the correct value for the plasmon energy. Let us now study the imaginary part of the dielectric function in detail for small energies. In this energy range it is due to absorption induced by scattering against imperfections and phonons. It should be reproduced quite nicely by the Drude expression:

$$\varepsilon_2(\omega) = \frac{4\pi}{\omega} \frac{\rho_0}{(\rho_0)^2 + (4\pi\omega/\omega_{pl}^2)^2} \approx \begin{cases} \dfrac{4\pi}{\omega\rho_0} & ; \ \omega \to 0 \\ \dfrac{1}{\omega^3} \dfrac{\rho_0 \omega_{pl}^4}{4\pi} & ; \ \omega \to \infty \end{cases} \quad (4.119)$$

In Physics Handbook [12] we find the value 98.4×10^{-8} Ωm for the static resistivity, ρ_0. In Figure 4.28 we show the low-frequency region on a log-log plot.

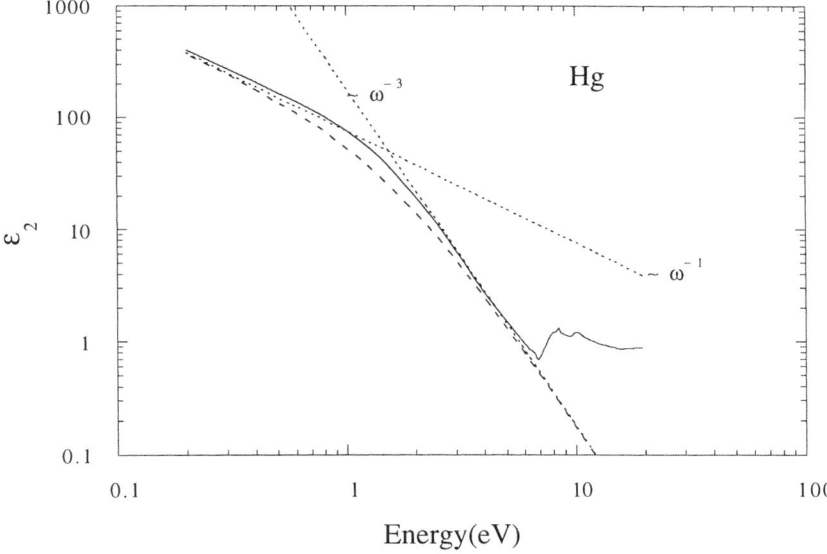

Figure 4.28: The imaginary part (solid curve) of the dielectric function of mercury as compared to the Drude result (dashed curve); asymptotes are also drawn.

The solid line is the experimental results and the dashed the Drude expression. The dotted straight lines are the two asymptotes. The results agree quite well but once again we see that there are additional contributions in the range 6 eV and upwards. These are believed to be due to excitations from d-bands. The mercury atoms have the configuration 5d^{10} 6s^2. Thus the *d*-electrons start to contribute before we reach the expected plasmon energy from the *s*-electrons. This is the reason for the discrepancy between the theoretical and experimental plasmon energies. We used the expected plasmon energy in the Drude expression and it worked quite well. This indicates that the problem is not from choosing the wrong number of contributing electrons.

In Figure 4.29 we compare the Drude result with both real and imaginary parts of the dielectric function for the whole experimental energy range. We have added a background dielectric constant of 2.4 to the real part. The Drude result is then

$$\varepsilon(\omega) = 2.4 - \left(4\pi/\omega_{pl}\right)^2 \frac{1}{\left(\rho_0\right)^2 + \left(4\pi\omega/\omega_{pl}^2\right)^2} + i\frac{4\pi}{\omega}\frac{\rho_0}{\left(\rho_0\right)^2 + \left(4\pi\omega/\omega_{pl}^2\right)^2} \quad (4.120)$$

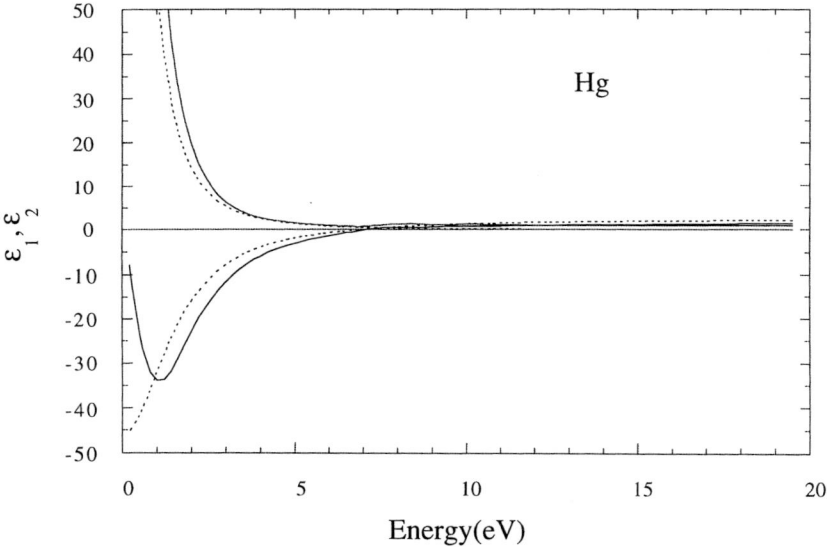

Figure 4.29: The theoretical results (dotted curves) for the real and imaginary parts of the dielectric function for mercury compared to the experimental results (solid curves).

The most striking difference between prediction and experiment is the upward turn of the experimental real part of the dielectric function towards the low energy side. This is due to additional excitations. We found in Figure 4.28 deviations in the imaginary part, as compared to the Drude result, in a limited energy range; there is an extra hump which as usual leads to a reduction in the real part for higher energies and an enhancement for lower energies.

4.4.3 Surface tension of mercury

Let us now try to determine the surface tension for mercury, using the same approach as in Section 4.4.1. The surface tension, or surface energy, is the energy per unit area needed to split a piece of material in two. One bulk plasmon is in this process changed into two surface plasmons and this causes the change in energy. The bulk plasmon energy is obtained from:

$$0 = \varepsilon_1(\omega_{pl}) = 2.4 - \left(\omega_{pl}^0/\omega_{pl}\right)^2 \qquad (4.121)$$

This gives the experimental value 7.0 eV. For the surface plasmon we get the energy by solving:

$$0 = \varepsilon_1(\omega_{spl}) + 1 = 3.4 - \left(\omega_{pl}^0/\omega_{spl}\right)^2 \qquad (4.122)$$

This gives the value 5.88 eV. The surface energy is, with the approximations used by Schmit and Lucas, given by

$$\begin{aligned}
\sigma_s &= \frac{1}{2}\left\{\frac{1}{A}\sum_q \frac{1}{2}\left[2\hbar\omega_{spl} - \hbar\omega_{pl}\right]\right\} \\
&= \frac{1}{2}\left\{\int_0^{k_c} \frac{d^2q}{(2\pi)^2} \frac{1}{2}\left[2\hbar\omega_{spl} - \hbar\omega_{pl}\right]\right\} \\
&= \frac{1}{2}\left\{\int_0^{k_c} \frac{dq\,q}{(2\pi)} \frac{1}{2}\left[2\hbar\omega_{spl} - \hbar\omega_{pl}\right]\right\} \\
&= \frac{(k_c)^2}{16\pi}\left[2\hbar\omega_{spl} - \hbar\omega_{pl}\right] \\
&= \frac{1}{16\pi}\left(\frac{\omega_{spl}}{v_F}\right)^2 \left[2\hbar\omega_{spl} - \hbar\omega_{pl}\right] \\
&= \frac{1}{16\pi}\left(\frac{\hbar\omega_{spl}}{\hbar^2 k_F/m}\right)^2 \left[2\hbar\omega_{spl} - \hbar\omega_{pl}\right] \\
&= 0.488 J/m^2 = 0.488 N/m
\end{aligned} \qquad (4.123)$$

This is to be compared to the value given in Physics Handbook [12]: 0.490 N/m. The agreement is quite good.

4.5 Quantum wells

In this section we will deal with a system that is easier to treat than that of two metallic half-spaces discussed in the previous sections. We will consider a model representing two parallel quantum wells. We will treat these as two-dimensional (2D) metallic sheets. In a 2D metallic sheet there are no bulk and surface. The collective modes are so-called 2D plasmons. Their dispersion does not, as in a 3D metal, start out at a constant frequency value; it starts out at zero and has a square root behavior for small momenta. The model very well represents a real system of quantum wells and these systems have attracted much attention during the last number of years. The reason is that one has managed to demonstrate experimentally current drag between the wells. If a current is forced to flow in one of the wells the carriers drag along the carriers in the other well so that also a current in that well flows. The electrons in the two wells scatter against each other.

We will use this model to calculate the force between the sheets as function of separation. One finds a very distinct division between three different separation ranges in which the distance dependence of the force is different. We will furthermore once again demonstrate that the interaction may be derived in two different ways; one way with many-body theory and the other from the zero-point energy of the modes.

4.5.1 Casimir and van der Waals forces between two 2D metallic sheets

The Casimir and van der Waals forces between objects can be approached in different ways. In one of the ways one studies the interactions between the carriers in the objects and how the interaction energy changes with separation between the objects. This energy is the correlation energy. Another approach is to study how the zero-point energy of the electromagnetic modes of the system changes with separation. It may be difficult to see any connection between these two approaches. However, the modes are determined by the boundary conditions at the surfaces of the objects, and these boundary conditions are the results of the adjustment of the charged carriers in the system to all electromagnetic fields. The two treatments are just the two sides of the same coin. Let us call the first approach, or method, the correlation energy method (CEM) and the second, the mode-summation method (MSM). In the MSM one can go along two different routes. Either one studies the change in the zero-point energy of the modes or one calculates the radiation pressure from the normal modes on the objects and thereby obtains the forces. We follow the first route.

Here, we will briefly describe the results from a calculation [13] on a model system of two strictly 2D metallic sheets embedded in a dielectric medium. This model system very well represents the "current-drag" system of Reference [14]. This system consists of two narrow electron quantum wells in GaAs. Each well can to a good approximation be treated as

4.5 Quantum Wells

strictly 2D as long as the following two conditions are fulfilled: the separation between the wells is big enough, so that the wave functions in the different wells are not overlapping; the wells are narrow enough so that in each well only one level is occupied and the closest unoccupied level is far enough up in energy for interband transitions to be negligible.

For the calculations one needs the dielectric properties of the system. The elements of the dielectric matrix, when retardation effects are neglected, are found to be:

$$\begin{cases} \text{Inlayer screening}: \varepsilon_{22}^{-1} = \varepsilon_{11}^{-1} = \dfrac{V_1}{V_{ext}^{(1)}} = \dfrac{1+\alpha_0(1-c_q^2)}{1+2\alpha_0 + \alpha_0^2(1-c_q^2)} \\ \qquad = \left[1+\alpha_0(1-c_q^2)\right]\left[1+\alpha_0(1+c_q)\right]^{-1}\left[1+\alpha_0(1-c_q)\right]^{-1} \\ \text{Interlayer screening}: \varepsilon_{21}^{-1} = \varepsilon_{12}^{-1} = \dfrac{V_2}{V_{ext}^{(2)}} = \dfrac{1}{1+2\alpha_0+\alpha_0^2(1-c_q^2)} \\ \qquad = \left[1+\alpha_0(1+c_q)\right]^{-1}\left[1+\alpha_0(1-c_q)\right]^{-1} \end{cases} \quad (4.124)$$

where $c_q = exp(-qd)$ and $\alpha_0(q,\omega)$ is the 2D polarizability. The potentials $V_{ext}^{(1)}$, $V_{ext}^{(2)}$, V_1, and V_2 are the external potential from an external charge distribution placed in layer *1*, the corresponding potential in layer *2*, the resulting potential in layer *1* when the potentials from all induced charges are taken into account and the resulting potential in layer *2*, respectively.

The dispersions of the collective modes of our system are obtained as the $\omega(q)$ for which the inverse dielectric functions diverge (or, equivalently, when the determinant of the dielectric matrix is zero). For this choice of ω and q we may have an induced electric field even in the absence of external perturbations. Thus, the dispersion curves for the collective modes are obtained as solutions when the denominator (the inter- and intra-layer versions of the inverse dielectric function have the same denominator) is put equal to zero. At infinite separation the denominator is just the 2D dielectric function squared. This function is zero on the plasmon dispersion curve. Thus we have two degenerate solutions in this limit. When the separation decreases the degeneracy is lifted. One curve moves upwards and the other downwards. This is illustrated in Figure 4.30 which is valid for $d = 500$ Å. At zero separation the lower one is pushed completely into the continuum and the upper one is the solution to the equation: $1+2\alpha_0(q,\omega) = 0$.

All the modes, for increasing q, sooner or later enter the continuum. The number of modes does not decrease with one each time a mode enters the continuum. The number of continuum modes increases with one at this point. We are facing a problem with this approach since we cannot determine where each mode goes. The energy of the collective mode is shared among the continuum modes. We use the approximation that the mode stays at the boundary of the continuum. With this approximation we get for the energy:

$$E(d) = \sum_{\mathbf{q}} \left\{ \left[\tfrac{\hbar}{2}\omega_1(q,d) + \tfrac{\hbar}{2}\omega_2(q,d) \right] - \left[\tfrac{\hbar}{2}\omega_1(q,\infty) + \tfrac{\hbar}{2}\omega_2(q,\infty) \right] \right\}$$
$$= A \int_0^\infty \frac{d^2q}{(2\pi)^2} \left\{ \left[\tfrac{\hbar}{2}\omega_1(q,d) + \tfrac{\hbar}{2}\omega_2(q,d) \right] - \left[\tfrac{\hbar}{2}\omega_1(q,\infty) + \tfrac{\hbar}{2}\omega_2(q,\infty) \right] \right\}$$
(4.125)

where $\omega_1(q,d)$ and $\omega_2(q,d)$ are the two collective modes and A is the area of each quantum well.

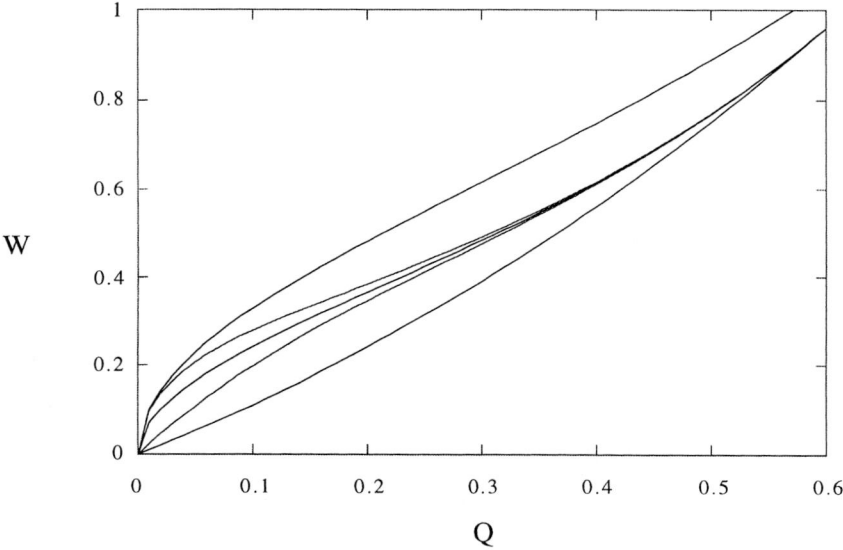

Figure 4.30: Dispersion of the collective modes for two 2D metallic sheets. The lowest curve is the upper boundary of the single-particle continuum. The middle curve is the two degenerate plasmon modes at infinite separation. The two curves closest to the middle curve are the two modes for finite separation. The upper curve is the mode that survives at zero separation.

We have used the energy at infinite separation as our reference energy. The result from this calculation is represented by curve 2 in Figure 4.31. Here, we have only included the plasmon modes, which are the only modes that are separated from the continuum. As we saw in Chapter 3 there are also modes inside the continuum that contribute to the energy and force. These can not be easily treated in the MSM. One way to take care of these would be to replace the dielectric function with that of the plasmon pole approximation. Then the contribution from the continuum modes are handed over to the only remaining collective mode. However, the result is then only approximate.

The main result of this work is presented in Figure 4.31. We present all results in the form of the interaction energy, E, per unit area. The force, F, is obtained from these results

4.5 Quantum Wells

just by performing the derivative with respect to separation. The asymptotes a1 and a2 represent the van der Waals and Casimir results, respectively. They are the large separation limits of the non-retarded and retarded treatments, respectively. The van der Waals result, a1, depends on the material parameters according to:

$$\begin{cases} E \approx -0.012562 e\hbar \sqrt{n/\kappa m^*} d^{-5/2} \\ F \approx 0.031406 e\hbar \sqrt{n/\kappa m^*} d^{-7/2} \end{cases} \quad (4.126)$$

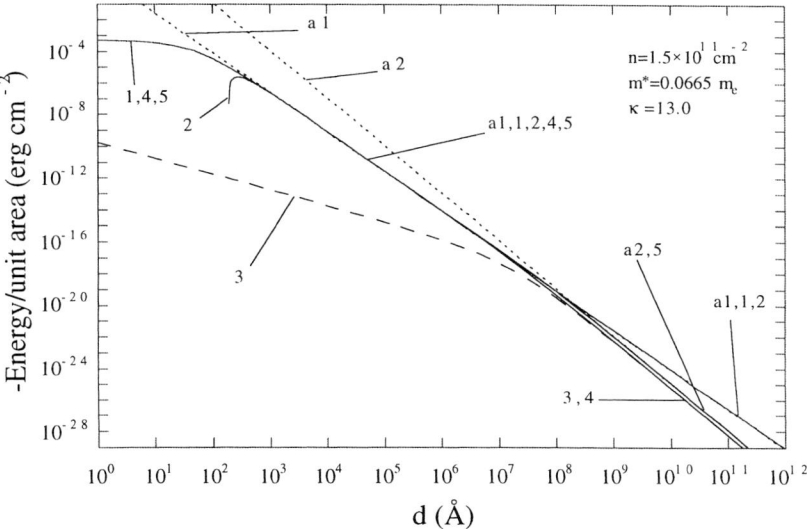

Figure 4.31: Interaction energy between two 2D metallic sheets as function of separation in different approximations.

This d-dependence is different from the behavior of two semi-infinite solids where E and F depend on d as d^{-2} and d^{-3}, respectively. It does not agree with the result for two thin non-metallic films either where the dependencies are d^{-4} and d^{-5}, respectively.

The Casimir result, a2, on the other hand agrees with the result for two semi-infinite metals. It is

$$\begin{cases} E = -\dfrac{\hbar \tilde{c} \pi^2}{720 d^3} \\ F = \dfrac{\hbar \tilde{c} \pi^2}{240 d^4} \end{cases} \quad (4.127)$$

The asymptote a2 has a steeper slope than a1 and the two asymptotes cross at the separation: $d \approx 1.1907 m^* c^2/ne^2$. To be noted is that a2 does not depend on the material parameters of the sheets.

Curve 1 represents our full non-retarded result from the CEM. It approaches the Van der Waals result for intermediate and large separations. In the correlation energy both single particle and collective excitations contribute. The result is obtained as an integral over the ωq-plane. In the small q-limit the contributions from the collective excitations dominate, and one can show that these contributions are just the zero-point energy of the modes. An exponential factor in the integrand guarantees that for large separations only the small momentum region contributes. Hence, in that limit the correlation energy approaches the zero-point energy of the modes. Curve 2 is the result from the calculation where the interaction energy comes entirely from the zero-point energy of the plasmon modes. We find that with this approach we actually get a repulsive force for small separations and a negative adhesion energy. As we mentioned earlier we let the modes stay at the boundary of the continuum instead of entering it. Another possibility would be to let the modes drop to zero energy as soon as they enter the continuum. We have also calculated the energy and force with this approximation. The results are quantitatively quite different but qualitatively the same, with a repulsive force at short range. For intermediate and large separations the result merges with curve 1 and the Van der Waals result, a1, and the result is independent of how we treat the modes when they merge with the single-particle continuum. Thus for large separations the approach, where only the zero-point energy of the plasmons is calculated, works well. Since this means a simpler calculation this approach may be preferable. This is no longer true for very large separations where the retardation effects become dominant. It turns out that in this regime the force caused by the Coulomb interaction is suppressed and replaced by photon interactions.

The result in the CEM is obtained in the form of the following integral:

$$E_c^{RPA}(d) - E_c^{RPA}(\infty) = \frac{\hbar}{4\pi^2} \int_0^\infty \int_0^\infty d\omega dq\, q \ln\left[1 - \frac{c_q^2 \alpha'_0{}^2(q,\omega)}{(1+\alpha'_0(q,\omega))^2}\right] \qquad (4.128)$$

where the primes indicate that the polarizabilities are calculated on the imaginary frequency axes. In the CEM both the longitudinal collective excitations and transverse single-particle excitations contribute to the energy. For large separations and neglect of retardation effects the collective part dominates the force. When retardation effects are included the collective part of the force is suppressed and another contribution appears. For large separations when retardation effects are important one has to include interactions between the electrons derived from the vector potential and not just from the scalar potential. This means that also transverse fields or photons contribute. It turns out that the longitudinal fields and the p-polarized transverse fields combine into one contribution and the s-polarized transverse fields give a separate contribution. The result from the first contribution is

$$E_{l+p}(d) = \frac{\hbar}{4\pi^2} \int_0^\infty \int_0^\infty d\omega dq\, q \ln\left\{1 - e^{-2qd\gamma'(q,\omega)}\left[\frac{\gamma'(q,\omega)\alpha'_0(q,\omega)}{1+\gamma'(q,\omega)\alpha'_0(q,\omega)}\right]^2\right\} \qquad (4.129)$$

4.5 Quantum Wells

where

$$\gamma'(q,\omega) = \gamma(q,i\omega) = \sqrt{1+(\omega/\tilde{c}q)^2} \qquad (4.130)$$

The result from the second contribution is

$$E_s(d) = \frac{\hbar}{4\pi^2} \int_0^\infty \int_0^\infty d\omega dq\, q \ln\left\{1 - e^{-2qd\gamma'(q,\omega)}\left[\frac{2\pi e^2 n/qm^* c^2 \gamma'(q,\omega)}{1+2\pi e^2 n/qm^* c^2 \gamma'(q,\omega)}\right]^2\right\} \qquad (4.131)$$

Curve 3 is the contribution from *s*-polarized photons. Curve 4 represents the contribution from the longitudinal and *p*-polarized interactions. It reproduces the full non-retarded result for small and intermediate separations. When it approaches asymptote a2 it changes direction and follows a curve representing half of the Casimir result. The total retarded result, that is, the sum of curves 3 and 4 is represented by curve 5. This full result coincides with the non-retarded result for small and intermediate separations and with the Casimir result for large separations. For large separations the scalar-potential contribution is suppressed and partly cancelled by the vector-potential contribution. This is maybe most clearly seen in the dispersion curves of the plasmon modes. These are, when retardation is taken into account, in the long–wave-length–limit pushed down below the dispersion curve for the light.

One interesting thing to notice is that the full result does not follow the upper one of the two asymptotes a1, and a2. Instead, it follows the lower one. This is contrary to what one would imagine since the result is the sum of the plasmon and photon contributions. The reason is that when the photons start to contribute the plasmon contribution drops in size, and vice versa.

The full non-retarded and full retarded results are the same for small and intermediate separations but split up for larger separations. In Figure 4.32 we have expanded the bifurcation region.

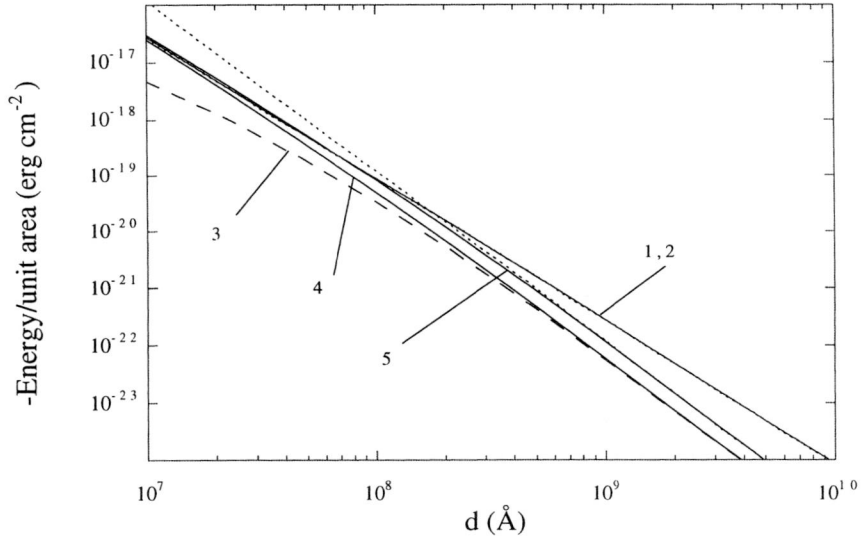

Figure 4.32: Figure 4.31 expanded around the bifurcation region.

Figure 4.33 shows the full retarded results for different carrier concentrations. The curves are shifted towards higher energies with increasing densities but they all approach the same density-independent asymptote for large separations. In the figure we have given the results for the densities 1×10^{10}, 1×10^{11}, 1×10^{12} and 1×10^{13} cm^{-2}, respectively. These examples span the reasonable density range for the real system we have chosen to model. The vertical bar in the figure indicates the 2D Thomas-Fermi screening length which is density independent, in contrast to the case in 3D. There is a clear connection between this screening length and the separation where the short range, single-particle correlations become important.

Let us now return to Figure 4.31 and study the full result, curve 4. This represents a general behavior for the energy and force between objects. There is an intermediate region where the interaction is of the van der Waals type. For larger separations retardation effects set in and there is a rather distinct change in separation dependence; the interaction drops off faster with separation; the interaction is of Casimir type. For small separations non-locality effects set in and the interaction saturates. For a metallic system, like the one treated here, this occurs at distances comparable to the Thomas-Fermi screening length. For semiconductors or insulators this happens at separations of the order of the size of the unit cell.

4.5 Quantum Wells

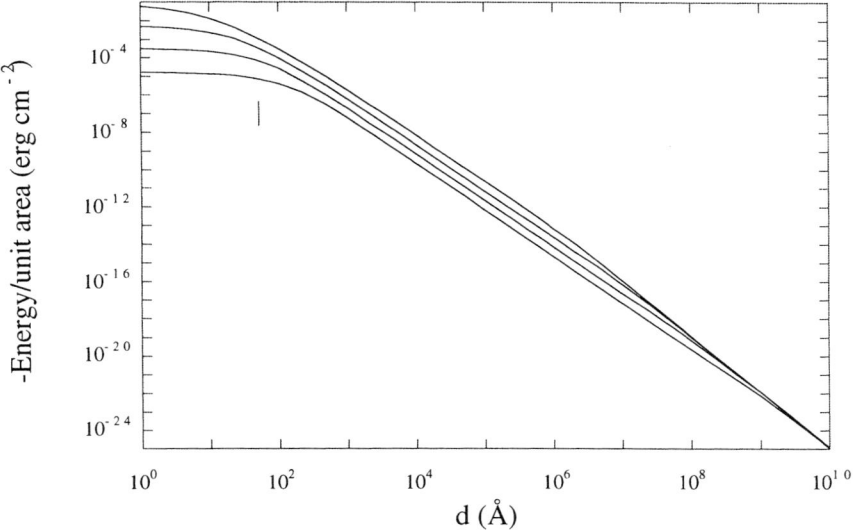

Figure 4.33: The full result for different carrier concentrations. The vertical line indicates the position of the Thomas-Fermi screening length. Se the test for details.

4.5.2 Plasmon-pole approximation

In the plasmon-pole approximation the single-particle continuum and the plasmon line are combined into one single collective mode. This means that the MSM works for all separations and the result is identical with that from the CEM. The plasmon-pole dielectric-function [15] is constructed in the following way: Peaks in the dynamical structure factor are interpreted as excitations of the system. If the full Random-Phase-Approximation (RPA) dielectric-function is used the structure factor has two peaks for small Q values (long wave lengths), one from e-h excitations (single-particle excitations) and one from plasmon excitations (collective excitations). The plasmon peak is for zero temperature a δ-function and is the dominating of the two peaks. The peaks come closer for higher momentum, eventually merge and only the e-h peak survives. The e-h peak is centered around $W = Q^2$, where we have used the scaled frequency and momentum as defined in Equation (4.133). The plasmon-pole dielectric-function is constructed in such a way that the structure factor consists of a single δ-function for all Q-values. For small Q-values it starts out like the plasmon peak and for large Q-values it follows the $W = Q^2$ curve. We demand that the weight of the δ-function is such that the f-sum rule is fulfilled and that the real part of the dielectric function has the correct static both long– and short– wave-length–limits as well as the correct high-frequency limit, that is,

$$\varepsilon(Q,0) \underset{Q \to \infty}{=} 1 + \frac{y}{2Q^3} \; ;$$

$$\varepsilon(Q,0) \underset{Q \to 0}{=} 1 + \frac{y}{Q} \; ; \qquad (4.132)$$

$$\varepsilon(Q,W) \underset{W \to \infty}{=} 1 - \frac{yQ}{2W^2}$$

where the functions have been expressed in the following dimensionless variables:

$$W = \frac{\hbar\omega}{4E_F} \; ; \quad E_F = \frac{\hbar^2 k_F^2}{2m^*}$$

$$Q = q/2k_F \; ; \quad P = p/2\hbar k_F \qquad (4.133)$$

$$y = \frac{m^* e^2}{\hbar^2 \kappa k_F} \; ; \quad D = d 2k_F$$

These demands result in

$$\varepsilon_B(Q,W) = 1 + \frac{yQ}{2\left(Q^2/2 + Q^4 - W^2 - i\delta\right)}$$

$$= 1 + \frac{yQ}{2\left(Q^2/2 + Q^4 - W^2\right)}$$

$$+ i \frac{yQ}{2\sqrt{Q^4 + Q^2/2}} \frac{\pi}{2} \left\{ \delta\left[W - \sqrt{Q^4 + Q^2/2}\right] + \delta\left[W + \sqrt{Q^4 + Q^2/2}\right] \right\}$$

$$(4.134)$$

From this follows

$$\alpha'_{0,B}(Q,W) = \alpha_{0,B}(Q,iW) = \frac{yQ}{2\left(W^2 + Q^2/2 + Q^4\right)} \qquad (4.135)$$

and

4.5 Quantum Wells

$$\varepsilon_B^{-1}(Q,W) = \frac{W^2 - (Q^2/2 + Q^4)}{W^2 - (yQ/2 + Q^2/2 + Q^4) + i\delta}$$

$$= \frac{W^2 - (Q^2/2 + Q^4)}{W^2 - (yQ/2 + Q^2/2 + Q^4)} \quad (4.136)$$

$$- i\frac{yQ/2}{W_{pl,B}(Q)} \frac{\pi}{2} \left\{ \delta\left[W - W_{pl,B}(Q)\right] + \delta\left[W + W_{pl,B}(Q)\right] \right\}$$

where

$$W_{pl,B}(Q) = \sqrt{yQ/2 + Q^2/2 + Q^4} \quad (4.137)$$

The results [16] from the calculation with the CEM for the plasmon-pole approximation and the full RPA dielectric function are presented in Figure 4.34.

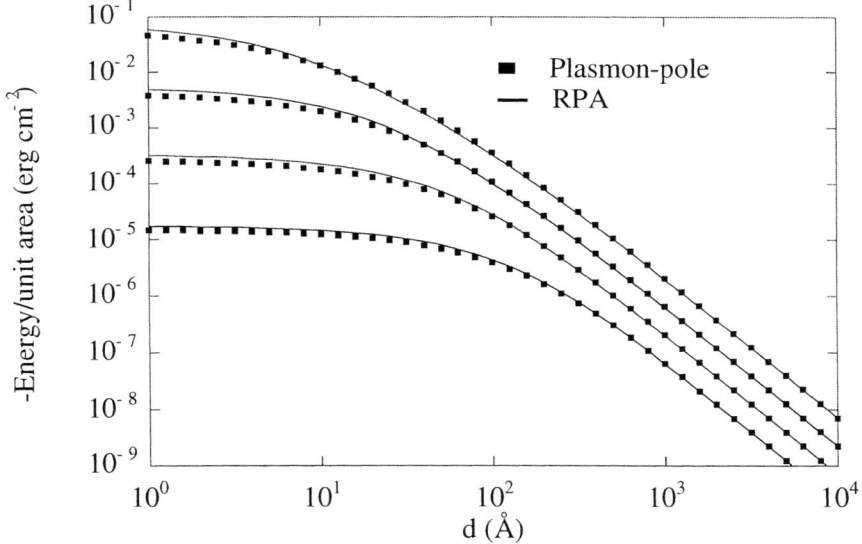

Figure 4.34: The interaction energy calculated with the plasmon-pole approximation (squares) and with full RPA (solid curves) for four different carrier concentrations.

The full result is represented by the full curve and the result from plasmon-pole approximation with squares. The results are for carrier densities in each sheet of 1×10^{10}, 1×10^{11}, 1×10^{12}, and 1×10^{13} cm^{-2}, respectively, counting the curves from below. The result from the plasmon-pole approximation (filled squares) is rather good for all densities.

In the MSM one attributes the interaction energy to the change in the sum of the zero-point energies of all the collective modes in the system. The dispersion curves for these modes are obtained as the frequency where the determinant of the dielectric matrix is zero, or equivalently where the denominators of the elements of the inverse dielectric matrix, below, are zero

$$\begin{cases} \varepsilon_{22}^{-1} = \varepsilon_{11}^{-1} = \dfrac{1+\alpha_0\left(1-c_q^2\right)}{1+2\alpha_0+\alpha_0^2\left(1-c_q^2\right)} \\ \varepsilon_{21}^{-1} = \varepsilon_{12}^{-1} = \dfrac{1}{1+2\alpha_0+\alpha_0^2\left(1-c_q^2\right)} \end{cases} \quad (4.138)$$

Thus, the modes are found as solutions to the equation

$$1+2\alpha_0(q,\omega)+\alpha_0^2(q,\omega)\left(1-c_q^2\right)=0 \quad (4.139)$$

There are two modes, $\omega_1(q)$ and $\omega_2(q)$. The interaction energy is given by

$$\begin{aligned} E(d) &= \frac{1}{\Omega}\sum_{\mathbf{q}}\left\{\left[\tfrac{\hbar}{2}\omega_1(q,d)+\tfrac{\hbar}{2}\omega_2(q,d)\right]-\left[\tfrac{\hbar}{2}\omega_1(q,\infty)+\tfrac{\hbar}{2}\omega_2(q,\infty)\right]\right\} \\ &= \int\frac{d^2q}{(2\pi)^2}\left\{\left[\tfrac{\hbar}{2}\omega_1(q,d)+\tfrac{\hbar}{2}\omega_2(q,d)\right]-\left[\tfrac{\hbar}{2}\omega_1(q,\infty)+\tfrac{\hbar}{2}\omega_2(q,\infty)\right]\right\} \end{aligned} \quad (4.140)$$

The energies of the two modes are found to be

$$W_{1,2}(Q) = \sqrt{Q^4+Q^2/2+yQ\left(1\pm e^{-DQ}\right)/2} \quad (4.141)$$

where D is the dimensionless separation introduced earlier in Equation (4.133). The integrals cannot be solved analytically. However we can easily find the asymptotic D-dependence for large D-values. In this limit only long wavelengths, that is, small Q-values, contribute. The plasmon dispersion for small Q, is

$$W_{1,2}(Q) \approx \sqrt{yQ\left(1\pm e^{-DQ}\right)/2} \quad (4.142)$$

and the asymptotic result is

$$E \approx -0.012562e\hbar\sqrt{n/\kappa m^*}\,d^{-5/2} \quad (4.143)$$

The results from the full calculations are identical to the results from the CEM. The results from the MSM and CEM are identical.

References

[1] S. Q. Wang and G. D. Mahan, Phys. Rev. **B6**, 4517 (1972).
[2] M. J. Sparnaay, Physica (Utrecht) **24**, 751 (1958); *Physics in the Making*, edited by A. Sarlemijn and M. J. Sparnaay (North-Holland, Amsterdam, 1989).
[3] S. K. Lamoreaux, Phys. Rev. Lett. **78**, 5 (1997).
[4] U. Mohideen and Anushree Roy, Phys. Rev. Lett. **81**,4549 (1998); B. W. Harris, F. Chen, and U. Mohideen, Phys. Rev. **A62**, 052109 (2000).
[5] M. Abramowitz, and I. A. Stegun, Handbook of Mathematical functions, formula 3.6.28.
[6] J. Schmit and A.A. Lucas, Solid State Commun. **11**, 415 (1972).
[7] R. A. Craig, Phys. Rev. **B6**, 1134 (1972).
[8] N. D. Lang and W. Kohn, Phys. Rev. **B1**, 4555 (1970).
[9] H. B. G. Casimir, D. Polder, Phys. Rev. **73**, 360 (1948); H. B. G. Casimir, J. Chim. Phys. **47**, 407 (1949).
[10] D. Kleppner, Physics Today, October 1990, p. 9.
[11] E. T. Arakawa, and T. Inagaki, *Handbook of optical constants of solids II* (Academic Press 1991), p. 461.
[12] C. Nordling and J. Österman, *Physics Handbook for Science and Engineering*, (Studentlitteratur, Lund 1980), **T-1.5**, p 36.
[13] Bo. E. Sernelius and Patrik Björk, Phys. Rev. **B57**, 6592 (1998).
[14] T. J. Gramila, J. P. Eisenstein, A. H. MacDonald, L. N. Pfeiffer, and K. W. West, Phys. Rev. Lett. **66**, 1216 (1991).
[15] L. Hedin, and S. Lundqvist, Solid State Physics **23**, 1 (1969).
[16] M. Boström and Bo. E. Sernelius, Physica Scripta **T79**, 89 (1999).

5 Forces

Why do thunderstorms bring rain?

There are four fundamental forces in nature. Some properties of these forces are summarized in Table 5.1. All these forces are necessary for the existence of the world we live in. The strongest of these forces are the strong and weak nuclear forces. They are responsible for the stability of the atomic nucleus. They are extremely short range and are therefore not directly observed in our every day life. The gravitational force is the weakest of the four. The gravitation between two electrons, for example, can be neglected as compared to the electromagnetic repulsion. It is weaker by a factor of 2.4×10^{-43}. Both the gravitational and electromagnetic forces are long range and have the same distance dependence. They vary as r^{-2} and the potentials as r^{-1}. Still, the gravitational force dominates between planets and is responsible for the fact that objects on earth are not thrown out into space. The cause of this advantage for the gravitation is that big objects are basically charge neutral and that positive and negative charges have opposite and cancelling effects on the electromagnetic force. It would not take a very large unbalance in charge to out balance the gravitational force between earth and the moon, for example. It would be enough to have a surplus of electrons, say, on both objects of an amount equal to the number of electrons normally contained in 1000 m^3 of a material.

Table 5.1: Comparison of the four fundamental forces in nature.

Interaction	Field Quanta	Coupling Constant	Range (m)
Electromagnetic	γ	$\alpha \approx 1/137$	∞
Strong Nuclear	Gluons	$\alpha_S \approx 100\,\alpha$	10^{-15}
Weak Nuclear	$W^{\pm}, Z^0$	$\alpha_W \approx 4\,\alpha$	10^{-18}
Gravitational	graviton?	$\alpha_G \approx 10^{-39}$	∞

Thus, it is not trivial to characterize the forces by their strength, since the strongest forces have the shortest range and the weakest is strongest at large separations. Furthermore, the potential from a charge in a polarizable medium with free carriers is short range, it varies as $v(r) = q\exp(-r/\lambda_{TF})/r$ instead of $v(r) = q/r$ as in free space. The Thomas-Fermi screening length, λ_{TF}, is of the order of the atomic spacing for ordinary metals while for doped semiconductors it can be from 10 to 100 Å. The nuclear forces are often modeled by a potential of the same type, a Yukawa potential, where now λ_{TF} is replaced by the range of the potential. The short range of the nuclear forces are probably due to an effect similar to the screening effect.

In our every day life the gravitational force makes us stay on the ground and objects we drop fall to the ground. The gravitational force from the moon induces the tide. Apart from these examples gravitation goes more or less unnoticed. The most important force is the electromagnetic one. This force prevents us from falling to the center of earth and it allows us to pick up things with our hands. The electromagnetic interaction is furthermore responsible for us being able to see things; it causes the reflection and refraction of light.

The forces and optical effects we treat in this book are all of electromagnetic character. The forces are always present. We will treat larger objects but let us start with atoms or molecules. The forces are always larger if the atoms or molecules are charged or have permanent dipole moments. However, in a system with many atoms or molecules there are screening effects that reduce the effective interaction between the charged particles and thermal fluctuations prevent the permanent dipoles to align. This means that the forces we are concerned with here are in most cases as important as those derived from the charges or permanent dipoles.

The force between two isolated ions or charged molecules are dominated by the direct Coulomb interaction, but the charges also induce dipoles and higher order multipoles in the two particles that lead to other forces. An ion and a neutral atom, for example are attracted to each other because of this induction. For neutral atoms or molecules one distinguishes between three types of forces: orientation, induction and dispersion forces. We will go through some examples below.

5.1 Two molecules with permanent dipole moments

In this section we derive the interaction energy and force between two molecules with permanent dipole moments.

The electric field from an electric dipole $\mathbf{p}_1$ at the origin is given by

$$\mathbf{E} = -\left[\frac{\mathbf{p}_1}{r^3} - \frac{3(\mathbf{p}_1 \cdot \mathbf{r})\mathbf{r}}{r^5}\right] \tag{5.1}$$

This can alternatively be expressed as

$$\mathbf{E} = -\tilde{\phi}\mathbf{p}_1 \; ; \; E_\mu = -\sum_\nu \phi_{\mu\nu} p_{1,\nu} \tag{5.2}$$

where the dipole-dipole matrix or tensor is

5.1 Two Molecules with Permanent Dipole Moments

$$\phi_{\mu\nu} = \frac{\delta_{\mu\nu}}{r^3} - \frac{3r_\mu r_\nu}{r^5} \tag{5.3}$$

The energy of a dipole in an electric field is

$$U = -\mathbf{p} \cdot \mathbf{E} = -\sum_\mu p_\mu E_\mu \tag{5.4}$$

Thus, the interaction energy between two dipoles is

$$U_{12} = -\mathbf{p}_1 \cdot \mathbf{E}_2 = -\mathbf{p}_2 \cdot \mathbf{E}_1 = \left[\frac{\mathbf{p}_2 \cdot \mathbf{p}_1}{r_{12}^3} - \frac{3(\mathbf{p}_1 \cdot \mathbf{r}_{12})(\mathbf{p}_2 \cdot \mathbf{r}_{12})}{r_{12}^5}\right] = \sum_\mu \sum_\nu p_{1,\mu} \phi_{\mu\nu}^{12} p_{2,\nu} \tag{5.5}$$

where the superscript of the dipole-dipole tensor indicates that its r-value is the separation between the two dipoles 1 and 2. It can also be rewritten on matrix form as

$$U_{12} = \mathbf{p}_2^* \cdot \tilde{\phi} \cdot \mathbf{p}_1 \tag{5.6}$$

where the dipole on the left is represented by a row vector, indicated by the star, and that on the right by a column vector. This can be written as

$$U_{12} = \frac{p_2 p_1}{r_{12}^3}\left[\cos(\theta_{12}) - 3\cos(\theta_1)\cos(\theta_2)\right] \tag{5.7}$$

where θ_i is the angle between dipole i and the line joining the two dipoles and θ_{12} is the angle between the dipoles.

If the orientation of the molecules are at random the interaction energy vanishes. At zero temperature, at least classically, the molecules are oriented so that their dipoles are aligned. This gives the lowest energy:

$$T = 0: U_{12} = -\frac{2 p_2 p_1}{r_{12}^3} \tag{5.8}$$

For finite temperatures the size of the energy is lower and vanishes at infinite temperature. For finite temperatures the results are obtained as the weighted average of the energy over all orientations. The weighting factor is the Boltzmann weighting factor, $\exp(-\beta U_{12})$. The result for high temperatures is

$$U_{12} \approx -\frac{2}{3}\left(\frac{p_2 p_1}{r_{12}^3}\right)^2 \beta \quad ; \quad \beta \ll \left(\frac{p_2 p_1}{r_{12}^3}\right)^{-1} \tag{5.9}$$

The separation dependence, r^{-6}, is the same as for the van der Waals interaction, which will be treated later. We mention this because in the beginning the van der Waals interactions was mistaken for the interaction between permanent dipoles.

Note that for finite temperatures the force is obtained as the derivative of Helmholtz' free energy with respect to separation and in the high temperature limit this quantity is half as big as the energy.

5.2 One ion and one molecule with permanent dipole moment

In this section we derive the interaction energy and force between an ion of charge q and a molecule with permanent dipole moment $\mathbf{p}$. This problem is important for ions in polar solvents like water.

The interaction energy is in this case

$$U_{12} = -\mathbf{p} \cdot \mathbf{E} = -\mathbf{p} \cdot \mathbf{r}\frac{q}{r^3} = -\frac{qp}{r^2}\cos\theta \tag{5.10}$$

The force between the objects is

$$\mathbf{F} = -\frac{dU_{12}}{dr}\hat{\mathbf{r}} = -\frac{2qp}{r^3}\cos\theta\,\hat{\mathbf{r}} \tag{5.11}$$

This force is attractive or repulsive, depending on the orientation of the dipole. The force is strong at the distance of contact between the ion and molecule and this explains the fact that it rains in connection with thunderstorms; the ions attract water molecules and act as nucleation centers. At very high temperatures the direction of the dipole is random and the average energy is zero. For zero temperature we have

$$U_{12} = -\frac{qp}{r^2} \quad ; \quad T = 0 \tag{5.12}$$

and for high temperatures:

$$U_{12} \approx -\frac{1}{3}\left(\frac{qp}{r^2}\right)^2 \beta \ ; \ \beta << \left(\frac{qp}{r^2}\right)^{-1} \qquad (5.13)$$

Note that for finite temperatures the force is obtained as the derivative of Helmholtz' free energy with respect to separation and in the high temperature limit this quantity is half as big as the energy.

5.3 Two molecules one with and one without permanent dipole moment

In this section we derive the interaction energy and force between two molecules where one has a permanent dipole moment and the other has not. If only one of the molecules has a permanent dipole moment we will still have a force between the molecules. This is also true if the second particle is an atom instead of a molecule. The electric field from the permanent dipole moment induces a dipole moment in the other molecule:

$$\mathbf{p}_2 = \alpha^{mol}\mathbf{E} \qquad (5.14)$$

and the interaction energy is

$$\begin{aligned} U_{12} &= -\tfrac{1}{2}\mathbf{p}_2 \cdot \mathbf{E} = \tfrac{1}{2}\mathbf{p}_2 \cdot \left[\frac{\mathbf{p}_1}{r^3} - \frac{3(\mathbf{p}_1 \cdot \mathbf{r})\mathbf{r}}{r^5}\right] \\ &= -\tfrac{1}{2}\alpha^{mol}\left[\frac{\mathbf{p}_1}{r^3} - \frac{3(\mathbf{p}_1 \cdot \mathbf{r})\mathbf{r}}{r^5}\right] \cdot \left[\frac{\mathbf{p}_1}{r^3} - \frac{3(\mathbf{p}_1 \cdot \mathbf{r})\mathbf{r}}{r^5}\right] \\ &= -\tfrac{1}{2}\alpha^{mol}\frac{(p_1)^2}{r^6}\left[1 + 3(\hat{\mathbf{p}}_1 \cdot \hat{\mathbf{r}})^2\right] = -\tfrac{1}{2}\alpha^{mol}\frac{(p_1)^2}{r^6}\left[1 + 3\cos^2(\theta)\right] \end{aligned} \qquad (5.15)$$

where α^{mol} is the molecular polarizability. The factor 1/2 is present because the interaction is induced. We have assumed that the molecular polarizability is a constant. For molecules that are not spherically symmetric or for atoms with partially filled electron shells the polarizability is actually a tensor and the induced polarization is not in the direction of the electric field from the permanent dipole.

The force between the molecules is

$$\mathbf{F} = -\frac{dU_{12}}{dr}\hat{\mathbf{r}} = -3\alpha^{mol}\frac{(p_1)^2}{r^7}\left[1 + 3\cos^2(\theta)\right]\hat{\mathbf{r}} \tag{5.16}$$

Thus the energy and the attractive force vary with separation as r^{-6} and r^{-7}, respectively. Thus, this interaction, having the same variation as the van der Waals interaction, means a possible source of misinterpretation.

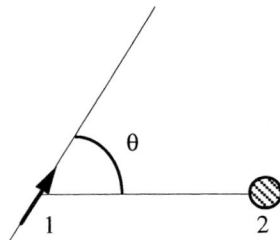

Figure 5.1: One polarizable atom or molecule (2) and one with permanent dipole moment (1).

On matrix form we have

$$U_{12} = -\frac{1}{2}\mathbf{p}_1^* \cdot \tilde{\phi} \cdot \tilde{\alpha}^{mol} \cdot \tilde{\phi} \cdot \mathbf{p}_1 \tag{5.17}$$

If the molecule or atom is spherically symmetric the polarizability tensor becomes a scalar function times the unit tensor, and

$$U_{12} = -\frac{1}{2}\alpha^{mol}\mathbf{p}_1^* \cdot \tilde{\phi} \cdot \tilde{\phi} \cdot \mathbf{p}_1 = -\frac{1}{2}\alpha^{mol}\mathbf{p}_1^* \cdot \tilde{\phi}^2 \cdot \mathbf{p}_1 \tag{5.18}$$

where

$$\tilde{\phi}^2{}_{\mu\nu} = \frac{\delta_{\mu\nu}}{r^6} + \frac{3r_\mu r_\nu}{r^8} \tag{5.19}$$

5.4 Two molecules without permanent dipole moments

In this section we will derive the interaction energy and force between two molecules without permanent dipole moments. We will neglect retardation effects; this is allowed except for very large separations between the atoms. In Chapter 6 we will return to this problem and include retardation effects; the treatment becomes more complicated but we feel that the picture would not be complete without it.

Johan Diderik van der Waals (1837-1923) graduated on his thesis: *On the continuity of the gaseous and liquid states*, in 1873. He found deviations for real gases to the ideal gas equation of state. He found empirically

$$\left(p + \frac{a}{V^2}\right)(V - b) = RT \tag{5.20}$$

instead of the ideal equation:

$$pV = RT \tag{5.21}$$

for a mole of gas. The correction constant b is due to the fact that the gas atoms take up a finite fraction of the volume, thus reducing the free volume. The factor a, which is of interest here, is due to the attractive force between the atoms, reducing the pressure exerted on the walls of the container. He was awarded the Nobel Price in 1910 for this and similar work on the equations of state for gases and fluids. We will derive the van der Waals equation of state in Section 5.13.

Van der Waals' force was found on empirical grounds and it was not until 1930 that London [1] gave a realistic explanation for this force. We will go through a similar derivation here. There will be an attractive force between two atoms or molecules even if none of them carry a permanent dipole moment. This is still true if the particles are spherically symmetric. This is a bit counter intuitive. One would imagine that two spherically symmetric, neutral atoms would not interact were they so far apart that their electron wave functions were not overlapping.

Let us return to the Lorentz model and study two atoms with one bound electron in each. The binding energies may be different. The driving fields for one atom are the ones produced by the dipole moments from the other. Thus from Equations (2.2) and (5.2) we have the two coupled differential equations

$$\begin{cases} \tilde{m}_1 \cdot \ddot{\mathbf{r}}_1 + \tilde{m}_1 \cdot \tilde{\Gamma}_1 \cdot \dot{\mathbf{r}}_1 + \omega_1^2 \tilde{m}_1 \cdot \mathbf{r}_1 = e\tilde{\phi}^{12} \cdot \mathbf{p}_2 \\ \tilde{m}_2 \cdot \ddot{\mathbf{r}}_2 + \tilde{m}_2 \cdot \tilde{\Gamma}_2 \cdot \dot{\mathbf{r}}_2 + \omega_2^2 \tilde{m}_2 \cdot \mathbf{r}_2 = e\tilde{\phi}^{21} \cdot \mathbf{p}_1 \end{cases} \tag{5.22}$$

The first (second) equation is the equation of motion for the electron on atom number *1* (*2*) and the external force on the right hand side is that produced by the dipole moment of atom number *2* (*1*). We have generalized Equation (2.2) to be valid also for anisotropic systems by letting the mass be represented by a tensor.

We note that

$$\mathbf{p} = -e\mathbf{r} \tag{5.23}$$

which gives

$$\begin{cases} \tilde{1} \cdot \ddot{\mathbf{p}}_1 + \tilde{\Gamma}_1 \cdot \dot{\mathbf{p}}_1 + \omega_1^2 \tilde{1} \cdot \mathbf{p}_1 = -e^2 \tilde{m}_1^{-1} \cdot \tilde{\phi}^{12} \cdot \mathbf{p}_2 \\ \tilde{1} \cdot \ddot{\mathbf{p}}_2 + \tilde{\Gamma}_2 \cdot \dot{\mathbf{p}}_2 + \omega_2^2 \tilde{1} \cdot \mathbf{p}_2 = -e^2 \tilde{m}_2^{-1} \cdot \tilde{\phi}^{21} \cdot \mathbf{p}_1 \end{cases} \tag{5.24}$$

Fourier transforming (see Appendix 2) these differential equations gives

$$\begin{cases} \tilde{1} \cdot \mathbf{p}_1(-i\omega)^2 + \tilde{\Gamma}_1 \cdot \mathbf{p}_1(-i\omega) + \omega_1^2 \tilde{1} \cdot \mathbf{p}_1 = -e^2 \tilde{m}_1^{-1} \cdot \tilde{\phi}^{12} \cdot \mathbf{p}_2 \\ \tilde{1} \cdot \mathbf{p}_2(-i\omega)^2 + \tilde{\Gamma}_2 \cdot \mathbf{p}_2(-i\omega) + \omega_2^2 \tilde{1} \cdot \mathbf{p}_2 = -e^2 \tilde{m}_2^{-1} \cdot \tilde{\phi}^{21} \cdot \mathbf{p}_1 \end{cases} \tag{5.25}$$

Let us now neglect the friction, and rearrange our equations:

$$\begin{cases} \left[\omega^2 - \omega_1^2\right] \tilde{1} \cdot \mathbf{p}_1 - e^2 \tilde{m}_1^{-1} \cdot \tilde{\phi}^{12} \cdot \mathbf{p}_2 = 0 \\ -e^2 \tilde{m}_2^{-1} \cdot \tilde{\phi}^{21} \cdot \mathbf{p}_1 + \left[\omega^2 - \omega_2^2\right] \tilde{1} \cdot \mathbf{p}_2 = 0 \end{cases} \tag{5.26}$$

Eliminating $\mathbf{p}_2$ gives

$$\left[\omega^2 - \omega_1^2\right]\left[\omega^2 - \omega_2^2\right]\tilde{1} \cdot \mathbf{p}_1 - e^2 \tilde{m}_1^{-1} \cdot \tilde{\phi}^{12} \cdot e^2 \tilde{m}_2^{-1} \cdot \tilde{\phi}^{21} \cdot \mathbf{p}_1 = 0 \tag{5.27}$$

Rearrangement leads to

$$\left\{ e^2 \tilde{m}_1^{-1} \cdot \tilde{\phi}^{12} \cdot e^2 \tilde{m}_2^{-1} \cdot \tilde{\phi}^{21} - \left[\omega^2 - \omega_1^2\right]\left[\omega^2 - \omega_2^2\right]\tilde{1} \right\} \cdot \mathbf{p}_1 = 0 \tag{5.28}$$

We look for the normal modes of the coupled system, that is, the ω values that solve this equation. These are found by letting the determinant of the matrix, multiplying $\mathbf{p}_1$ in Equation (5.28), be zero. This leads to an equation of the form:

5.4 Two Molecules without Permanent Dipole Moments

$$\left[\left(\omega^2 - \omega_1^2\right)\left(\omega^2 - \omega_2^2\right) - \lambda_1\right]\left[\left(\omega^2 - \omega_1^2\right)\left(\omega^2 - \omega_2^2\right) - \lambda_2\right]$$
$$\times \left[\left(\omega^2 - \omega_1^2\right)\left(\omega^2 - \omega_2^2\right) - \lambda_3\right] = 0 \quad (5.29)$$

where the λ-values are complicated combinations of matrix elements. We are interested in forces for large separations where dipole-dipole interactions dominate. It turns out that we do not need the full results in this case, we only need the sum of the λ-values. We note that Equation (5.28) is on a form similar to an eigenvalue equation:

$$\tilde{A} \cdot \mathbf{X} = \lambda \mathbf{X} = \lambda \tilde{1} \cdot \mathbf{X} \quad (5.30)$$

where now

$$\begin{cases} \tilde{A} = e^2 \tilde{m}_1^{-1} \cdot \tilde{\phi}^{12} \cdot e^2 \tilde{m}_2^{-1} \cdot \tilde{\phi}^{21} \\ \lambda = \left[\omega^2 - \omega_1^2\right]\left[\omega^2 - \omega_2^2\right] \end{cases} \quad (5.31)$$

The matrices are 3×3 matrices which means that there are three λ values. The sum of these values is equal to the trace of $\mathbf{A}$, that is, the sum of the diagonal elements of $\mathbf{A}$.

Now, the determinant in Equation (5.29) consists of three factors. Each factor leads to two solutions with positive frequency:

$$\Omega^i{}_{1,2} = \sqrt{\tfrac{1}{2}\left(\omega_1^2 + \omega_2^2\right) \pm \sqrt{\tfrac{1}{4}\left(\omega_1^2 - \omega_2^2\right)^2 + \lambda_i}} \quad ; \quad i = 1,2,3 \quad (5.32)$$

The interaction energy is

$$\Delta E = \sum_{i=1}^{3}\left[\tfrac{1}{2}\hbar\left(\Omega^i{}_1 + \Omega^i{}_2\right) - \tfrac{1}{2}\hbar(\omega_1 + \omega_2)\right]$$
$$\approx -\tfrac{1}{4}\hbar \frac{1}{\omega_1\omega_2(\omega_1 + \omega_2)} \sum_{i=1}^{3} \lambda_i \quad (5.33)$$

where the last relation is valid for large separations. Thus we have

$$\Delta E = -\tfrac{1}{4}\hbar \frac{1}{\omega_1\omega_2(\omega_1 + \omega_2)} Tr\left[e^2 \tilde{m}_1^{-1} \cdot \tilde{\phi}^{12} \cdot e^2 \tilde{m}_2^{-1} \cdot \tilde{\phi}^{21}\right] \quad (5.34)$$

Now, we may write

$$\frac{1}{\omega_1\omega_2(\omega_1+\omega_2)} = \frac{1}{\pi}\int_{-\infty}^{\infty} d\omega \frac{1}{\omega^2+\omega_1^2}\frac{1}{\omega^2+\omega_2^2} \tag{5.35}$$

and this means that

$$\Delta E = -\frac{\hbar}{4\pi}\int_{-\infty}^{\infty} d\omega Tr\left[\tilde{\alpha}_1(i\omega)\cdot\tilde{\phi}^{12}\cdot\tilde{\alpha}_2(i\omega)\cdot\tilde{\phi}^{21}\right] \tag{5.36}$$

This is equal to the result obtained from a rigorous many-particle derivation. The polarizabilities contain, in general, contributions from excitations of many different electrons with different excitation energies.

In the isotropic case we have

$$\begin{aligned}\Delta E &= -\frac{\hbar}{4\pi}\int_{-\infty}^{\infty} d\omega\alpha_1(i\omega)\alpha_2(i\omega)Tr\left[\tilde{\phi}^{12}\cdot\tilde{\phi}^{21}\right]\\ &= -\frac{\hbar}{4\pi}\int_{-\infty}^{\infty} d\omega\alpha_1(i\omega)\alpha_2(i\omega)\frac{1}{r^6}\left[1+3\frac{x^2}{r^2}+1+3\frac{y^2}{r^2}+1+3\frac{z^2}{r^2}\right]\\ &= -\frac{3\hbar}{2\pi}\frac{1}{r^6}\int_{-\infty}^{\infty} d\omega\alpha_1(i\omega)\alpha_2(i\omega)\\ &= -\frac{3\hbar}{\pi}\frac{1}{r^6}\int_{0}^{\infty} d\omega\alpha_1(i\omega)\alpha_2(i\omega)\end{aligned} \tag{5.37}$$

This is the van der Waals interaction. Often one of the excitation frequencies dominates and one can write [2]

$$\alpha_i(i\omega) \approx \frac{\alpha_i(0)}{1+(\omega/\omega_i)^2} \tag{5.38}$$

This is the so-called London approximation and it leads to

$$\begin{aligned}\Delta E &\approx -\frac{3\hbar}{\pi}\frac{\alpha_1(0)\alpha_2(0)}{r^6}\int_0^{\infty} d\omega\frac{1}{1+(\omega/\omega_1)^2}\frac{1}{1+(\omega/\omega_2)^2}\\ &= -\frac{3\hbar}{\pi}\frac{\alpha_1(0)\alpha_2(0)}{r^6}\frac{\pi}{2}\frac{\omega_1\omega_2}{\omega_1+\omega_2}\\ &= -\frac{3}{2}\frac{\alpha_1(0)\alpha_2(0)}{r^6}\frac{E_1 E_2}{E_1+E_2}\end{aligned} \tag{5.39}$$

If the two atoms are the same we have

$$\Delta E \approx -\frac{3}{4} \frac{\alpha^2(0)}{r^6} E_1 \qquad (5.40)$$

This interaction is responsible for the parameter a in Equation (5.20). Several other interactions, treated in the previous sections, have the same distance dependence. This is a latent source of misinterpretations.

5.5 Two ions

In the case of a polarizable atom near an ion also the charge of the ion induces a polarization of the atom. For two ions we have

$$V_{ij} = Z_i Z_j \frac{e^2}{r} - \frac{e^2}{2r^4}\left[Z_i^2 \alpha_{2,j} + Z_j^2 \alpha_{2,i}\right] - \frac{e^2}{2r^6}\left[Z_i^2 \alpha_{4,j} + Z_j^2 \alpha_{4,i}\right] - \ldots \qquad (5.41)$$

The first term is the direct Coulomb interaction, the second due to induced dipoles and the third to induced quadrupoles. The third term has the same separation dependence as the dipole-dipole contribution for neutral atoms, leading to the van der Waals interaction.

5.6 Three or more polarizable atoms

The interaction between three polarizable ions or atoms is usually known as the Axilrod-Teller [3] interaction. It is

$$\Delta E = -\frac{\hbar}{4\pi} \int_{-\infty}^{\infty} d\omega \Big\{ Tr\left[\tilde{\alpha}_1(i\omega) \cdot \tilde{\phi}^{12} \cdot \tilde{\alpha}_2(i\omega) \cdot \tilde{\phi}^{23} \cdot \tilde{\alpha}_3(i\omega) \cdot \tilde{\phi}^{31}\right] \\ + Tr\left[\tilde{\alpha}_1(i\omega) \cdot \tilde{\phi}^{13} \cdot \tilde{\alpha}_3(i\omega) \cdot \tilde{\phi}^{32} \cdot \tilde{\alpha}_2(i\omega) \cdot \tilde{\phi}^{21}\right] \Big\} \qquad (5.42)$$

For isotropic atoms we have

$$\Delta E = -\frac{\hbar}{2\pi} \int_{-\infty}^{\infty} d\omega \alpha_1(i\omega)\alpha_2(i\omega)\alpha_3(i\omega) Tr\left[\tilde{\phi}^{12} \cdot \tilde{\phi}^{23} \cdot \tilde{\phi}^{31}\right] \qquad (5.43)$$

This follows from the cyclic properties of the trace and the fact that the dipole-dipole tensor is symmetric. The product of the three tensors is

$$\left[\tilde{\phi}^{12} \cdot \tilde{\phi}^{23} \cdot \tilde{\phi}^{31}\right]_{\mu\nu} = \frac{1}{r_{12}^3 r_{23}^3 r_{31}^3}\left[\delta_{\mu\nu} - 3\frac{r_{12,\mu} r_{12,\nu}}{r_{12}^2} - 3\frac{r_{23,\mu} r_{23,\nu}}{r_{23}^2} - 3\frac{r_{31,\mu} r_{31,\nu}}{r_{31}^2}\right.$$
$$+ 9\frac{r_{12,\mu} r_{23,\nu}}{r_{12}^2 r_{23}^2}(\mathbf{r}_{12} \cdot \mathbf{r}_{23}) + 9\frac{r_{23,\mu} r_{31,\nu}}{r_{23}^2 r_{31}^2}(\mathbf{r}_{23} \cdot \mathbf{r}_{31})$$
$$+ 9\frac{r_{31,\mu} r_{12,\nu}}{r_{31}^2 r_{12}^2}(\mathbf{r}_{31} \cdot \mathbf{r}_{12})$$
$$\left. + 27\frac{r_{23,\mu} r_{31,\nu}}{r_{12}^2 r_{23}^2 r_{31}^2}(\mathbf{r}_{12} \cdot \mathbf{r}_{23})(\mathbf{r}_{31} \cdot \mathbf{r}_{12})\right]$$

$$(5.44)$$

To be noted is that this tensor is not symmetric with respect to the interchange of two of the three **r**-vectors. However the trace is and we have

$$Tr\left[\tilde{\phi}^{12} \cdot \tilde{\phi}^{23} \cdot \tilde{\phi}^{31}\right] = \frac{1}{r_{12}^3 r_{23}^3 r_{31}^3}\left[-6 + 9\frac{1}{r_{12}^2 r_{23}^2}(\mathbf{r}_{12} \cdot \mathbf{r}_{23})^2 + 9\frac{1}{r_{23}^2 r_{31}^2}(\mathbf{r}_{23} \cdot \mathbf{r}_{31})^2\right.$$
$$+ 9\frac{1}{r_{31}^2 r_{12}^2}(\mathbf{r}_{31} \cdot \mathbf{r}_{12})^2$$
$$\left. + 27\frac{1}{r_{12}^2 r_{23}^2 r_{31}^2}(\mathbf{r}_{12} \cdot \mathbf{r}_{23})(\mathbf{r}_{23} \cdot \mathbf{r}_{31})(\mathbf{r}_{31} \cdot \mathbf{r}_{12})\right]$$
$$= \frac{3}{r_{12}^3 r_{23}^3 r_{31}^3}\left[-2 + 3\cos^2(\theta_2) + 3\cos^2(\theta_3) + 3\cos^2(\theta_1)\right.$$
$$\left. + 9\cos(\theta_1)\cos(\theta_2)\cos(\theta_3)\right]$$
$$= \frac{3}{r_{12}^3 r_{23}^3 r_{31}^3}\left[1 + 3\cos(\theta_1)\cos(\theta_2)\cos(\theta_3)\right]$$

$$(5.45)$$

We find

5.6 Three or More Polarizable Atoms

$$\Delta E = -\frac{3\hbar}{2\pi r_{12}^3 r_{23}^3 r_{31}^3}[1+3\cos(\theta_1)\cos(\theta_2)\cos(\theta_3)]\int_{-\infty}^{\infty}d\omega\alpha_1(i\omega)\alpha_2(i\omega)\alpha_3(i\omega)$$

$$= -\frac{3\hbar}{2\pi r_{12}^3 r_{23}^3 r_{31}^3}[1+3\cos(\theta_1)\cos(\theta_2)\cos(\theta_3)]\int_{-\infty}^{\infty}d\omega\alpha^3(i\omega)$$

(5.46)

where the last line is valid if the three atoms are identical. Using the London approximation we find

$$\Delta E = -\frac{9\alpha_{d,i}(0)^3 E_i}{16 r_{12}^3 r_{23}^3 r_{31}^3}[1+3\cos(\theta_1)\cos(\theta_2)\cos(\theta_3)]$$

(5.47)

Thus we find that the introduction of one additional atom modifies the interaction between two original atoms. This can lead to *catalytic effects*.

The derivation above can be extended to more than three atoms. For four atoms we have

$$\Delta E = -\frac{\hbar}{4\pi}\frac{1}{4}\sum_{\substack{ijkl=1\\ \text{all different}}}^{4}\int_{-\infty}^{\infty}d\omega Tr\left[\tilde{\alpha}_i(i\omega)\cdot\tilde{\phi}^{ij}\cdot\tilde{\alpha}_j(i\omega)\cdot\tilde{\phi}^{jk}\cdot\tilde{\alpha}_k(i\omega)\cdot\tilde{\phi}^{kl}\cdot\tilde{\alpha}_l(i\omega)\cdot\tilde{\phi}^{li}\right]$$

(5.48)

By always keeping the same atom first, atom *1* say, we can drop the factor 1/4 in front of the summation and get

$$\Delta E = -\frac{\hbar}{4\pi}\sum_{\substack{jkl=2\\ \text{all different}}}^{4}\int_{-\infty}^{\infty}d\omega Tr\left[\tilde{\alpha}_1(i\omega)\cdot\tilde{\phi}^{1j}\cdot\tilde{\alpha}_j(i\omega)\cdot\tilde{\phi}^{jk}\cdot\tilde{\alpha}_k(i\omega)\cdot\tilde{\phi}^{kl}\cdot\tilde{\alpha}_l(i\omega)\cdot\tilde{\phi}^{l1}\right]$$

(5.49)

For a general *n* number of atoms we have

$$\Delta E = -\frac{\hbar}{4\pi}\frac{1}{n}\sum_{\substack{j_1 j_2\ldots j_n=1\\ \text{all different}}}^{n}\int_{-\infty}^{\infty}d\omega Tr\left[\tilde{\alpha}_{j_1}(i\omega)\cdot\tilde{\phi}^{j_1 j_2}\cdot\tilde{\alpha}_{j_2}(i\omega)\cdots\tilde{\phi}^{j_n j_1}\right]$$

$$= -\frac{\hbar}{4\pi}\sum_{\substack{j_2\ldots j_n=2\\ \text{all different}}}^{n}\int_{-\infty}^{\infty}d\omega Tr\left[\tilde{\alpha}_1(i\omega)\cdot\tilde{\phi}^{1 j_2}\cdot\tilde{\alpha}_{j_2}(i\omega)\cdots\tilde{\phi}^{j_n 1}\right]$$

(5.50)

5.7 Interaction between macroscopic objects

When we want to determine the force between macroscopic objects we can in principle go on as before and sum the contribution to the energy from two-particle, three-particle... interactions between all the atoms constituting the objects. In some of the contributions all atoms are within the same object. These contributions are part of the binding energy of that object and do not contribute to the force between the objects.

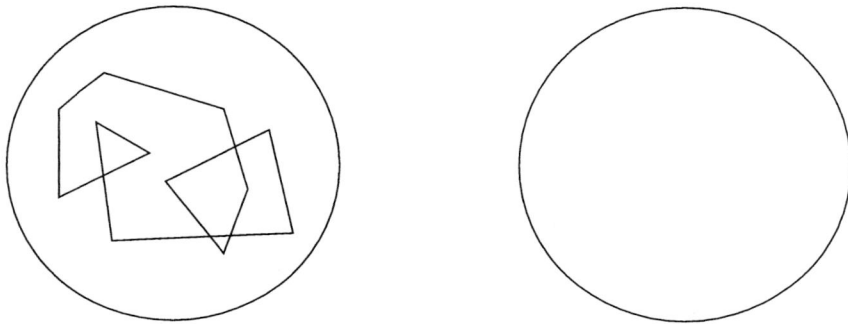

Figure 5.2: Interactions that contribute to the binding energy of the object to the left, but not to the interaction energy and force between the objects.

The contribution in the Figure 5.3 belong to the contribution making up the force to the lowest order. Next order contribution has four lines connecting the two objects.

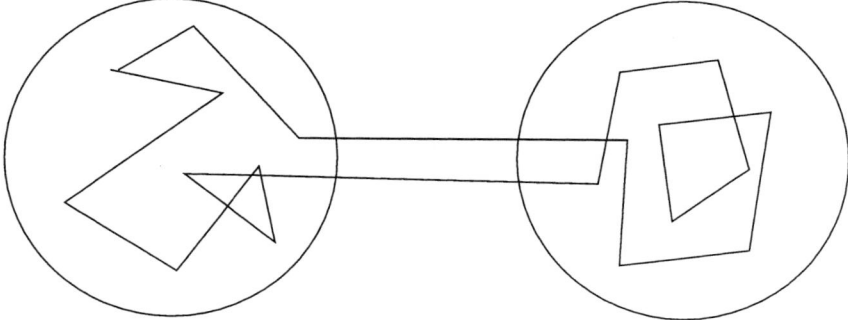

Figure 5.3: Interactions that contribute to the lowest order to the interaction energy and force between the objects.

The loops inside the objects can be summed separately and make a screening of the potentials. When the objects contain many atoms it is much better to treat the whole object as a continuous particle where the atoms respond collectively to the potentials and there are induced dipoles and higher order multipoles. In principle we solve the Maxwell's equations with the the proper boundary conditions at the surfaces of the objects and thereby find the normal modes of the system. Then from this we calculate the energy as a function of separation and find the force.

5.8 Interaction between two spheres: limiting results

To find the normal modes for a system of two spheres is in general very complicated. However, in the two limiting cases of large and small separations the problem is feasible. We start by considering these limits since this often is enough.

For large separations only dipole-dipole interactions are important and the energy variation with separation is the same as for two atoms:

$$\Delta E = -\frac{3\hbar}{\pi}\frac{1}{r^6}\int_0^\infty d\omega \alpha_1(i\omega)\alpha_2(i\omega) \qquad (5.51)$$

The polarizabilities of the spheres are

$$\alpha_i(\omega) = R_i^3 \frac{\varepsilon_i(\omega)-1}{\varepsilon_i(\omega)+2} \qquad (5.52)$$

For metal spheres we have

$$\alpha_i(0) = R_i^3 \quad ; \quad E_i = \frac{\hbar\omega_{pl,i}}{\sqrt{3}} \qquad (5.53)$$

which leads to

$$\Delta E = -\frac{\sqrt{3}}{2}\frac{R_1^3 R_2^3}{r^6}\frac{\hbar\omega_{pl,1}\hbar\omega_{pl,2}}{\hbar\omega_{pl,1}+\hbar\omega_{pl,2}} \qquad (5.54)$$

and for two identical metallic spheres:

$$\Delta E = -\frac{\sqrt{3}}{4}\frac{R^6}{r^6}\hbar\omega_{pl} \qquad (5.55)$$

For small separations the surfaces at closest distance appear flat and the result is quite reasonable. It is

$$\Delta E \approx -\frac{\hbar}{16\pi^2 \ell^2}\left(\frac{2\pi R_1 R_2 \ell}{R_1 + R_2}\right)\int_0^\infty d\omega \frac{[\varepsilon_1(i\omega)-1][\varepsilon_2(i\omega)-1]}{[\varepsilon_1(i\omega)+1][\varepsilon_2(i\omega)+1]} \quad ; \quad \ell = r - R_1 - R_2 \tag{5.56}$$

which can be viewed as the energy between two flat surfaces with a finite area, an effective area:

$$A_{eff} = \left(\frac{2\pi R_1 R_2 \ell}{R_1 + R_2}\right) \tag{5.57}$$

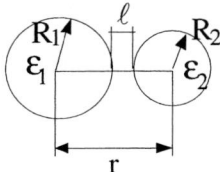

Figure 5.4: The notation used in case of two interacting spheres.

5.9 Interaction between two spheres: general results

The general result [4] for the interaction energy between two spheres is

$$\Delta E_{12} = -(\hbar/4\pi)\int_{-\infty}^{\infty} d\omega \sum_{h=1}^{\infty} h^{-1} \sum_{l_{11},l_{12},\ldots l_{1h}=1}^{\infty} \sum_{l_{21},l_{22},\ldots l_{2h}=1}^{\infty} C(\mathbf{l}_1,\mathbf{l}_2)$$
$$\cdot \Delta(l_{11},1)(R_1/r_{21})^{2l_{11}+1} \Delta(l_{21},2)(R_2/r_{21})^{2l_{21}+1} \ldots \Delta(l_{2h},2)(R_2/r_{21})^{2l_{2h}+1} \tag{5.58}$$

where

$$C(\mathbf{l}_1,\mathbf{l}_2) = \sum_m \binom{l_{11}+l_{21}}{l_{11}+m}\binom{l_{21}+l_{12}}{l_{21}+m}\cdots\binom{l_{1h}+l_{2h}}{l_{1h}+m}\binom{l_{2h}+l_{11}}{l_{2h}+m} \tag{5.59}$$

5.9 Interaction Between two Spheres: General Results

The result is for zero temperature and the integral is performed along the imaginary frequency axis. The multipole susceptibilities for a sphere surrounded by a medium with dielectric function ε are

$$\Delta(l,j) = \frac{l(\varepsilon_j - \varepsilon)}{l\varepsilon_j + (l+1)\varepsilon} \tag{5.60}$$

For large separations we keep the lowest order contribution ($h=1$):

$$\Delta E_{12} = -(\hbar/4\pi) \int_{-\infty}^{\infty} d\omega \cdot \Delta(1,1)(R_1/r_{21})^3 \Delta(1,2)(R_2/r_{21})^3 6$$

$$= -(\hbar 6/4\pi)(R_1/r_{21})^3 (R_2/r_{21})^3 \int_{-\infty}^{\infty} d\omega \cdot \frac{(\varepsilon_1 - \varepsilon)}{\varepsilon_1 + 2\varepsilon} \frac{(\varepsilon_2 - \varepsilon)}{\varepsilon_2 + 2\varepsilon} \tag{5.61}$$

$$= -(3\hbar/2\pi)\frac{1}{r^6} \int_{-\infty}^{\infty} d\omega \cdot \alpha_1(i\omega)\alpha_2(i\omega)$$

where we have used:

$$C(1,1) = \sum_m \binom{1+1}{1+m}\binom{1+1}{1+m} = \binom{1+1}{1-1}\binom{1+1}{1-1} + \binom{1+1}{1+0}\binom{1+1}{1+0} + \binom{1+1}{1+1}\binom{1+1}{1+1} = 6 \tag{5.62}$$

The largest contribution goes as r^{-6} and agrees with what we presented in the previous section. Next largest contribution goes as r^{-8} and is

$$\Delta E_{12} = -\frac{\hbar}{2} \int_{-\infty}^{\infty} \frac{d\omega}{2\pi} \left[C(1,2)\Delta(1,1)\left(\frac{R_1}{r_{21}}\right)^3 \Delta(2,2)\left(\frac{R_2}{r_{21}}\right)^5 \right.$$

$$\left. + C(2,1)\Delta(2,1)\left(\frac{R_1}{r_{21}}\right)^5 \Delta(1,2)\left(\frac{R_2}{r_{21}}\right)^3 \right]$$

$$= -\frac{15}{(r_{21})^8} \frac{\hbar}{2} \int_{-\infty}^{\infty} \frac{d\omega}{2\pi} \left[\Delta(1,1)(R_1)^3 \Delta(2,2)(R_2)^5 \right. \tag{5.63}$$

$$\left. + \Delta(2,1)(R_1)^5 \Delta(1,2)(R_2)^3 \right]$$

$$= -\frac{15}{(r_{21})^8} \frac{\hbar}{2} \int_{-\infty}^{\infty} \frac{d\omega}{2\pi} \left[\alpha_{1,1}(i\omega)\alpha_{2,2}(i\omega) + \alpha_{2,1}(i\omega)\alpha_{1,2}(i\omega) \right]$$

where in the last line the first index represents the multipole order and the second the object number. These higher multipole interactions also occur for molecules. We only considered dipoles earlier. Thus the spheres behave as molecules.

For finite temperatures the integration along the imaginary axis is replaced by a summation over discrete frequencies:

$$\int_{-\infty}^{\infty} \frac{d\omega}{2\pi} f(i\omega) \to \frac{1}{\hbar\beta} \sum_{n=-\infty}^{\infty} f(i\omega_n) \; ; \; \omega_n = \frac{n 2\pi}{\hbar\beta} \tag{5.64}$$

This will be shown in Section 6.2.

If we now expand the van der Waals interaction between two molecules or between two spheres at large separations in powers of the separation we have

$$V_{vdW}(r) = -\frac{C_6}{r^6} - \frac{C_8}{r^8} - \frac{C_{10}}{r^{10}} - \cdots \tag{5.65}$$

The first term is the result of dipole-dipole interactions, the second of dipole-quadrupole-interactions and the third term has contributions from dipole-octupole and quadrupole-quadrupole interactions.

Thus we can divide the C constants into the different contributions:

$$\begin{cases} C_{6,i,j} = C_{i,j}(1,1) \\ C_{8,i,j} = C_{i,j}(1,2) + C_{i,j}(2,1) \\ C_{10,i,j} = C_{i,j}(1,3) + C_{i,j}(2,2) + C_{i,j}(3,1) \\ \vdots \end{cases} \tag{5.66}$$

where

$$C_{i,j}(M,L) = \frac{(2M+2L)!}{(2M)!(2L)!} \int_{-\infty}^{\infty} \frac{d\omega}{2\pi} \left[\alpha_{M,i}(i\omega)\alpha_{L,j}(i\omega)\right] \tag{5.67}$$

The 2^M pole polarizability is

$$\alpha_{M,i}(i\omega) = R_i^{2M+1} \frac{M[\varepsilon_i(i\omega) - \varepsilon(i\omega)]}{[M\varepsilon_i(i\omega) + (M+1)\varepsilon(i\omega)]} \tag{5.68}$$

where $\varepsilon(i\omega)$ is the dielectric constant, or function, of the surrounding medium.

5.9.1 Radially varying dielectric functions

In the case of two spheres with radially varying dielectric functions the result is [4]

$$\Delta E_{12} = -\frac{\hbar}{32\pi} \int_{-\infty}^{\infty} d\omega \int_0^{R_1} dr_i \, d\ln\varepsilon_1(i\omega)/dr_i \int_0^{R_2} dr_j \, d\ln\varepsilon_2(i\omega)/dr_j$$

$$\times \left\{ \frac{r_i r_j}{r_{12}^2 - (r_i + r_j)^2} + \frac{r_i r_j}{r_{12}^2 - (r_i - r_j)^2} + \frac{1}{2} \ln \frac{r_{12}^2 - (r_i + r_j)^2}{r_{12}^2 - (r_i - r_j)^2} \right\}$$

(5.69)

If now the dielectric functions are constant throughout the spheres the derivatives of the logarithms of the dielectric functions become δ-functions,

$$d\ln\varepsilon_1(i\omega)/dr_i \to -\delta(r_i - R_1) \ln \frac{\varepsilon_1(i\omega)}{\varepsilon(i\omega)}$$

(5.70)

and we get

$$\Delta E_{12} = -\frac{\hbar}{32\pi} \int_{-\infty}^{\infty} d\omega \ln \frac{\varepsilon_1(i\omega)\varepsilon_2(i\omega)}{\varepsilon^2(i\omega)}$$

$$\times \left\{ \frac{R_1 R_2}{r_{12}^2 - (R_1 + R_2)^2} + \frac{R_1 R_2}{r_{12}^2 - (R_1 - R_2)^2} + \frac{1}{2} \ln \frac{r_{12}^2 - (R_1 + R_2)^2}{r_{12}^2 - (R_1 - R_2)^2} \right\}$$

(5.71)

If the spheres have the same radius, R, and the separation between the spheres is ℓ, we have

$$\Delta E_{12} = -\frac{\hbar}{64} \int_{-\infty}^{\infty} \frac{d\omega}{2\pi} \ln \frac{\varepsilon_1(i\omega)\varepsilon_2(i\omega)}{\varepsilon^2(i\omega)} \left\{ \frac{1}{x(x+2)} + \frac{1}{(x+1)^2} + 2\ln \frac{x(x+2)}{(x+1)^2} \right\}$$

(5.72)

where $x = \ell/2R$.

5.10 General expression for small separations

For small separations one can find a general expression for the interaction energy between spheres, cylinders and half-spaces [4]:

$$\Delta E_{12} = -\frac{\hbar}{32\pi^2}\left[\frac{2\pi R_1 R_2}{R_1+R_2}\right]^{1-n/2}\Gamma(1+n/2)L^n d^{-(1+n/2)}\sum_{l=1}^{\infty}\frac{\langle\omega_l\rangle}{l^3} \quad (5.73)$$

where $n = 0$ is for spheres, $n = 1$ for cylinders and $n = 2$ for half-spaces. The characteristic dielectric integrals are

$$\langle\omega_l\rangle = \int_{-\infty}^{\infty} d\omega \left[\frac{\varepsilon_1(i\omega)-\varepsilon(i\omega)}{\varepsilon_1(i\omega)+\varepsilon(i\omega)}\frac{\varepsilon_2(i\omega)-\varepsilon(i\omega)}{\varepsilon_2(i\omega)+\varepsilon(i\omega)}\right]^l \quad (5.74)$$

for zero temperature. For finite temperatures they are replaced by

$$\langle\omega_l\rangle = \frac{2\pi}{\hbar\beta}\sum_{i\omega_n}\left[\frac{\varepsilon_1(i\omega_n)-\varepsilon(i\omega_n)}{\varepsilon_1(i\omega_n)+\varepsilon(i\omega_n)}\frac{\varepsilon_2(i\omega_n)-\varepsilon(i\omega_n)}{\varepsilon_2(i\omega_n)+\varepsilon(i\omega_n)}\right]^l \quad (5.75)$$

where $\varepsilon(i\omega_n)$ is the dielectric function of the surrounding medium.

5.11 Cylinders and half-spaces

In the case of two cylinders at large separations we have [4]

$$\Delta E_{12} = -\left[\frac{\hbar\langle\omega_1\rangle L}{4\pi r_{21}}\right]\sum_{m,n=1}^{\infty}\frac{\Gamma^2\left(m+n+\frac{1}{2}\right)}{m!(m-1)!n!(n-1)!}\left(\frac{R_1}{r_{21}}\right)^{2m}\left(\frac{R_2}{r_{21}}\right)^{2n} \quad (5.76)$$

For cylinders the multipole susceptibilities are

5.11 Cylinders and Half-Spaces

$$\Delta(m,i) = \frac{(\varepsilon_i - \varepsilon)I_m(kR_i)(d/dR_i)I_m(kR_i)}{\varepsilon_i K_m(kR_i)(d/dR_i)I_m(kR_i) - \varepsilon I_m(kR_i)(d/dR_i)K_m(kR_i)}$$

$$= \frac{(\varepsilon_i - \varepsilon)R_i(d/dR_i)I_m^2(kR_i)}{(\varepsilon_i + \varepsilon) + (\varepsilon_i - \varepsilon)R_i(d/dR_i)I_m(kR_i)K_m(kR_i)} \quad (5.77)$$

where the I_m and K_m functions are the so-called modified Bessel functions [5].
For two half spaces we have

$$\Delta(i) = \frac{(\varepsilon_i - \varepsilon)}{(\varepsilon_i + \varepsilon)} \quad (5.78)$$

and

$$\Delta E_{12} = -\left(\frac{L^2}{2\pi d^2}\right) \frac{\hbar}{2} \int_{-\infty}^{\infty} \frac{d\omega}{2\pi} \int_0^{\infty} dxx \ln\left[1 - \Delta(1)\Delta(2)e^{-2x}\right]$$

$$= -\left(\frac{\hbar L^2}{32\pi^2 d^2}\right) \sum_{l=1}^{\infty} \frac{\langle \omega_l \rangle}{l^3} \quad (5.79)$$

This expression is valid for all separations (neglecting retardation).

To get the result for small separations between a sphere and a plane we can start from the expression for two spheres and let the radius of one go to infinity. The result is

$$\Delta E_{12} = -\frac{\hbar}{32\pi^2}\left[\frac{2\pi R_1 R_2}{R_1 + R_2}\right]d^{-1}\sum_{l=1}^{\infty}\frac{\langle \omega_l \rangle}{l^3} \to -\frac{\hbar R_1}{16\pi}d^{-1}\sum_{l=1}^{\infty}\frac{\langle \omega_l \rangle}{l^3} \quad (5.80)$$

It is the same as for two planes but with an effective area:

$$\Delta E_{12} = \frac{\hbar}{32\pi^2 d^2}[2\pi R_1 d]\sum_{l=1}^{\infty}\frac{\langle \omega_l \rangle}{l^3} \quad (L^2 \to 2\pi R_1 d) \quad (5.81)$$

where the expression within the brackets is the effective area.

For small separations between a cylinder and a plane we may start from the expression for two cylinders and let the radius of one go to infinity. The result is

$$\Delta E_{12} = -\frac{\hbar}{32\pi^2}\left[\frac{2\pi R_1 R_2}{R_1+R_2}\right]^{\frac{1}{2}}\Gamma\left(\frac{3}{2}\right)Ld^{-\frac{3}{2}}\sum_{l=1}^{\infty}\frac{\langle\omega_l\rangle}{l^3}$$

$$\rightarrow -\frac{\hbar}{32\pi^2}\sqrt{2\pi R_1}\,\Gamma\left(\frac{3}{2}\right)Ld^{-\frac{3}{2}}\sum_{l=1}^{\infty}\frac{\langle\omega_l\rangle}{l^3}$$

(5.82)

It is the same as for two planes but with an effective area:

$$\Delta E_{12} = -\frac{\hbar}{32\pi^2 d^2}\left[\Gamma\left(\frac{3}{2}\right)L\sqrt{2\pi R_1 d}\right]\sum_{l=1}^{\infty}\frac{\langle\omega_l\rangle}{l^3} \tag{5.83}$$

where the expression within the brackets is the effective area.

5.12 Summation of pair interactions

A much faster way to obtain results for the interaction energy is to just sum over pair interactions [4]. One treats the systems as composed of polarizable atoms or molecules and sums the contribution from all pairs. This gives, in general, results that have the right distance dependence but the strength is not quite right. The forces are not strictly additive. The result is asymptotically correct for diluted systems.

Let us perform such a calculation for two planes of thickness δ and separation d. We perform a general calculation that is valid both in the non-retarded and in the retarded limits. Let the interaction energy between two molecules be

$$V = -Br^{-\gamma} \tag{5.84}$$

In the non-retarded limit we have

$$B = \frac{3\hbar\omega_0\alpha(0)^2}{4}\,;\,\gamma = 6 \tag{5.85}$$

and in the retarded:

$$B = \frac{23\hbar c\alpha(0)^2}{4\pi}\,;\,\gamma = 7 \tag{5.86}$$

We have derived Equation (5.85) (see Equation (5.40)) but not yet Equation (5.86). We

5.12 Summation of Pair Interactions

will do that in Section 6.1.2 in a way analogous to the derivation of the result in the non-retarded limit. We will furthermore in Section 6.10 derive both these limiting results, the non-retarded and the retarded, in an alternative way.

Let us now return to the derivation. First we consider one atom a distance d from a layer of thickness δ:

$$V_{al}(d) = -Bn_2 \int_d^{d+\delta} dz \int d^2r \left(r^2 + z^2\right)^{-\gamma/2}$$
$$= -\frac{B2\pi n_2}{(\gamma-2)(\gamma-3)}\left[d^{-(\gamma-3)} - (d+\delta)^{-(\gamma-3)}\right] \tag{5.87}$$

This result can be used to find the results for two layers. We sum the contributions from all atoms in one layer interacting with the other layer. The energy between the planes is now

$$V_{ll}(d) = n_1 \text{Area} \int_0^{\delta} dz \left[-\frac{B2\pi n_2}{(\gamma-2)(\gamma-3)} \right]\left[(d+z)^{-(\gamma-3)} - (d+\delta+z)^{-(\gamma-3)}\right]$$
$$= -\frac{B2\pi n_2 n_1 \text{Area}}{(\gamma-2)(\gamma-3)} \frac{-1}{(\gamma-4)}\left[(d+z)^{-(\gamma-4)} - (d+\delta+z)^{-(\gamma-4)}\right]_0^{\delta}$$
$$= -\frac{B2\pi n_2 n_1 \text{Area}}{(\gamma-2)(\gamma-3)(\gamma-4)}\left[(d)^{-(\gamma-4)} - 2(d+\delta)^{-(\gamma-4)} + (d+2\delta)^{-(\gamma-4)}\right] \tag{5.88}$$

and the energy per unit area:

$$\frac{V_{ll}(d)}{\text{Area}} = -\frac{B2\pi n_2 n_1}{(\gamma-2)(\gamma-3)(\gamma-4)}\left[(d)^{-(\gamma-4)} - 2(d+\delta)^{-(\gamma-4)} + (d+2\delta)^{-(\gamma-4)}\right] \tag{5.89}$$

In the non-retarded van der Waals limit we have

$$\frac{V_{ll}(d)}{\text{Area}} = -\frac{\hbar\omega_0 \alpha(0)^2 \pi n_2 n_1}{16}\left[(d)^{-2} + (d+2\delta)^{-2} - 2(d+\delta)^{-2}\right] \tag{5.90}$$

or

$$\frac{V_{ll}(d)}{\text{Area}} = -\frac{A}{12\pi}\left[(d)^{-2} + (d+2\delta)^{-2} - 2(d+\delta)^{-2}\right] \tag{5.91}$$

where

$$A = \frac{3\pi^2 n^2 \alpha(0)^2 \hbar\omega_0}{4} \tag{5.92}$$

is the so-called *Boer-Hamaker* constant of the material [6]. Under certain conditions the force between two objects can be written as the product of two factors; one factor depends on the geometrical shapes of the two objects and is the same for all material combinations; the other depends on the materials of the objects and the surrounding medium only and not on the shapes. The separation dependence is to be found in the first factor and the second is just the Boer-Hamaker constant or just Hamaker constant. This division can only safely be made in diluted systems. We will return to this constant several times in the book.

The force per unit area is

$$F = -\frac{A}{6\pi}\left[(d)^{-3} + (d+2\delta)^{-3} - 2(d+\delta)^{-3}\right] \tag{5.93}$$

In the retarded or Casimir limit we have for the potential per unit area

$$\frac{V_{ll}(d)}{Area} = -\frac{23\hbar c n^2 \alpha(0)^2}{120}\left[(d)^{-3} - 2(d+\delta)^{-3} + (d+2\delta)^{-3}\right] \tag{5.94}$$

and for the force per unit area

$$F = -\frac{23\hbar c n^2 \alpha(0)^2}{40}\left[(d)^{-4} - 2(d+\delta)^{-4} + (d+2\delta)^{-4}\right] \tag{5.95}$$

A similar summation over pairs can be performed for two spheres. It is more complicated to perform so we just give the results [4]. They are for the non-retarded van der Waals limit

$$V = -\frac{A}{6}\left[\frac{2}{s^2-4} + \frac{2}{s^2} + \ln\left(\frac{s^2-4}{s^2}\right)\right]; \; s = r/R \tag{5.96}$$

or

$$V = -\frac{A}{12}\left[\frac{1}{x(2+x)} + \frac{1}{(1+x)^2} + 2\ln\left(\frac{x(2+x)}{(1+x)^2}\right)\right]; \; x = \frac{\ell}{2R} \tag{5.97}$$

For large separations we have

$$V = -\frac{16A}{9}\left[\frac{1}{s^6} + \frac{6}{s^8} + \cdots\right] \tag{5.98}$$

5.12 Summation of Pair Interactions

and for small separations:

$$V \approx -\frac{A}{24}\frac{1}{x} = -\frac{AR}{12}\frac{1}{\ell} \qquad (5.99)$$

For the retarded case we only need the result in the large separation limit. From Equation (5.98) and comparison between Equations (5.85) and (5.86) we get

$$V \approx \frac{23c}{3\pi\omega_0}\frac{1}{r}\left(-\frac{16AR^6}{r^6}\right) = -\frac{92\pi\hbar c R^6 n^2 \alpha(0)^2}{9}\frac{1}{r^7} \qquad (5.100)$$

or more directly from using Equation (5.86) for each pair and realizing that we have N^2 interacting pairs contributing to the interaction, each giving the same contribution:

$$\begin{aligned}V &\approx -\frac{23\hbar c}{4\pi}N^2\alpha(0)^2\frac{1}{r^7} = -\frac{23\hbar c}{4\pi}n^2\left(\frac{4\pi R^3}{3}\right)^2\alpha(0)^2\frac{1}{r^7} \\ &= -\frac{92\pi\hbar c R^6 n^2 \alpha(0)^2}{9}\frac{1}{r^7}\end{aligned} \qquad (5.101)$$

where n is the density of polarizable entities in the spheres. Alternatively we may start with a more exact treatment and then take the diluted limit. We treat each sphere as a polarizable entity with polarizability given by Equation (5.68) with $M = 1$ and the dielectric function of the surrounding medium being unity. We then have

$$\begin{aligned}V &= -\frac{23\hbar c}{4\pi}\alpha_{sphere}(0)^2\frac{1}{r^7} = -\frac{23\hbar c}{4\pi}\left[R^3\frac{\varepsilon(0)-1}{\varepsilon(0)+2}\right]^2\frac{1}{r^7} \\ &\approx [\text{diluted limit}] \approx -\frac{23\hbar c}{4\pi}\left[R^3\frac{4\pi n\alpha(0)}{3}\right]^2\frac{1}{r^7} \\ &= -\frac{92\pi\hbar c R^6 n^2 \alpha(0)^2}{9}\frac{1}{r^7}\end{aligned} \qquad (5.102)$$

where we have used the following relation for the dielectric function of a diluted system with n polarizable entities per unit volume, each with polarizability $\alpha(\omega)$:

$$\varepsilon(\omega) = 1 + 4\pi n\alpha(\omega) \qquad (5.103)$$

5.13 Derivation of the van der Waals equation of state

In this section we derive the van der Waals equation of state for a non-ideal gas:

$$\left(p + \frac{an^2}{V^2}\right)(V - nb) = nRT \tag{5.104}$$

or

$$\left(p + \frac{an^2}{V^2}\right)(V - nb) = Nk_BT = \frac{N}{\beta} \tag{5.105}$$

where k_B is the Boltzmann constant and $\beta = 1/k_BT$.

Let the number density of gas molecules be ρ (=N/V). The interaction between the molecules leads to an energy shift for each molecule according to:

$$\Delta E = \int_{r_0}^{\infty} 4\pi r^2 \rho \, dr \left(-Br^{-\gamma}\right)$$
$$= -\frac{4\pi B\rho}{(\gamma - 3)r_0^{\gamma - 3}} = -C\rho \tag{5.106}$$

where we have summed over pair interactions, $V = -Br^{-\gamma}$, as discussed in the previous section. The free volume in which the molecules can move is

$$\Delta \tilde{V} = V - N\,4\pi r_0^3/3 = V\left(1 - \rho\,4\pi r_0^3/3\right) = V(1 - D\rho) \tag{5.107}$$

where we have subtracted the volume taken up by the molecules themselves. We have introduced the constants C and D to make the derivation that follows more transparent. We will now determine the chemical potential for the gas and from this obtain the equation of state. The chemical potential is determined from the relation

$$\frac{N}{\tilde{V}} = \int \frac{d^3k}{(2\pi)^3} n_B(k)$$
$$= \int_0^{\infty} dk \, \frac{4\pi k^2}{(2\pi)^3} \exp\left[-\beta\left(\frac{\hbar^2 k^2}{2m} + \Delta E - \mu\right)\right] \tag{5.108}$$

5.13 van der Waals Equation of State

where on the right-hand side we have summed over all states weighted by the Boltzmann distribution function. The above equation can be rewritten as

$$\frac{\rho}{1-\rho D} = \frac{1}{2\pi^2} \exp[-\beta(\Delta E - \mu)] \int_0^\infty dk\, k^2 \exp\left[-\beta\left(\frac{\hbar^2 k^2}{2m}\right)\right] \tag{5.109}$$

and rearrangement gives

$$\frac{2\pi^2 \rho \exp[\beta(\Delta E - \mu)]}{1-\rho D} = \int_0^\infty dk\, k^2 \exp\left[-\left(\sqrt{\frac{\hbar^2 \beta}{2m}}\, k\right)^2\right]$$

$$= \left(\sqrt{\frac{\hbar^2 \beta}{2m}}\right)^{-3} \int_0^\infty dk\, k^2 \exp[-k^2] \tag{5.110}$$

$$= \left(\sqrt{\frac{\hbar^2 \beta}{2m}}\right)^{-3} \frac{\sqrt{\pi}}{4}$$

Thus we have

$$\exp[-\beta(\Delta E - \mu)] = \left(\sqrt{\frac{2\pi\hbar^2 \beta}{m}}\right)^3 \frac{\rho}{1-\rho D} \tag{5.111}$$

$$= \lambda_T^3 \frac{\rho}{1-\rho D}$$

where λ_T is the so-called thermal wave length. The final result for the chemical potential becomes

$$\mu = -C\rho + \frac{1}{\beta} \ln\left(\frac{\rho \lambda_T^3}{1-\rho D}\right) \tag{5.112}$$

Now, we know from thermodynamics that Gibb's free energy or potential is

$$G = \mu N \tag{5.113}$$

and that

$$V = \left[\frac{\partial G(p,N,T)}{\partial p}\right]_{N,T} = N\left[\frac{\partial \mu}{\partial p}\right]_{N,T} \tag{5.114}$$

From this follows that

$$\left[\frac{\partial p}{\partial \mu}\right]_{N,T} = \rho \tag{5.115}$$

and

$$\left[\frac{\partial p}{\partial \rho}\right]_{N,T} = \left[\frac{\partial p}{\partial \mu}\right]_{N,T}\left[\frac{\partial \mu}{\partial \rho}\right]_{N,T} = \rho\left[\frac{\partial \mu}{\partial \rho}\right]_{N,T} \tag{5.116}$$

Thus

$$\begin{aligned}
p &= \int_0^\rho d\rho\, \rho \left[\frac{\partial \mu}{\partial \rho}\right]_{N,T} = \int_0^\rho d\rho\, \rho \frac{\partial}{\partial \rho}\left[-C\rho + \frac{1}{\beta}\ln\left(\frac{\rho \lambda_T^3}{1-\rho D}\right)\right]_{N,T} \\
&= \int_0^\rho d\rho\, \rho\left[-C + \frac{1}{\beta \rho}\left(\frac{1}{1-\rho D}\right)\right] = \int_0^\rho d\rho\left[-C\rho + \frac{1}{\beta}\left(\frac{1}{1-\rho D}\right)\right] \\
&= -C\frac{\rho^2}{2} - \frac{1}{\beta D}\ln(1-\rho D)
\end{aligned} \tag{5.117}$$

To get further we have to make use of the fact that ρD is much smaller than unity, which is fulfilled for a gas. We expand the logarithm and keep the two lowest order terms:

$$\begin{aligned}
p &\approx -C\frac{\rho^2}{2} - \frac{1}{\beta D}\left[-\rho D - \frac{1}{2}(\rho D)^2\right] \\
&= -C\frac{\rho^2}{2} + \frac{\rho D}{\beta D}\left[1 + \frac{1}{2}(\rho D)\right] \\
&\approx -C\frac{\rho^2}{2} + \frac{\rho}{\beta\left(1 - \frac{1}{2}\rho D\right)}
\end{aligned} \tag{5.118}$$

and rewrite this expression as

5.13 van der Waals Equation of State

$$\left(p + C\frac{\rho^2}{2}\right)\left(\frac{1}{\rho} - \frac{1}{2}D\right) = \frac{1}{\beta} \tag{5.119}$$

or

$$\left(p + C\frac{\rho^2}{2}\right)\left(\frac{N}{\rho} - \frac{N}{2}D\right) = \frac{N}{\beta} \tag{5.120}$$

or

$$\left(p + C\frac{N^2}{2V^2}\right)\left(V - \frac{N}{2}D\right) = \frac{N}{\beta} = Nk_BT \tag{5.121}$$

Now we may identify the parameters in the van der Waals equation of state. We find

$$an^2 = \frac{CN^2}{2} = \frac{2\pi BN^2}{(\gamma-3)r_0^{\gamma-3}} \tag{5.122}$$

and

$$nb = \frac{N}{2}D = \frac{N2\pi r_0^3}{3} \tag{5.123}$$

This completes the derivation of the van der Waals equation of state for non-ideal gases.

References

[1] F. London, Z. Phys. Chem. B **11**, 222 (1930); Z. Phys. **63**, 245 (1930).
[2] S. Q. Wang and G. D. Mahan, Phys. Rev. **B6**, 517 (1972).
[3] B. M. Axilrod, and E. Teller, J. Chem. Phys. **11**,299 (1943).
[4] D. Langbein, *Theory of Van der Waals Attraction*, (Springer, New York, 1974,) in the series Springer Tracts in Modern Physics.
[5] *Handbook of Mathematical Functions*, edited by M. Abramowitz and I. A. Stegun, National Bureau of Standards Applied Mathematical Series, No 55 (U. S. G. P. O., Washington, D.C., 10th printing, 1972), formula 9.6.1.
[6] J. H. Boer, Trans. Faraday Soc. **32**, 10 (1936); H. C. Hamaker, Physica **4**, 1058 (1937).

6 Energy and force

Why do birds of a feather flock together? Why does liquid helium climb the walls of a beaker?

We have many times in this book demonstrated that the interaction energy of a system can be expressed as the change in the zero-point energy of all electromagnetic modes of the system. In Chapter 3 we showed that for the bulk of a metal the plasmon mode contributed but also the single-particle continuum did. We furthermore showed that in principle we could, by keeping the discreteness, find modes inside the continuum. The zero-point energy of these modes makes up the rest of the interaction energy. This method is not practical, however, since there are so many modes in the continuum; it was only used for demonstrating purposes. In Section 4.3 we expressed the Casimir force between two metal plates in terms of the change in the zero-point energy of wave-guide type modes between the plates; also here there are too many modes involved to make it feasible to keep track of the change in the zero-point energy of each individual mode. In Section 4.4 we used the surface plasmon modes to calculate the surface energy of metals. In Section 4.5 we discussed the force between two quantum wells. For large separations we could use the zero-point energy of the 2D plasmons of the system to calculate the force. For smaller separations also the continuum contributes and we could not use this method – it becomes impractical. We could circumvent the difficulties by introducing the plasmon-pole approximation in which the continuum disappears and we only have plasmon modes. Alternatively we used the correlation-energy method where the interaction is obtained from many-body theory and is identified as interlayer correlation energy. In Section 5.4 we expressed the force between two polarizable atoms as the change of the zero-point energy of the normal modes of the system.

Thus in some situations we may study the individual modes of the system but in some we may not. Sometimes it can be difficult to find the modes; sometimes the modes form a continuum; when damping is present maybe there is no distinct mode. For water (see Figure 2.27) the real part of the dielectric function does not turn negative neither in the regions of vibration modes, in the infra red, nor in the region of intra-atomic electronic transitions, in the ultraviolet range. The modes are still there and contribute but they are not stable. In this chapter we will go through methods that can be used when we have problems finding the modes or if they form a continuum. Sometimes it is easier to use these methods even though it is possible to find the individual modes.

We begin with the method for zero temperature and then treat finite temperatures. In the demonstration of the method we use the interaction between two polarizable atoms as an illustration. We treat the problem first neglecting retardation effects and demonstrate that we get the same result as we got before in Chapter 5. Then we repeat the calculations including retardation.

Then we apply the method to planar geometries. We derive the surface energy of a planar surface with and without retardation effects included. We also derive the van der Waals and Casimir interactions between two planar surfaces. We calculate the surface energy in two different ways: in one way we use the change in energy when one bulk mode is transformed into two surface modes as the material is split into two pieces; in the other approach we start with the modes in the gap between two half-spaces of the material and calculate the change in energy when the two halves are brought from zero to infinite separation. In this last method also the force between the surfaces as function of separation is obtained as a side result.

6.1 Interaction energy at zero temperature

We will repeatedly make use of an extension to the so-called *Argument Principle* familiar to most of us from under-graduate mathematical courses on analytical functions. Let us study a region in the complex frequency plane. We have two functions defined in this region; one, $\varphi(z)$, is analytical in the whole region; one, $f(z)$, has poles and zeros inside the region. The following relation holds for an integration path around the region:

$$\frac{1}{2\pi i}\oint dz \varphi(z) \frac{d}{dz} \ln f(z) = \sum \varphi(z_o) - \sum \varphi(z_\infty) \tag{6.1}$$

where z_0 and z_∞ are the zeros and poles, respectively, of function $f(z)$. In the Argument Principle the function φ is replaced by unity and the right hand side then equals the number of zeros minus the number of poles for function $f(z)$ inside the integration path.

Let us return briefly to Equation (3.52). There we found that for each **q**-value the contribution to the Exchange-Correlation energy is the frequency at the zeros of the dielectric function minus the frequency at the poles all multiplied by $\hbar/2$. Thus to reproduce this result we choose the two functions to be:

$$\varphi(z) = \hbar z/2 \quad ; \quad f(z) = \varepsilon(\mathbf{q}, z) \tag{6.2}$$

The contour should include the whole of the positive real frequency axis. The function $f(z)$ is the function in the defining equation for the normal modes of the system:

$$f(\omega) = 0 \tag{6.3}$$

By using this theorem we end up with an integration along a closed contour in the complex frequency plane. In most cases it is fruitful to choose the contour shown in Figure 6.1.

6.1 Interaction Energy at Zero Temperature

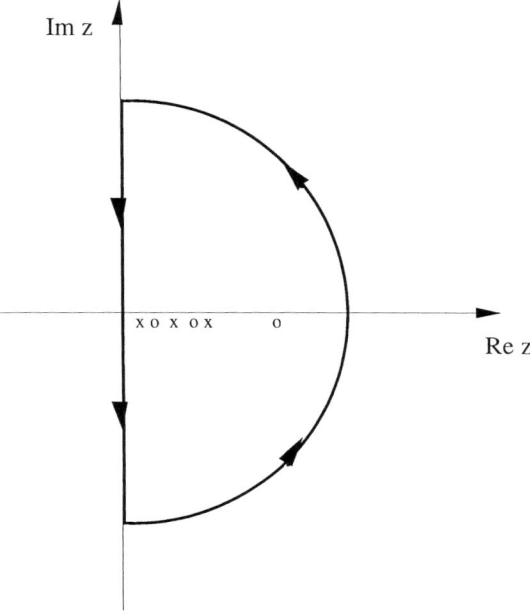

Figure 6.1: Integration contour in the complex z-plane. Crosses and circles are poles and zeros, respectively, of the function $f(z)$. The radius of the circle is let to go to infinity.

We have the freedom to multiply the function $f(z)$ with an arbitrary constant without changing the result on the right hand side of Equation (6.1). If we choose it carefully we can make the contribution from the curved part of the contour vanish and we are only left with an integration along the imaginary frequency axis.

6.1.1 Interaction between two polarizable atoms revisited: no retardation

We start from Equation (5.28), that is,

$$\left\{ e^2 \tilde{m}_1^{-1} \cdot \tilde{\phi}^{12} \cdot e^2 \tilde{m}_2^{-1} \cdot \tilde{\phi}^{21} - \left[\omega^2 - \omega_1^2\right]\left[\omega^2 - \omega_2^2\right]\tilde{1} \right\} \cdot \mathbf{p}_1 = 0 \tag{6.4}$$

and limit ourselves to the isotropic case where we have

$$\left\{\frac{e^4}{m_1 m_2}\tilde{\phi}^2 - \left[\omega^2 - \omega_1^2\right]\left[\omega^2 - \omega_2^2\right]\tilde{1}\right\} \cdot \mathbf{p}_1 = 0 \tag{6.5}$$

The tensors in Equation (6.4) are both symmetric with respect to interchange of the two particle indices so we may drop these without any loss of generality. The tensor in the first term of Equation (6.5) is one of these tensors squared. It was introduced in Equation (5.19).

If we now choose the z-axis to be along the line joining the two atoms the tensor $\tilde{\phi}^2$ becomes diagonal:

$$\tilde{\phi}^2 = \frac{1}{r^6}\begin{pmatrix} 1 & 0 & 0 \\ 0 & 1 & 0 \\ 0 & 0 & 4 \end{pmatrix} \tag{6.6}$$

and Equation (6.5) can be written as

$$\mathbf{A} \cdot \mathbf{p}_1 = 0 \tag{6.7}$$

where the matrix $\mathbf{A}$ is diagonal. The condition for normal modes is

$$|\mathbf{A}| = \left[\frac{e^4}{m_1 m_2 r^6} - \left(\omega^2 - \omega_1^2\right)\left(\omega^2 - \omega_2^2\right)\right]^2 \left[\frac{4e^4}{m_1 m_2 r^6} - \left(\omega^2 - \omega_1^2\right)\left(\omega^2 - \omega_2^2\right)\right] = 0 \tag{6.8}$$

Now we are ready to calculate the energy of the system. We chose the following functions:

$$\varphi(z) = \hbar z/2 \tag{6.9}$$

and

$$f(z) = \frac{\left[e^4/m_1 m_2 r^6 - \left(z^2 - \omega_1^2\right)\left(z^2 - \omega_2^2\right)\right]^2 \left[4e^4/m_1 m_2 r^6 - \left(z^2 - \omega_1^2\right)\left(z^2 - \omega_2^2\right)\right]}{\left[-\left(z^2 - \omega_1^2\right)\left(z^2 - \omega_2^2\right)\right]^2 \left[-\left(z^2 - \omega_1^2\right)\left(z^2 - \omega_2^2\right)\right]}$$

$$= \left[1 - \frac{e^4/m_1 m_2 r^6}{\left(z^2 - \omega_1^2\right)\left(z^2 - \omega_2^2\right)}\right]^2 \left[1 - \frac{4e^4/m_1 m_2 r^6}{\left(z^2 - \omega_1^2\right)\left(z^2 - \omega_2^2\right)}\right] \tag{6.10}$$

where we have chosen the determinant of $\mathbf{A}$ at the separation r divided by the determinant at infinite separation. In doing so we have determined our zero of energy to be

6.1 Interaction Energy at Zero Temperature

the energy at infinite separation. We further have that $f(z)$ tends to unity at the arched part of the contour as the radius tends to infinity and the logarithm tends to zero fast enough for the integral to be zero.

Thus we have

$$V(r) = \frac{1}{2\pi i}\oint dz\, \hbar z/2 \frac{d}{dz}\ln\left\{\left[1 - \frac{e^4/m_1 m_2 r^6}{(z^2 - \omega_1^2)(z^2 - \omega_2^2)}\right]^2 \left[1 - \frac{4e^4/m_1 m_2 r^6}{(z^2 - \omega_1^2)(z^2 - \omega_2^2)}\right]\right\}$$

$$= -\frac{1}{2\pi i}\int_{-\infty}^{\infty} d(i\omega)\,\hbar(i\omega)/2\,\frac{d}{d(i\omega)}\ln\left\{\left[1 - \frac{e^4/m_1 m_2 r^6}{((i\omega)^2 - \omega_1^2)((i\omega)^2 - \omega_2^2)}\right]^2\right.$$

$$\left.\times\left[1 - \frac{4e^4/m_1 m_2 r^6}{((i\omega)^2 - \omega_1^2)((i\omega)^2 - \omega_2^2)}\right]\right\}$$

(6.11)

Rearrangement gives

$$V(r) = -\frac{\hbar}{4\pi}\int_{-\infty}^{\infty} d\omega\,\omega\, \frac{d}{d\omega}\ln\left\{\left[1 - \frac{e^4/m_1 m_2 r^6}{(\omega^2 + \omega_1^2)(\omega^2 + \omega_2^2)}\right]^2\right.$$

$$\left.\times\left[1 - \frac{4e^4/m_1 m_2 r^6}{(\omega^2 + \omega_1^2)(\omega^2 + \omega_2^2)}\right]\right\}$$

(6.12)

Identifying the polarizabilities gives

$$V(r) = -\frac{\hbar}{4\pi}\int_{-\infty}^{\infty} d\omega\,\omega\, \frac{d}{d\omega}\ln\left\{[1 - \alpha_1'(\omega)\alpha_2'(\omega)/r^6]^2 [1 - 4\alpha_1'(\omega)\alpha_2'(\omega)/r^6]\right\} \quad (6.13)$$

where the primes on the polarizabilities denote that they are to be calculated on the imaginary frequency axis. Performing a partial integration leads to

$$V(r) = \frac{\hbar}{4\pi}\int_{-\infty}^{\infty} d\omega\, \ln\left\{[1 - \alpha_1'(\omega)\alpha_2'(\omega)/r^6]^2 [1 - 4\alpha_1'(\omega)\alpha_2'(\omega)/r^6]\right\} \quad (6.14)$$

For large separation we may expand the logarithm and obtain the van der Waals result in agreement with Equation (5.37):

$$V_{vdW}(r) = -\frac{3\hbar}{2\pi}\frac{1}{r^6}\int_{-\infty}^{\infty}d\omega\alpha_1'(\omega)\alpha_2'(\omega) = -\frac{3\hbar}{\pi}\frac{1}{r^6}\int_{0}^{\infty}d\omega\alpha_1'(\omega)\alpha_2'(\omega) \qquad (6.15)$$

6.1.2 Interaction between two polarizable atoms revisited: retardation

When retardation is included we find a more general force that agrees with the van der Waals force for short and intermediate distances but drops off faster with distance at larger distances. The finite speed of light has two effects: The electric field at the position of one of the atoms, as generated by the dipole on the other atom, is delayed by the time it takes for an electromagnetic wave to travel the distance between the atoms; this is the true retardation effect. The other effect is that a time dependent dipole moment generates a more general electric field than the one we have discussed so far. The field no longer depends on the dipole moment only but also on its first time derivative and its second time-derivative. Let the time dependent dipole moment be defined by:

$$\mathbf{p}(t) = \hat{\mathbf{p}}p(t) \qquad (6.16)$$

Note that the length of the vector is changing with time and the direction is unchanged. The produced electric field is

$$\mathbf{E}(\mathbf{r}) = -[\hat{\mathbf{p}} - (\hat{\mathbf{p}}\cdot\hat{\mathbf{r}})\hat{\mathbf{r}}]\frac{1}{c^2 r}\ddot{p}\left(t-\frac{r}{c}\right) - [\hat{\mathbf{p}} - 3(\hat{\mathbf{p}}\cdot\hat{\mathbf{r}})\hat{\mathbf{r}}]\left[\frac{1}{cr^2}\dot{p}\left(t-\frac{r}{c}\right) + \frac{1}{r^3}p\left(t-\frac{r}{c}\right)\right] \qquad (6.17)$$

In neglecting the retardation effects we let the speed of light go to infinity and regain the original expression for the field, used before in Equation (5.1). Including retardation we find the more general equations of motion for the two electrons:

$$\begin{cases} \tilde{1}\cdot\ddot{\mathbf{p}}_1(t) + \omega_1^2\tilde{1}\cdot\mathbf{p}_1(t) = -e^2\tilde{m}_1^{-1}\cdot\left\{\tilde{\phi}^{12}\cdot\left[\mathbf{p}_2\left(t-\frac{r}{c}\right) + \left(\frac{r}{c}\right)\dot{\mathbf{p}}_2\left(t-\frac{r}{c}\right)\right] + \left(\frac{r}{c}\right)^2\tilde{\vartheta}^{12}\cdot\ddot{\mathbf{p}}_2\left(t-\frac{r}{c}\right)\right\} \\ \tilde{1}\cdot\ddot{\mathbf{p}}_2(t) + \omega_2^2\tilde{1}\cdot\mathbf{p}_2(t) = -e^2\tilde{m}_2^{-1}\cdot\left\{\tilde{\phi}^{21}\cdot\left[\mathbf{p}_1\left(t-\frac{r}{c}\right) + \left(\frac{r}{c}\right)\dot{\mathbf{p}}_1\left(t-\frac{r}{c}\right)\right] + \left(\frac{r}{c}\right)^2\tilde{\vartheta}^{21}\cdot\ddot{\mathbf{p}}_1\left(t-\frac{r}{c}\right)\right\} \end{cases}$$
$$(6.18)$$

Note the delay in the time arguments on the right hand side of the equations. We have also introduced a new tensor:

6.1 Interaction Energy at Zero Temperature

$$\tilde{\vartheta}^{12}{}_{\mu\nu} = \frac{\delta_{\mu\nu}}{r^3} - \frac{r_\mu r_\mu}{r^5} \tag{6.19}$$

After Fourier transforming we have

$$\begin{cases} \tilde{\mathbf{1}} \cdot \mathbf{p}_1(-i\omega)^2 + \omega_1^2 \tilde{\mathbf{1}} \cdot \mathbf{p}_1 = -e^2 \tilde{m}_1^{-1} \cdot \left[\tilde{\phi}^{12}(1 - i\omega r/c) + \tilde{\vartheta}^{12}(-i\omega r/c)^2\right] \cdot \mathbf{p}_2 e^{i\omega r/c} \\ \tilde{\mathbf{1}} \cdot \mathbf{p}_2(-i\omega)^2 + \omega_2^2 \tilde{\mathbf{1}} \cdot \mathbf{p}_2 = -e^2 \tilde{m}_2^{-1} \cdot \left[\tilde{\phi}^{21}(1 - i\omega r/c) + \tilde{\vartheta}^{21}(-i\omega r/c)^2\right] \cdot \mathbf{p}_1 e^{i\omega r/c} \end{cases}$$
$$\tag{6.20}$$

Eliminating $\mathbf{p}_2$ and rearrangement leads to

$$\left\{ e^4 \tilde{m}_1^{-1} \cdot \left[\tilde{\phi}^{12}(1 - i\omega r/c) + \tilde{\vartheta}^{12}(i\omega r/c)^2\right] \cdot \tilde{m}_2^{-1} \cdot \left[\tilde{\phi}^{21}(1 - i\omega r/c) + \tilde{\vartheta}^{21}(i\omega r/c)^2\right] e^{i2\omega r/c} \right.$$
$$\left. - \left[\omega_1^2 - \omega^2\right]\left[\omega_2^2 - \omega^2\right]\tilde{\mathbf{1}} \right\} \cdot \mathbf{p}_1 = 0 \tag{6.21}$$

Assuming isotropic atoms gives

$$\left\{ \frac{e^4}{m_1 m_2} \left[(\tilde{\phi})^2 (1 - i\omega r/c)^2 + (\tilde{\vartheta})^2 (i\omega r/c)^4 + 2(\tilde{\phi} \cdot \tilde{\vartheta})(1 - i\omega r/c)(i\omega r/c)^2\right] e^{i2\omega r/c} \right.$$
$$\left. - \left[\omega_1^2 - \omega^2\right]\left[\omega_2^2 - \omega^2\right]\tilde{\mathbf{1}} \right\} \cdot \mathbf{p}_1 = 0 \tag{6.22}$$

and after rearrangements we have

$$\left\{ \frac{e^4}{m_1 m_2} \left[(\tilde{\phi})^2 \left(1 - \left(\frac{\omega r}{c}\right)^2 - 2i\left(\frac{\omega r}{c}\right)\right) + (\tilde{\vartheta})^2 \left(\frac{\omega r}{c}\right)^4 - 2(\tilde{\phi} \cdot \tilde{\vartheta})\left(\left(\frac{\omega r}{c}\right)^2 - i\left(\frac{\omega r}{c}\right)^3\right)\right] e^{i2\omega r/c} \right.$$
$$\left. - \left[\omega_1^2 - \omega^2\right]\left[\omega_2^2 - \omega^2\right]\tilde{\mathbf{1}} \right\} \cdot \mathbf{p}_1 = 0 \tag{6.23}$$

The two tensors are both symmetric with respect to interchange of the two particle indices so we may drop these without any loss of generality. It furthermore turns out that

$$\left(\tilde{\vartheta}\right)^2 = \frac{1}{r^3}\tilde{\vartheta}$$
$$\tilde{\phi}\cdot\tilde{\vartheta} = \frac{1}{r^3}\tilde{\vartheta} \qquad (6.24)$$

For simplicity let us choose the z-direction to be along the line joining the two atoms. Then the tensors are

$$\tilde{\phi}^2 = \frac{1}{r^6}\begin{pmatrix} 1 & 0 & 0 \\ 0 & 1 & 0 \\ 0 & 0 & 4 \end{pmatrix} \qquad (6.25)$$

$$\tilde{\vartheta} = \frac{1}{r^3}\begin{pmatrix} 1 & 0 & 0 \\ 0 & 1 & 0 \\ 0 & 0 & 0 \end{pmatrix} \qquad (6.26)$$

$$\tilde{\phi}\cdot\tilde{\vartheta} = \tilde{\vartheta}^2 = \frac{1}{r^6}\begin{pmatrix} 1 & 0 & 0 \\ 0 & 1 & 0 \\ 0 & 0 & 0 \end{pmatrix} \qquad (6.27)$$

We define the matrix $\mathbf{A}$ as the matrix multiplying $\mathbf{p}_1$ in Equation (6.23) and find the normal modes as the solutions to the equation $|\mathbf{A}| = 0$.

Now,

$$|\mathbf{A}(\omega)| = \left\{\frac{e^4}{m_1 m_2 r^6}\left[1 - i2\left(\frac{\omega r}{c}\right) - 3\left(\frac{\omega r}{c}\right)^2 + i2\left(\frac{\omega r}{c}\right)^3 + \left(\frac{\omega r}{c}\right)^4\right]e^{i2\omega r/c}\right.$$
$$\left. - \left(\omega_1^2 - \omega^2\right)\left(\omega_2^2 - \omega^2\right)\right\}^2$$
$$\times \left\{\frac{4e^4}{m_1 m_2 r^6}\left[1 - i2\left(\frac{\omega r}{c}\right) - \left(\frac{\omega r}{c}\right)^2\right]e^{i2\omega r/c} - \left(\omega_1^2 - \omega^2\right)\left(\omega_2^2 - \omega^2\right)\right\}$$
$$(6.28)$$

In this situation it is not so trivial to find the normal mode solutions and proceed along the lines in Section 5.4. The solutions will be complex valued and are difficult to find. This is a situation where the method introduced in Section 6.1 is favorable.

We are almost ready to write down the expression for the potential. Our expression for the determinant is valid in the upper half of the complex frequency plane. We also need the

6.1 Interaction Energy at Zero Temperature

corresponding expression in the lower half. We have to make an analytical continuation. The function we have is a retarded function. We need the time-ordered version. It is even in z. We see that the real (imaginary) part of the expression we have is even (odd) in the real part of z. Thus if we let

$$|\mathbf{A}(z)| = |\mathbf{A}(a+ib)| \tag{6.29}$$

we have

$$|\mathbf{A}(-a+ib)| = |\mathbf{A}(z)|^* \tag{6.30}$$

If we now make the analytical continuation into the lower half plane we see that

$$|\mathbf{A}(a-ib)| = |\mathbf{A}(-a+ib)| = |\mathbf{A}(a+ib)|^* \tag{6.31}$$

In the first equality we used the inversion symmetry and in the second we used Equation (6.30). Thus, the function $f(z)$ we choose is

$$f_u(z) = \left[1 - e^4 e^{i2zr/c} \frac{1 - i2\left(\frac{zr}{c}\right) - 3\left(\frac{zr}{c}\right)^2 + i2\left(\frac{zr}{c}\right)^3 + \left(\frac{zr}{c}\right)^4}{m_1 m_2 r^6 \left(\omega_1^2 - z^2\right)\left(\omega_2^2 - z^2\right)}\right]^2$$

$$\times \left[1 - 4e^4 e^{i2zr/c} \frac{\left[1 - i2\left(\frac{zr}{c}\right) - \left(\frac{zr}{c}\right)^2\right]}{m_1 m_2 r^6 \left(\omega_1^2 - z^2\right)\left(\omega_2^2 - z^2\right)}\right] \tag{6.32}$$

for z in the upper half plane and

$$f_l(z) = f_u\left(z^*\right)^* =$$

$$= \left[1 - e^4 e^{-i2zr/c} \frac{1 + i2\left(\frac{zr}{c}\right) - 3\left(\frac{zr}{c}\right)^2 - i2\left(\frac{zr}{c}\right)^3 + \left(\frac{zr}{c}\right)^4}{m_1 m_2 r^6 \left(\omega_1^2 - z^2\right)\left(\omega_2^2 - z^2\right)}\right]^2$$

$$\times \left[1 - 4e^4 e^{-i2zr/c} \frac{\left[1 + i2\left(\frac{zr}{c}\right) - \left(\frac{zr}{c}\right)^2\right]}{m_1 m_2 r^6 \left(\omega_1^2 - z^2\right)\left(\omega_2^2 - z^2\right)}\right] \tag{6.33}$$

in the lower half plane. This means that on the imaginary frequency axis we have

$$f(i\omega) = \left[1 - e^4 e^{-2|\omega|r/c} \frac{1 + 2\left(\frac{|\omega|r}{c}\right) + 3\left(\frac{\omega r}{c}\right)^2 + 2\left(\frac{|\omega|r}{c}\right)^3 + \left(\frac{\omega r}{c}\right)^4}{m_1 m_2 r^6 \left(\omega_1^2 + \omega^2\right)\left(\omega_2^2 + \omega^2\right)} \right.$$

$$\left. \times \left[1 - 4e^4 e^{-2|\omega|r/c} \frac{\left[1 + 2\left(\frac{|\omega|r}{c}\right) + \left(\frac{\omega r}{c}\right)^2\right]}{m_1 m_2 r^6 \left(\omega_1^2 + \omega^2\right)\left(\omega_2^2 + \omega^2\right)} \right] \right]^2 \quad (6.34)$$

We have chosen the function so that the integration along the curved part of the path in Figure 6.1 vanishes and the reference level for the potential is at infinite separation. We also find that the function is even with respect to frequency on the imaginary frequency axis, as it should be. We may now identify the polarizabilities for the two atoms as we have done several times before and find

$$f(i\omega) = \left[1 - e^{-2|\omega|r/c} \alpha_1'(\omega)\alpha_2'(\omega) \frac{1}{r^6} \left[1 + 2\left(\frac{|\omega|r}{c}\right) + 3\left(\frac{\omega r}{c}\right)^2 + 2\left(\frac{|\omega|r}{c}\right)^3 + \left(\frac{\omega r}{c}\right)^4 \right] \right.$$

$$\left. \times \left[1 - 4e^{-2|\omega|r/c} \alpha_1'(\omega)\alpha_2'(\omega) \frac{1}{r^6} \left[1 + 2\left(\frac{|\omega|r}{c}\right) + \left(\frac{\omega r}{c}\right)^2 \right] \right] \right]^2 \quad (6.35)$$

We may now write down the fully retarded potential between two polarizable atoms. In analogy with Equation (6.14) it is

$$V(r) = \frac{\hbar}{4\pi} \int_{-\infty}^{\infty} d\omega \ln \left\{ \left[1 - \frac{4e^{-2|\omega|r/c}}{r^6} \alpha_1'(\omega)\alpha_2'(\omega) \left[1 + 2\left(\frac{|\omega|r}{c}\right) + \left(\frac{\omega r}{c}\right)^2 \right] \right] \right.$$

$$\left. \times \left[1 - \frac{e^{-2|\omega|r/c}}{r^6} \alpha_1'(\omega)\alpha_2'(\omega) \left[1 + 2\left(\frac{|\omega|r}{c}\right) + 3\left(\frac{\omega r}{c}\right)^2 + 2\left(\frac{|\omega|r}{c}\right)^3 + \left(\frac{\omega r}{c}\right)^4 \right] \right]^2 \right\}$$

$$(6.36)$$

Along the imaginary axis the integrand is even and we can limit ourselves to the upper part and multiply by two:

6.1 Interaction Energy at Zero Temperature

$$V(r) = \frac{\hbar}{2\pi} \int_0^\infty d\omega \ln\left\{\left[1 - \frac{4e^{-2\omega r/c}}{r^6}\alpha_1'(\omega)\alpha_2'(\omega)\left[1 + 2\left(\frac{\omega r}{c}\right) + \left(\frac{\omega r}{c}\right)^2\right]\right]\right.$$

$$\left.\times\left[1 - \frac{e^{-2\omega r/c}}{r^6}\alpha_1'(\omega)\alpha_2'(\omega)\left[1 + 2\left(\frac{\omega r}{c}\right) + 3\left(\frac{\omega r}{c}\right)^2 + 2\left(\frac{\omega r}{c}\right)^3 + \left(\frac{\omega r}{c}\right)^4\right]\right]^2\right\}$$

(6.37)

For large distances the logarithm may be expanded and only the lowest order term be kept. With large distances we here mean that they are large enough for the interaction to be weak. Then we find:

$$V_{CP}(r) = -\frac{\hbar}{\pi r^6}\int_0^\infty d\omega \alpha_1(i\omega)\alpha_2(i\omega) e^{-2\omega r/c}\left[3 + 6(\omega r/c) + 5(\omega r/c)^2\right.$$

$$\left. + 2(\omega r/c)^3 + (\omega r/c)^4\right]$$

(6.38)

This is the Casimir-Polder interaction and it gives the van der Waals result for intermediate separations and the retarded result for large separations. We will now demonstrate that this is the case. Let us first start with the van der Waals limit. Assume that $\omega r/c$ is small compared to unity. The expression in the square brackets reduces to 3 and the exponential prefactor to unity:

$$V_{CP}(r) \underset{\omega r/c \to 0}{\approx} -\frac{3\hbar}{\pi}\frac{1}{r^6}\int_0^\infty d\omega \alpha_1(i\omega)\alpha_2(i\omega)$$

(6.39)

This is the van der Waals result.
To find the other limiting result we make the substitution $u=\omega r/c$. Then we have:

$$V_{CP}(r) = -\frac{\hbar c}{\pi r^7}\int_0^\infty du\, \alpha_1\left(i\frac{uc}{r}\right)\alpha_2\left(i\frac{uc}{r}\right) e^{-2u}\left[3 + 6u + 5u^2 + 2u^3 + u^4\right]$$

(6.40)

The exponential factor guarantees that only small u values contribute to the integral. If r is big enough we can replace the polarizabilities with the static ones and move them outside the integral. Then we have:

$$V_{CP}(r) \underset{r\to\infty}{\approx} -\frac{\hbar c \alpha_1(0)\alpha_2(0)}{\pi r^7}\int_0^\infty du\, e^{-2u}\left[3 + 6u + 5u^2 + 2u^3 + u^4\right]$$

$$= -\frac{\hbar c \alpha_1(0)\alpha_2(0)}{\pi r^7}\frac{23}{4} = -\frac{23\hbar c}{4\pi}[\alpha_1(0)\alpha_2(0)]\frac{1}{r^7}$$

(6.41)

which is the Casimir result. Thus we see that for intermediate separations the potential goes as r^{-6} and for large separations as r^{-7}.

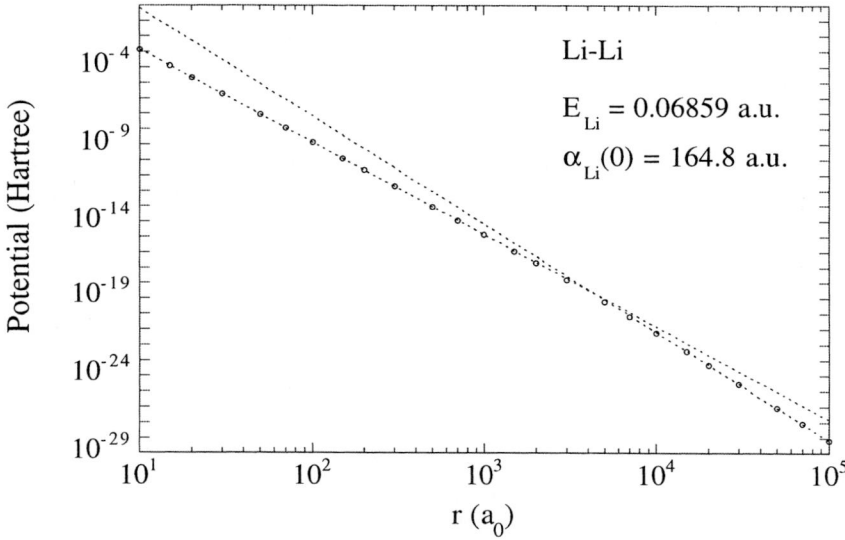

Figure 6.2: The Casimir-Polder potential for a pair of lithium atoms (circles) and the Casimir and van der Waals asymptotes (dotted straight lines). Atomic units are used on both axes.

In Figure 6.2 we demonstrate the results for a pair of lithium atoms. The circles are the result from Equation (6.40). The steepest asymptote, the Casimir asymptote, is the result from Equation (6.41). The remaining asymptote, the van der Waals asymptote, is the result from Equation (6.39). The two parameters needed in the calculation, the static polarizability and a dominating excitation energy, are given in the figure.

6.2 Interaction energy at finite temperature

We have so far concentrated our treatment to zero temperature. For finite temperatures things are more complicated. When we calculate the force between two objects at finite temperatures, from how the energy changes with separation, it is important to know how the objects are thermodynamically connected to the rest of the world. In most cases the temperature effects are very weak. However, there is an exception: liquids like water with permanent dipole moments have important contributions to their dielectric functions for very small frequencies. This leads to important temperature effects. When we gradually increase the temperature from 0 K the temperature effects first appear at large separations only, and then at smaller and smaller separations.

If the objects are completely isolated with no heat exchange, no particle exchange and if they do not perform any work, except on each other, all work performed by the force between the objects goes to the change of the internal energies of the objects; thus in this case we should calculate the *internal energy* as function of separation and from this variation obtain the force. It should be noted that in this case the temperature will change with separation.

If the objects are in contact with a heat bath, so that heat is freely exchanged with the surroundings, some of the work that otherwise would go into the change of the internal energy now leaks out (or in) in the form of heat; in this case we should calculate the *Helmholtz' free energy* as function of separation and from this obtain the force; this is the most common situation.

If the objects also can exchange particles with the surroundings, energy is leaking out of the objects also via this channel. In this case we should calculate the *thermodynamic potential*. This situation occurs when two particles are in equilibrium with their solution or when two liquid droplets are in equilibrium with their vapor phase.

Often the volume of the objects stays unchanged as well as the surrounding pressure. In this case we can use the internal energy and the *enthalpy* interchangeably; this also holds for the Helmholtz' and *Gibbs'* free energies.

We will here derive the interaction energy in the form of Helmholtz' free energy, $\mathfrak{F}$. We derived it in Section 4.3.2 in terms of the energies of the normal modes. Here we will make a more general derivation using the *Argument Principle*. In terms of the normal modes we found for the free energy

$$\mathfrak{F} = -\frac{1}{\beta}\sum_{\mathbf{k}}\left[-\beta\frac{1}{2}\hbar\omega_{\mathbf{k}} - \ln\left(1 - e^{-\beta\hbar\omega_{\mathbf{k}}}\right)\right] = \sum_{\mathbf{k}}\left[\frac{1}{2}\hbar\omega_{\mathbf{k}} + \frac{1}{\beta}\ln\left(1 - e^{-\beta\hbar\omega_{\mathbf{k}}}\right)\right]$$
$$= \sum_{\mathbf{k}}\frac{1}{\beta}\ln\left(2\sinh\tfrac{1}{2}\beta\hbar\omega_{\mathbf{k}}\right)$$
(6.42)

and from this we get the change in free energy as

$$\Delta \mathfrak{F} = \sum_{\mathbf{k}} \left[\frac{1}{\beta} \ln\left(2\sinh\tfrac{1}{2}\beta\hbar\omega_{\mathbf{k}}\right) - \frac{1}{\beta}\ln\left(2\sinh\tfrac{1}{2}\beta\hbar\omega_{\mathbf{k}}^0\right) \right] \qquad (6.43)$$

We make the observation that we may use the *Argument Principle* but now with $\ln[2\sinh(1/2\beta\hbar z)]/\beta$ instead of $\hbar z/2$ for $\varphi(z)$ in the integrand. There is however one complication. This new function has poles of its own in the complex frequency plane. We have to choose our contour so that it includes all poles and zeros from the function $f(z)$ but excludes those of $\varphi(z)$. The poles of function $\varphi(z)$ all are on the imaginary frequency axis. We use the same contour as in Section 6.1 but let the straight part of the contour lie just to the right of, and infinitesimally close to, the imaginary axis. We have

$$\begin{aligned}
\Delta \mathfrak{F} &= \frac{1}{2\pi i} \int_{\infty}^{-\infty} d(i\omega) \frac{1}{\beta} \ln\left(2\sinh\tfrac{1}{2}\beta\hbar i\omega\right) \frac{d}{d(i\omega)} \ln f(i\omega) \\
&= -\frac{1}{2\pi i} \int_{-\infty}^{+\infty} d\omega \frac{1}{\beta} \ln\left(2\sinh\tfrac{1}{2}\beta\hbar i\omega\right) \frac{d}{d\omega} \ln f(i\omega) \qquad (6.44) \\
&= \frac{1}{2\pi i} \int_{-\infty}^{+\infty} d\omega \frac{\hbar}{2} \coth\left(\tfrac{1}{2}\beta\hbar i\omega\right) \ln f(i\omega) = \frac{\hbar}{4\pi i} \int_{-\infty}^{+\infty} d\omega \coth\left(\tfrac{1}{2}\beta\hbar i\omega\right) \ln f(i\omega)
\end{aligned}$$

The coth function has poles on the imaginary z-axis and they should not be inside the semi circle. The poles are at

$$z_n = i\frac{2\pi n}{\hbar\beta} \quad ; \quad n = 0,\pm 1,\pm 2,\ldots \qquad (6.45)$$

and all residues are the same, equal to $2/\hbar\beta$. We integrate along the imaginary axis and deform the path along small semi circles around each pole. The integration path is illustrated in Figure 6.3.

6.2 Interaction Energy at Finite Temperature

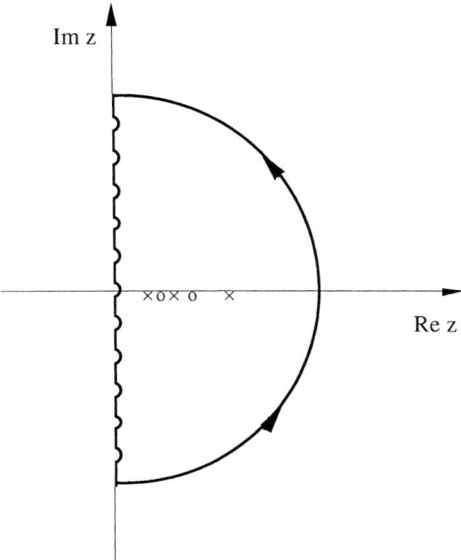

Figure 6.3: Integration contour in the complex z-plane. Crosses and circles are poles and zeros, respectively, of the function $f(z)$. The small semi circles are centered around the poles of the coth function in the integrand.

The integration along the axis results in zero since the integrand is odd with respect to ω. The only surviving contributions are the ones from the small semi circles. The result is

$$\Delta\tilde{\mathcal{F}} = \frac{\hbar}{4\pi i}\sum_{\omega_n}\frac{i2\pi}{2}\frac{2}{\hbar\beta}\ln f(i\omega_n) = \frac{1}{2\beta}\sum_{\omega_n}\ln f(i\omega_n) \ ; \ \omega_n = \frac{2\pi n}{\hbar\beta} \ ; \ n = 0,\pm 1,\pm 2,\ldots \tag{6.46}$$

Since the summand is even in n we can write this as

$$\Delta\tilde{\mathcal{F}} = \frac{1}{\beta}{\sum_{\omega_n}}'\ln f(i\omega_n) \ ; \ \omega_n = \frac{2\pi n}{\hbar\beta} \ ; \ n = 0,1,2,\ldots \tag{6.47}$$

where the prime on the summation sign indicates that the $n = 0$ term should be multiplied by a factor of one half. This factor of one half is because there is only one term with $|n| = 0$ in the original summation but two for all other integers.

When the temperature goes to zero the spacing between the discrete frequencies goes to zero and the summation may be replaced by an integration:

$$\Delta \mathfrak{F} = \frac{1}{\beta} \sum_{\omega_n}{}' \ln f(i\omega_n) \xrightarrow{T \to 0} \frac{\hbar \beta}{2\pi} \frac{1}{\beta} \int_0^\infty d\omega \ln f(i\omega_n) = \hbar \int_0^\infty \frac{d\omega}{2\pi} \ln f(i\omega_n) = \Delta E$$

(6.48)

and we regain the contribution to the internal energy from the interactions, the change in zero-point energy of the modes.

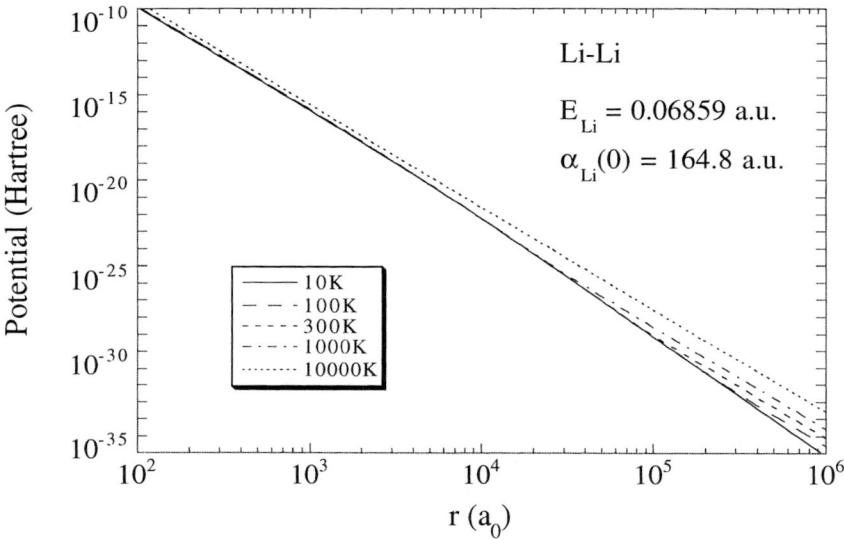

Figure 6.4: Temperature effects for the potential between two lithium atoms.

In Figure 6.4 we show the result for the temperature dependence of the potential between two lithium atoms. The result, counting from below, is for the temperatures 10, 100, 300, 1000, and 10000 K, respectively. We see that the effects first appear at large separations, in the Casimir region; the distance dependence returns to that of the van der Waals asymptote. Very high temperatures are needed to change the potential in the van der Waals region. In the hight-temperature limit only the first term, the $n = 0$ term, contributes and the result is linear in temperature. This is consistent with the result in Equation (4.110) for the Casimir effect. The results in Figure 6.4 are calculated neglecting any temperature dependence in the polarizabilities themselves; for very high temperatures we may expect temperature dependence in the function $f(z)$. At what temperature this becomes important depends on the size of the dominating characteristic energies of the system in question. In the case of larger objects there are also effects from the thermal expansion which modifies their dielectric properties.

6.3 Surface energy, method 1: no retardation

The bulk modes are given by the solutions to

$$\varepsilon(q,\omega) = 0 \tag{6.49}$$

and we know that the interaction energy is the change in the zero point energy of these solutions when the interaction is turned on. The energy is basically the zeros of the dielectric function minus the poles.

When the material is split up one bulk mode is replaced by two surface modes, one on each created surface. We want the energy of one of the surfaces. Thus we will calculate the contribution from one of the surface modes minus one half of the contribution from the bulk mode.

The surface modes are given by the solution to

$$\frac{\varepsilon(k,\omega)+1}{2} = 0 \tag{6.50}$$

if vacuum surrounds the sample.

We showed in Section 6.1 and subsections that the contribution to the interaction energy from one normal mode, found as solution to the equation $f(\omega) = 0$, is given by the expression

$$\Delta E(r) = \hbar \int_0^\infty \frac{d\omega}{2\pi} \ln[f(i\omega)] \tag{6.51}$$

At the surface we have one mode for each wave vector, **k**, in the plane of the surface. Thus the surface energy is

$$\begin{aligned} E &= \sum_{\mathbf{k}} \hbar \int_0^\infty \frac{d\omega}{2\pi} \left\{ \ln\left[\frac{\varepsilon(\mathbf{k},i\omega)+1}{2}\right] - \frac{1}{2}\ln[\varepsilon(\mathbf{k},i\omega)] \right\} \\ &= \Omega \int_0^\infty \frac{d^2k}{(2\pi)^2} \hbar \int_0^\infty \frac{d\omega}{2\pi} \frac{1}{2} \ln\left\{ \frac{[\varepsilon'(\mathbf{k},\omega)+1]^2}{4\varepsilon'(\mathbf{k},\omega)} \right\} \\ &= \frac{\hbar\Omega}{8\pi^2} \int_0^\infty \int_0^\infty d\omega dk k \ln\left\{ \frac{[\varepsilon'(k,\omega)+1]^2}{4\varepsilon'(k,\omega)} \right\} \end{aligned} \tag{6.52}$$

Remember that the prime on the dielectric function just denotes that it is the function on the imaginary frequency axis. Thus we find the following surface energy per unit area:

$$\sigma = \frac{\hbar}{8\pi^2} \int_0^\infty \int_0^\infty d\omega dk k \ln\left\{\frac{[1+\varepsilon'(k,\omega)]^2}{4\varepsilon'(k,\omega)}\right\} \tag{6.53}$$

We see that if the k-dependence of the dielectric function is neglected the result diverges. One way to simplify the treatment is to neglect the k-dependence but introduce a cutoff, k_c, in the momentum integration. Then we find

$$\sigma = \frac{\hbar k_c^2}{16\pi^2} \int_0^\infty d\omega \ln\left\{\frac{[1+\varepsilon'(\omega)]^2}{4\varepsilon'(\omega)}\right\} \tag{6.54}$$

For a metal the cutoff should be related to the inverse Thomas-Fermi screening length, q_{TF}, and for a semiconductor, where the interaction is caused by phonons, it should be such that the total number of modes equals the number of unit cells at the surface. The physical reason for the cutoff is that the coupling in general decreases with momentum and in the phonon case there are no modes outside the Brillouin zone. For a metal we have

$$\sigma = \frac{\hbar k_c^2}{16\pi^2} \int_0^\infty d\omega \ln\left\{\frac{\left[2+(\omega_{pl}/\omega)^2\right]^2}{4\left[1+(\omega_{pl}/\omega)^2\right]}\right\} = \frac{\hbar \omega_{pl} k_c^2}{16\pi^2} \int_0^\infty d\omega \ln\left\{\frac{\left[2+(1/\omega)^2\right]^2}{4\left[1+(1/\omega)^2\right]}\right\}$$

$$= \frac{\hbar \omega_{pl} k_c^2}{16\pi^2} \int_0^\infty d\omega \ln\left\{\frac{[1+2\omega^2]^2}{4\omega^2[1+\omega^2]}\right\} = \frac{(\sqrt{2}-1)}{16\pi} \hbar \omega_{pl} k_c^2 \tag{6.55}$$

This is in accordance with the result of Schmit and Lucas, which was discussed before in Section 4.4.1. They used the cutoff $k_c = q_{TF}/\sqrt{6}$.

In Section 6.5 we will use another method to calculate the surface energy. If we have two surfaces at infinite separation and bring them together there is an attractive force between them. This means that one gains energy when bringing them together. This is the energy stored as surface energy in the surfaces. When bringing the surfaces very close together one also has effects from the repulsion when the atoms come in direct contact. We furthermore found, in the Section 4.5 on quantum wells, that the van der Waals interaction saturates for metallic surfaces when the separation is smaller than the Thomas-Fermi screening-length. Thus, by limiting the energy contribution to the gain in bringing the surfaces from infinity to a finite separation, we may include, in an approximate way, other effects to the surface energy than obtained in our idealized van der Waals treatment.

6.4 Surface energy, method 1: retardation

When including retardation we use the full solutions of Maxwell's equations at the surface and we also have to take the transverse bulk modes into account. The bulk modes are given by the solutions to

$$\varepsilon(q,\omega) = 0 \tag{6.56}$$

and

$$\varepsilon(q,\omega)(\omega/c)^2 - q^2 = 0 \tag{6.57}$$

where the first equation gives the longitudinal modes and the second the transverse ones. Remember that there are two degenerate transverse modes. The contribution to the interaction energy from one of the set of bulk modes is

$$\frac{\hbar}{2}\left(\sum \omega_{i,o} - \sum \omega_{i,\infty}\right) = \hbar \int_0^\infty \frac{d\omega}{2\pi} \ln\left\{\frac{\varepsilon'(q,\omega)\left[\varepsilon'(q,\omega)(\omega/c)^2 + q^2\right]^2}{\left[(\omega/c)^2 + q^2\right]^2}\right\} \tag{6.58}$$

New, with retardation is that also transverse modes contribute. Taken together, all the transverse modes give a negligible contribution, as compared to the longitudinal modes. This statement also holds for the surface modes. We still have included this section to make the treatment complete. The transverse modes are unimportant for the surface energy. They are important for the forces between objects at very large separation, though. This is demonstrated later in this chapter. Now, the surface modes are determined by the relation

$$\varepsilon(k,\omega)\sqrt{k^2 - (\omega/c)^2} + \sqrt{k^2 - \varepsilon(k,\omega)(\omega/c)^2} = 0 \tag{6.59}$$

if vacuum surrounds the sample. This gives a **TM** mode. There is also a condition for **TE** modes, but it has no solution according to Section 4.1. We have found in Section 4.2 that between two surfaces also the **TE** mode contributes and if we assume that our surface is one of two, infinitely apart, we may include this mode.

It is not trivial in this retarded case to see which bulk modes go where when the surface is created. Besides, the vacuum should also be treated as a medium and vacuum modes should also turn into surface modes. We calculate the energy for two surfaces and divide the result by two. One longitudinal bulk mode is peeled off. So is one transverse mode from the medium and one transverse mode from the vacuum. We find the following surface energy per unit area:

$$\sigma = \frac{\hbar}{8\pi^2} \int_0^\infty \int_0^\infty d\omega dk k \ln \left\{ \frac{\left[\varepsilon'(k,\omega)\sqrt{k^2 + (\omega/c)^2} + \sqrt{k^2 + \varepsilon'(k,\omega)(\omega/c)^2} \right]^2}{4\varepsilon'(k,\omega)\left[k^2 + \varepsilon'(k,\omega)(\omega/c)^2 \right]} \right.$$

$$\left. \times \frac{\left[\sqrt{k^2 + (\omega/c)^2} + \sqrt{k^2 + \varepsilon'(k,\omega)(\omega/c)^2} \right]^2}{4\left[k^2 + (\omega/c)^2 \right]} \right\} \quad (6.60)$$

Note that we have been careful to rearrange the factors in the integrand so that the integration around the curved part of the path in the complex frequency plane does not contribute. In this case one can not obtain a simple general expression for the surface energy by introducing a cut off and neglecting the momentum dependence of the dielectric function.

That the above expression is correct can be verified from the energy calculated in Section 6.6. by letting the separation between the half-spaces go to infinity. It is much easier, more automatic, to use the method described there when the effects from retardation are included.

6.5 Surface energy, method 2: no retardation

In this section and the next we will determine the surface energy by calculating the zero-point energies of the modes between two half-spaces of the material; the surface energy is equal to the energy gain when the two surfaces are brought from infinite separation to zero separation, or rather to a finite but small separation.

We have earlier found that there are two surface modes when we have a gap in between two half-spaces of a material characterized by a dielectric function. These modes are, when retardation is neglected, determined by the following equation:

$$[\varepsilon(k,\omega)+1]^2 - e^{-2kd_0}[\varepsilon(k,\omega)-1]^2 = 0 \quad (6.61)$$

Continuing along the same lines as in the previous section we find

$$\sigma = \frac{\hbar}{8\pi^2} \int_0^\infty \int_0^\infty d\omega dk k \ln \left\{ \frac{[\varepsilon'(k,\omega)+1]^2}{[\varepsilon'(k,\omega)+1]^2 - e^{-2kd_0}[\varepsilon'(k,\omega)-1]^2} \right\} \quad (6.62)$$

6.5 Surface Energy, Method 1: no Retardation

In the numerator we have used the relation for infinite separation, where the second term vanishes and the two modes are degenerate. In the above expression we have as before divided the interaction energy by two to get the energy per surface area of one of the surfaces. If we neglect the k-dependence in the dielectric function, for a finite separation, this integral stays finite even without a momentum cutoff. Now neglecting the k-dependence we have

$$\begin{aligned}
\sigma &= \frac{\hbar}{8\pi^2}\int_0^\infty\int_0^\infty d\omega dk\, k \ln\left\{\frac{[\varepsilon'(\omega)+1]^2}{[\varepsilon'(\omega)+1]^2 - e^{-2kd_0}[\varepsilon'(\omega)-1]^2}\right\} \\
&= \frac{\hbar}{8\pi^2 d_0^2}\int_0^\infty\int_0^\infty d\omega dx\, x \ln\left\{\frac{[\varepsilon'(\omega)+1]^2}{[\varepsilon'(\omega)+1]^2 - e^{-2x}[\varepsilon'(\omega)-1]^2}\right\} \\
&= -\frac{\hbar}{8\pi^2 d_0^2}\int_0^\infty\int_0^\infty d\omega dx\, x \ln\left\{1 - e^{-2x}\left[\frac{\varepsilon'(\omega)-1}{\varepsilon'(\omega)+1}\right]^2\right\}
\end{aligned} \tag{6.63}$$

Thus the interaction energy between the planes is

$$\Delta E(d) = -2\sigma(d) = \frac{\hbar}{4\pi^2 d^2}\int_0^\infty\int_0^\infty d\omega dx\, x \ln\left\{1 - e^{-2x}\left[\frac{\varepsilon'(\omega)-1}{\varepsilon'(\omega)+1}\right]^2\right\} \tag{6.64}$$

Here we see that the separation dependence of the interaction energy between two half-spaces is d^{-2} and from this follows that the van der Waals force goes as d^{-3}.

Expanding the logarithm we have for the surface energy

$$\begin{aligned}
\sigma &= -\frac{\hbar}{8\pi^2 d_0^2}\int_0^\infty\int_0^\infty d\omega dx\, x \ln\left\{1 - e^{-2x}\left[\frac{\varepsilon'(\omega)-1}{\varepsilon'(\omega)+1}\right]^2\right\} \\
&= \frac{\hbar}{8\pi^2 d_0^2}\int_0^\infty\int_0^\infty d\omega dx\, x \sum_n \frac{1}{n} e^{-2xn}\left[\frac{\varepsilon'(\omega)-1}{\varepsilon'(\omega)+1}\right]^{2n} \\
&= \frac{\hbar}{8\pi^2 d_0^2}\int_0^\infty d\omega \sum_{n=1}^\infty \frac{1}{4n^3}\left[\frac{\varepsilon'(\omega)-1}{\varepsilon'(\omega)+1}\right]^{2n} \\
&= \frac{\hbar}{32\pi^2 d_0^2}\int_0^\infty d\omega \sum_{n=1}^\infty \frac{1}{n^3}\left[\frac{\varepsilon'(\omega)-1}{\varepsilon'(\omega)+1}\right]^{2n}
\end{aligned} \tag{6.65}$$

Here we find the characteristic integrals introduced in Equation (5.74). For a metal the result is reduced into

$$\sigma = \frac{\hbar}{32\pi^2 d_0^2} \int_0^\infty d\omega \sum_{n=1}^\infty \frac{1}{n^3} \left[\frac{(\omega_{pl}/\omega)^2}{2+(\omega_{pl}/\omega)^2} \right]^{2n}$$

$$= \frac{\hbar\omega_{pl}}{32\pi^2 d_0^2} \int_0^\infty dx \sum_{n=1}^\infty \frac{1}{n^3} \left[\frac{1}{1+2x^2} \right]^{2n}$$

$$= \frac{\hbar\omega_{pl}}{32\pi^2 d_0^2 \sqrt{2}} \int_0^\infty dx \sum_{n=1}^\infty \frac{1}{n^3} \left[\frac{1}{1+x^2} \right]^{2n}$$

$$= \frac{\hbar\omega_{pl}}{32\pi d_0^2 \sqrt{2}} \left[\frac{1}{4} + \frac{5}{32\cdot 2^3} + \frac{63}{512\cdot 3^3} + \frac{429}{4096\cdot 4^3} + \cdots \right]$$

(6.66)

6.6 Surface energy, method 2: retardation

In this section we repeat the derivations of the previous section but now we include retardation effects. When retardation is included the condition for having normal modes of **TM** type is

$$\left[\varepsilon(\omega) + \sqrt{k^2 - \varepsilon(\omega)(\omega/c)^2} \Big/ \sqrt{k^2 - (\omega/c)^2} \right]^2$$
$$- e^{-2\sqrt{k^2-(\omega/c)^2}\,d} \left[\varepsilon(\omega) - \sqrt{k^2 - \varepsilon(\omega)(\omega/c)^2} \Big/ \sqrt{k^2 - (\omega/c)^2} \right]^2 = 0$$

(6.67)

and for having modes of **TE** type:

$$\left[\sqrt{k^2 - \varepsilon(\omega)(\omega/c)^2} + \sqrt{k^2 - (\omega/c)^2} \right]^2$$
$$- e^{-2\sqrt{k^2-(\omega/c)^2}\,d} \left[\sqrt{k^2 - \varepsilon(\omega)(\omega/c)^2} - \sqrt{k^2 - (\omega/c)^2} \right]^2 = 0$$

(6.68)

This leads to

6.6 Surface Energy, Method 1: Retardation

$$\Delta E(d) = \frac{\hbar}{4\pi^2} \int_0^\infty \int_0^\infty d\omega dk\, k \ln\left\{1 - e^{-2\sqrt{k^2+\left(\frac{\omega}{c}\right)^2}d} \frac{\left[\varepsilon'(\omega)\sqrt{k^2+\left(\frac{\omega}{c}\right)^2} - \sqrt{k^2+\varepsilon'(\omega)\left(\frac{\omega}{c}\right)^2}\right]^2}{\left[\varepsilon'(\omega)\sqrt{k^2+\left(\frac{\omega}{c}\right)^2} + \sqrt{k^2+\varepsilon'(\omega)\left(\frac{\omega}{c}\right)^2}\right]^2}\right\}$$

$$+ \frac{\hbar}{4\pi^2} \int_0^\infty \int_0^\infty d\omega dk\, k \ln\left\{1 - e^{-2\sqrt{k^2+\left(\frac{\omega}{c}\right)^2}d} \frac{\left[\sqrt{k^2+\varepsilon'(\omega)\left(\frac{\omega}{c}\right)^2} - \sqrt{k^2+\left(\frac{\omega}{c}\right)^2}\right]^2}{\left[\sqrt{k^2+\varepsilon'(\omega)\left(\frac{\omega}{c}\right)^2} + \sqrt{k^2+\left(\frac{\omega}{c}\right)^2}\right]^2}\right\}$$

(6.69)

Here it is difficult to use a transformation to find the distance dependence since now both momentum and frequency appear in the exponent. For large separations however we can only have contributions for both k and ω/c small. Otherwise the exponential factor is zero and the energy vanishes. This means that the dielectric function may be replaced by its static value:

$$\Delta E(d)|_{d\text{ large}} = \frac{\hbar}{4\pi^2} \int_0^\infty \int_0^\infty d\omega dk\, k \ln\left\{1 - e^{-2\sqrt{k^2+\left(\frac{\omega}{c}\right)^2}d} \frac{\left[\varepsilon'(0)\sqrt{k^2+\left(\frac{\omega}{c}\right)^2} - \sqrt{k^2+\varepsilon'(0)\left(\frac{\omega}{c}\right)^2}\right]^2}{\left[\varepsilon'(0)\sqrt{k^2+\left(\frac{\omega}{c}\right)^2} + \sqrt{k^2+\varepsilon'(0)\left(\frac{\omega}{c}\right)^2}\right]^2}\right\}$$

$$+ \frac{\hbar}{4\pi^2} \int_0^\infty \int_0^\infty d\omega dk\, k \ln\left\{1 - e^{-2\sqrt{k^2+\left(\frac{\omega}{c}\right)^2}d} \frac{\left[\sqrt{k^2+\varepsilon'(0)\left(\frac{\omega}{c}\right)^2} - \sqrt{k^2+\left(\frac{\omega}{c}\right)^2}\right]^2}{\left[\sqrt{k^2+\varepsilon'(0)\left(\frac{\omega}{c}\right)^2} + \sqrt{k^2+\left(\frac{\omega}{c}\right)^2}\right]^2}\right\}$$

(6.70)

Now we may make the substitutions $k \to k/2d$; $\omega \to \omega/2d$ and find

$$\Delta E(d)|_{d\text{ large}} = \frac{\hbar}{32\pi^2 d^3} \int_0^\infty \int_0^\infty d\omega dk\, k \ln\left\{1 - e^{-\sqrt{k^2+\left(\frac{\omega}{c}\right)^2}} \frac{\left[\varepsilon'(0)\sqrt{k^2+\left(\frac{\omega}{c}\right)^2} - \sqrt{k^2+\varepsilon'(0)\left(\frac{\omega}{c}\right)^2}\right]^2}{\left[\varepsilon'(0)\sqrt{k^2+\left(\frac{\omega}{c}\right)^2} + \sqrt{k^2+\varepsilon'(0)\left(\frac{\omega}{c}\right)^2}\right]^2}\right\}$$

$$+ \frac{\hbar}{32\pi^2 d^3} \int_0^\infty \int_0^\infty d\omega dk\, k \ln\left\{1 - e^{-\sqrt{k^2+\left(\frac{\omega}{c}\right)^2}} \frac{\left[\sqrt{k^2+\varepsilon'(0)\left(\frac{\omega}{c}\right)^2} - \sqrt{k^2+\left(\frac{\omega}{c}\right)^2}\right]^2}{\left[\sqrt{k^2+\varepsilon'(0)\left(\frac{\omega}{c}\right)^2} + \sqrt{k^2+\left(\frac{\omega}{c}\right)^2}\right]^2}\right\}$$

(6.71)

The separation dependence of the interaction energy is now d^{-3} and from this follows that the Casimir force goes as d^{-4}. Thus the force at large separations drops off faster when retardation is included. For a metal the dielectric function diverges for zero frequency. This may be used to find the result:

$$\Delta E(d)\big|_{\text{d large}} = \frac{\hbar}{32\pi^2 d^3} \int_0^\infty \int_0^\infty d\omega dk\, k \ln\left\{1 - e^{-\sqrt{k^2 + \left(\frac{\omega}{c}\right)^2}}\right\}$$

$$+ \frac{\hbar}{32\pi^2 d^3} \int_0^\infty \int_0^\infty d\omega dk\, k \ln\left\{1 - e^{-\sqrt{k^2 + \left(\frac{\omega}{c}\right)^2}}\right\}$$

$$= \frac{\hbar}{16\pi^2 d^3} \int_0^\infty \int_0^\infty d\omega dk\, k \ln\left\{1 - e^{-\sqrt{k^2 + \left(\frac{\omega}{c}\right)^2}}\right\}$$

$$= \frac{\hbar c}{16\pi^2 d^3} \int_0^\infty \int_0^\infty dx dk\, k \ln\left\{1 - e^{-\sqrt{k^2 + x^2}}\right\} \qquad (6.72)$$

$$= \frac{\hbar c}{16\pi^2 d^3} \int_0^\infty \int_0^\infty dx dy\, y \ln\left\{1 - e^{-\sqrt{y^2 + x^2}}\right\}$$

$$= \frac{\hbar c}{16\pi^2 d^3} \int_0^\infty dr\, r \int_0^{\pi/2} d\theta\, r \cos\theta \ln\left\{1 - e^{-r}\right\}$$

$$= \frac{\hbar c}{16\pi^2 d^3} \int_0^\infty dr\, r^2 \ln\left\{1 - e^{-r}\right\} = -\frac{\hbar c \pi^2}{720 d^3}$$

for the energy and

$$F(d)\big|_{\text{d large}} = -\frac{\hbar c \pi^2}{240 d^4} \qquad (6.73)$$

for the force.

With this derivation of the Casimir energy and force we obtained the results much more easily than with the method described earlier in Section 4.3.1. We may furthermore notice that the **TE** and **TM** modes give the same contribution to the result at large separations. This is in full accordance with the results of Section 4.5.

6.7 Finite temperatures

We will in this section as an illustration go through the calculation of the Helmholtz' free energy for two water half-spaces. From the distance dependence of this result the van der Waals and Casimir forces may be obtained.

We showed in Section 6.2 that for finite temperature we get the Helmholtz free energy from the zero-temperature internal energy by making the following replacement:

$$\hbar \int_0^\infty \frac{d\omega}{2\pi} \rightarrow \frac{1}{\beta} {\sum_{\omega_n}}' \quad ; \quad \omega_n = \frac{2\pi n}{\hbar \beta} \quad ; \quad n = 0,1,2,\ldots \tag{6.74}$$

The integration along the upper part of the imaginary frequency axis is replaced by a discrete summation along the upper part of the imaginary frequency axis. The higher the temperature the larger the separation between the points of summation. For low temperatures the separation between the points becomes small and the summation can be replaced by an integration and we obtain the zero-temperature result. To be noted is that in the summation above the $n = 0$ term should be multiplied by one half. This is important to remember especially for high temperatures where only this term survives.

At what temperature we may replace the summation with an integration depends on how fast the integrand varies with frequency. Since different contributions appear in different energy scales and have different frequency dependence we may sometimes replace a part of the summation with an integration and keep the summation for the rest of the frequency range. It should furthermore be noted that the dielectric properties of the materials are temperature dependent; thus the temperature dependence comes both from the discrete summation points and from the summand.

To be able to calculate the energy we need the dielectric function of the system on the imaginary frequency axis; this is the case both for a zero-temperature calculation and for one at finite temperatures. If we use analytical model–dielectric functions there are no difficulties. If we want to do a detailed calculation for a real material it is more difficult. All measurements of the dielectric properties are done on the real axis. We showed in Equation (2.1) that we could get the imaginary part of the dielectric function from the real part and visa versa by using the Kramers-Kronig dispersion relations. These relations are valid on the real axis only. However we can with generalizations of these relations get the dielectric function in a general point of the complex frequency plane. It follows from the analytical properties of these functions. The dielectric function on the imaginary axis can be obtained from the real and imaginary parts of the dielectric function for real frequencies. It is actually enough to know one of the functions. In terms of the imaginary part it is obtained as

$$\varepsilon(i\omega) = 1 + \frac{2}{\pi} \int_0^\infty d\omega' \, \frac{\omega' \varepsilon_2(\omega')}{(\omega')^2 + \omega^2} \tag{6.75}$$

This expression has been used together with the experimental data for water, presented in Section 2.7, and the result is given in Figure 6.5. We have also, with circles, indicated the position of the discrete imaginary frequencies, $2\pi n/\beta$, entering the finite temperature calculations. The position of the points depends on the temperature and we have here assumed 300 K. They are equidistant on a linear plot but become very scarce for low frequencies and very crowded for high frequencies on a logarithmic plot. On the logarithmic plot the variation of the curve with frequency is similar in different frequency ranges. It is useful to use this plot to get a feeling for if the discrete summation may be replaced by an integration without loosing some temperature dependence of a contribution. We clearly see that to get the correct temperature dependence for energy contributions below 0.1 eV we need to use the discrete summation, but for higher frequencies we may replace the summation by an integration. To be noted is that we have put a circle at the left-most axis to indicate the first summation point. It should have been placed at zero frequency but this point can never be included on a logarithmic plot.

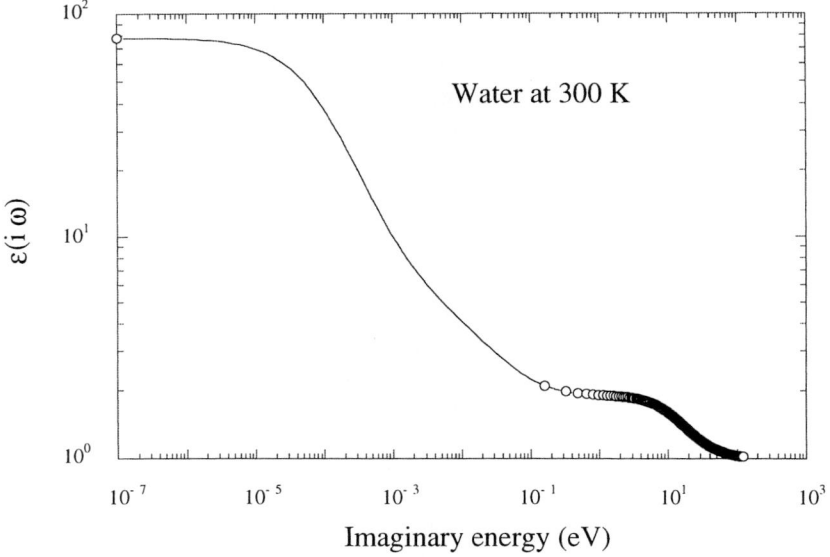

Figure 6.5: The dielectric function of water on the imaginary frequency axis. The circles indicate the position of the discrete summation frequencies.

Performing the summation leading to the free energy contribution from the interactions we find 1.36 eV per mode. The zero frequency contribution makes up 0.03 eV of this value. This may seem like a small relative contribution but in calculating the interaction between water and other substances their dielectric properties may be similar for high frequencies and different for low frequencies; this means that the contribution to their interaction from the high-frequency range is suppressed and the low-frequency contribution may have quite a large relative importance.

6.7 Finite Temperatures

When finding the dielectric function on the imaginary axis we could have used either of the following expressions instead of the one given above:

$$\varepsilon(i\omega) = 1 + \frac{2}{\pi} \int_0^\infty d\omega' \frac{\omega[\varepsilon_1(\omega')-1]}{(\omega')^2 + \omega^2} \quad;$$

$$\varepsilon(i\omega) = 1 + \frac{1}{\pi} \int_0^\infty d\omega' \frac{\omega[\varepsilon_1(\omega')-1] + \omega' \varepsilon_2(\omega')}{(\omega')^2 + \omega^2} \quad (6.76)$$

but as our data points are limited to a large but still finite frequency range we have chosen to use the imaginary part since it is small at both ends of the range. When using the real part one has to be careful and extrapolate the real part down to zero frequency in order not to get serious departure for low frequencies.

The dielectric function on the imaginary axis has some resemblance to the real part of the dielectric function along the real axis but it is monotonously decreasing with frequency and is much nicer to deal with. We compare the two in Figure 6.6.

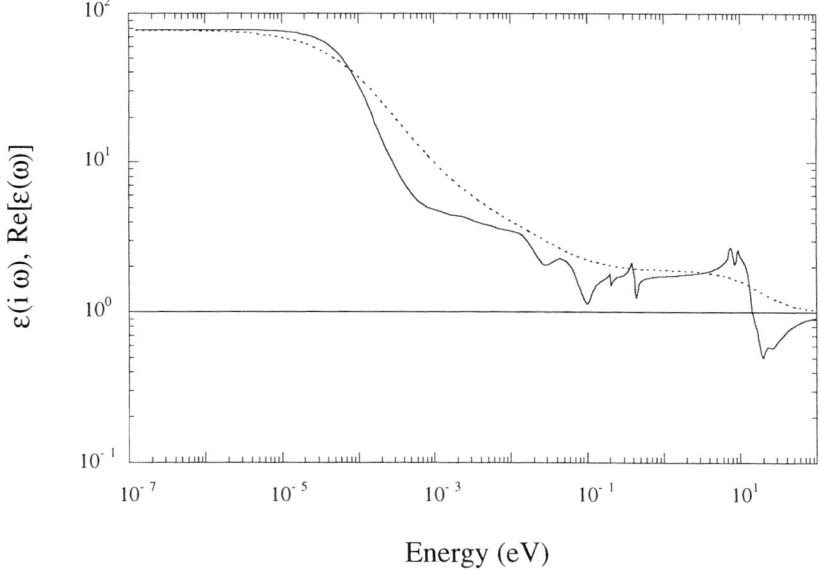

Figure 6.6: Comparison of the real part of the dielectric function on the real axis (solid curve) and the dielectric function on the imaginary axis (dotted curve).

We will end this section with performing the calculation of the Hamaker constant in two situations making use of the obtained results for water. The configurations we will study is two water half-spaces separated by air and the inverse problem with a water film in air; these situations give the same result. We will perform the calculations both at 300 K and at the boiling temperature 373 K to see if there are any important temperature effects. Much happens when the temperature is raised. The liquid expands and the various contributions to the dielectric function are temperature dependent. We will only include one temperature dependence – the dependence of the Debye rotational polarization which is responsible for the strong screening in the microwave range and whose temperature dependence we treated in Section 2.6. The difference in screening is demonstrated in figure 6.7. The dipolar contribution has decreased at the boiling point.

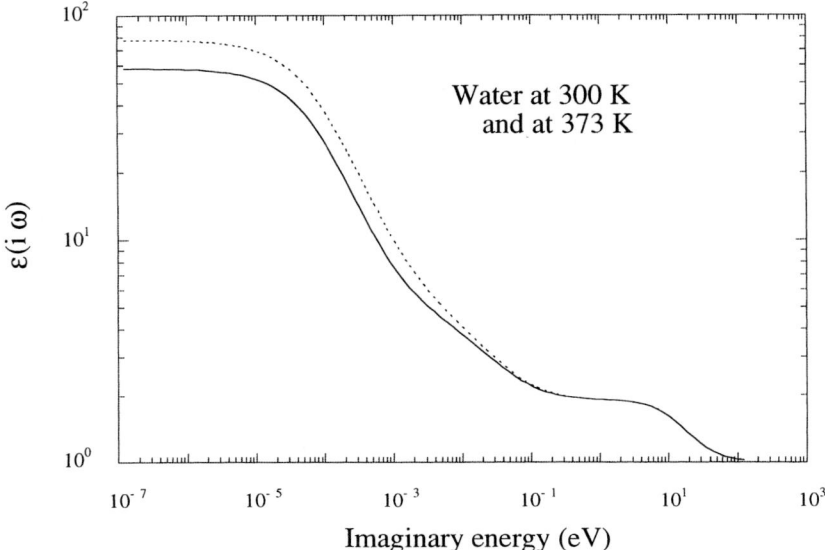

Figure 6.7: The dielectric function on the imaginary axis for water at 300 K (dotted curve) and at 373 K (solid curve).

Under certain assumptions the interaction energy between two objects may be written as a product of two factors; one material dependent factor, the so called Hamaker constant; one separation dependent factor whose form depends on the geometrical shapes of the objects. We introduced this constant in Section 6.12. In the case of two half-spaces the interaction energy is given by

$$\Delta E(d) = -\frac{A}{12\pi}\frac{1}{d^2} \tag{6.77}$$

where A is the *Hamaker constant*. From this follows that

$$A = \frac{6}{\beta}\sum_{\omega_n}\int_0^\infty dx\, x \ln\left\{1 - e^{-2x}\left[\frac{\varepsilon'(\omega_n)-1}{\varepsilon'(\omega_n)+1}\right]^2\right\} \tag{6.78}$$

6.7 Finite Temperatures

For this separation to be possible we have to neglect any momentum dependence in the dielectric functions. For large separations this is no mayor problem since the exponential factor in the integrand is an effective momentum cut off at a small momentum value. When retardation effects are taken into account this description is no longer as useful.

We find, for 300 K, the Hamaker constant is 0.3525 eV (5.65×10^{-20} J); the zero frequency contribution makes up 0.0218 eV (0.35×10^{-20} J) of this value. For 373 K we find the Hamaker constant is 0.3558 eV (5.70×10^{-20} J); the zero frequency contribution makes up 0.0265 eV (0.42×10^{-20} J) of this value. Thus the temperature dependence is very weak. The zero frequency contribution increases and the other contributions decrease. The net result is a weak increase.

From the Hamaker constant one may get an estimate of the surface tension. If we assume that the van der Waals force is active down to an effective separation d_0 we get the surface tension as

$$\gamma = \frac{A}{24\pi d_0^2} \tag{6.79}$$

For most materials it works remarkably well [1] with the choice of $d_0 = 1.65$ Å. With our result for water the use of this value would give a surface tension of 28 mJ/m^2. To obtain the experimental result we would need to use a 1 Å separation instead. This deviation from the rule of thumb is typical for H-bonded liquids like water. Probably there will be a short range orientational adjustment of the polar molecules when the two surfaces are near contact; this will give an important contribution to the surface tension but is not included in our derivation. This fact could also be understood from the experimental [2] values for the surface tension which decrease rather strongly with temperature.

For larger separations one needs to take retardation effects into account which also means inclusion of the **TE** modes between the surfaces.

6.7.1 Retarded interaction energy

For zero temperature we have derived the following expression in Equation (6.69):

$$\Delta E(d) = \frac{\hbar}{4\pi^2} \int_0^\infty \int_0^\infty d\omega dk\, k \ln\left\{1 - e^{-2\sqrt{k^2 + \left(\frac{\omega}{c}\right)^2}\, d} \frac{\left[\varepsilon'(\omega)\sqrt{k^2 + \left(\frac{\omega}{c}\right)^2} - \sqrt{k^2 + \varepsilon'(\omega)\left(\frac{\omega}{c}\right)^2}\right]^2}{\left[\varepsilon'(\omega)\sqrt{k^2 + \left(\frac{\omega}{c}\right)^2} + \sqrt{k^2 + \varepsilon'(\omega)\left(\frac{\omega}{c}\right)^2}\right]^2}\right\}$$

$$+ \frac{\hbar}{4\pi^2} \int_0^\infty \int_0^\infty d\omega dk\, k \ln\left\{1 - e^{-2\sqrt{k^2 + \left(\frac{\omega}{c}\right)^2}\, d} \frac{\left[\sqrt{k^2 + \varepsilon'(\omega)\left(\frac{\omega}{c}\right)^2} - \sqrt{k^2 + \left(\frac{\omega}{c}\right)^2}\right]^2}{\left[\sqrt{k^2 + \varepsilon'(\omega)\left(\frac{\omega}{c}\right)^2} + \sqrt{k^2 + \left(\frac{\omega}{c}\right)^2}\right]^2}\right\}$$

(6.80)

where the first term represents the contribution from the **TM** modes and the second the contribution from the **TE** modes.

For finite temperatures this internal energy change turns into the free energy change:

$$\Delta E(d) = \frac{1}{2\pi\beta} {\sum_{\omega_n}}' \int_0^\infty dk\, k \ln\left\{1 - e^{-2\sqrt{k^2 + \left(\frac{\omega_n}{c}\right)^2}\, d} \frac{\left[\varepsilon'(\omega_n)\sqrt{k^2 + \left(\frac{\omega_n}{c}\right)^2} - \sqrt{k^2 + \varepsilon'(\omega_n)\left(\frac{\omega_n}{c}\right)^2}\right]^2}{\left[\varepsilon'(\omega_n)\sqrt{k^2 + \left(\frac{\omega_n}{c}\right)^2} + \sqrt{k^2 + \varepsilon'(\omega_n)\left(\frac{\omega_n}{c}\right)^2}\right]^2}\right\}$$

$$+ \frac{1}{2\pi\beta} {\sum_{\omega_n}}' \int_0^\infty dk\, k \ln\left\{1 - e^{-2\sqrt{k^2 + \left(\frac{\omega_n}{c}\right)^2}\, d} \frac{\left[\sqrt{k^2 + \varepsilon'(\omega_n)\left(\frac{\omega_n}{c}\right)^2} - \sqrt{k^2 + \left(\frac{\omega_n}{c}\right)^2}\right]^2}{\left[\sqrt{k^2 + \varepsilon'(\omega_n)\left(\frac{\omega_n}{c}\right)^2} + \sqrt{k^2 + \left(\frac{\omega_n}{c}\right)^2}\right]^2}\right\}$$

(6.81)

One thing to notice is that the last term will not contribute to the zero-frequency contribution. The results for our system are shown in Figure 6.8.

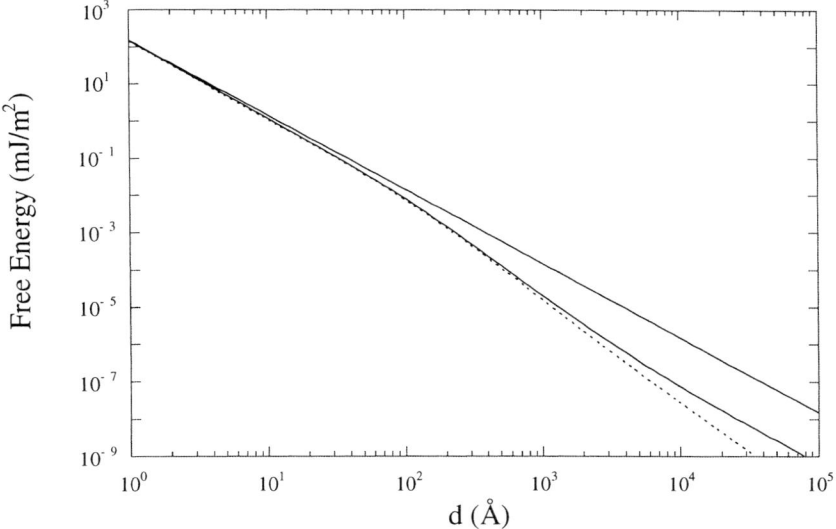

Figure 6.8: The free energy per unit area for a water film as function of thickness d. The upper curve is the non-retarded result and the lower solid curve includes retardation effects. Both solid curves are for 300 K. The results are also valid for two water half-spaces separated by d. The dotted curve is the retarded result for 10 K.

The upper curve represents the non-retarded van der Waals energy and the lower solid curve the full result including retardation effects. We note that the distance dependence is no longer the same as for the zero temperature case; for zero temperature the energy varies as d^{-2} for small separations and as d^{-3} for large separations. For finite temperature and large separations only the zero frequency contribution survives and the energy once again varies as d^{-2}. The results are valid for 300 K. The temperature effects are too small to make it meaningful to show also the result at the boiling point. The dotted curve is the result for 10 K; neglecting the fact that water actually freezes. Here the deviation from the d^{-3} dependence occurs at larger separations. Thus we see that the temperature and separation dependencies are rather complicated for a system like this with important dipolar contributions to the dielectric function.

6.8 Recent results for metals

Recently one has been able to measure the van der Waals and Casimir energies between metal plates with such high an accuracy that a comparison with theory is meaningful [3-5].

In the two first references the force is measured for separations in the range 0.062-0.350 µm; this is the van der Waals range. The agreement between theory and experiments is good. The agreement is better than 1%. One has to be careful here and specify what one means with this number. The agreement is better than 1% for the smallest separations where the force is largest and the relative uncertainties are smallest. The root-mean-square deviation for the experimental force from the theoretical one is also smaller than 1% of the measured force at the smallest separation. However, the accuracy for each individual separation is much worse at the high separation end of the measurement range. The other experiment, by Lamoreaux, is more interesting. It is extended to larger separations – up to 3.4 µm; this is clearly into the Casimir range. Here the forces are much smaller and more difficult to measure with good accuracy. With the same way to estimate the accuracy, the experiments were done within an accuracy of 5%. What makes these experiments more interesting is that in this range one starts to see temperature effects. The experiments were performed at 300 K. The experimental results with error bars are shown in Figure 6.9.

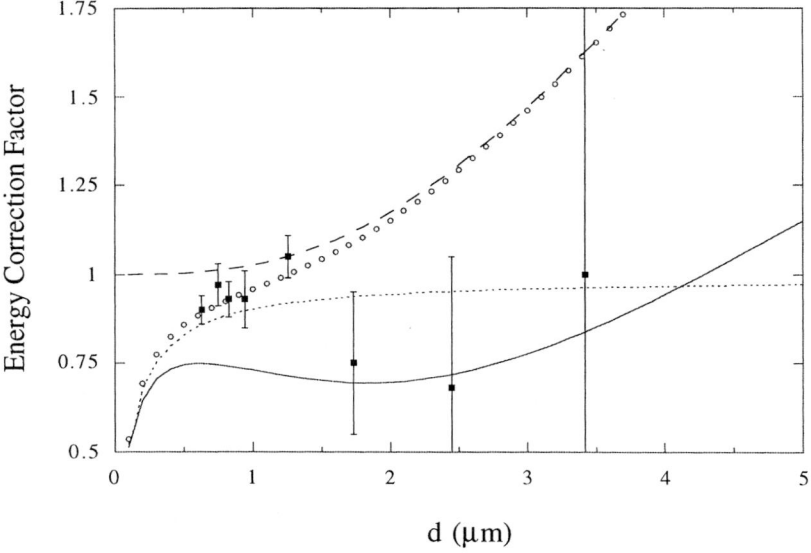

Figure 6.9: Energy Correction Factor as function of separation between two gold plates. The solid curve is the full result at 300 K to be compared to the experimental data points with error bars. The dotted curve is the full result at zero-temperature. The dashed curve is the temperature dependent Casimir result for a perfect metal. The circles are the result from a Drude model neglecting dissipation. All curves are the interaction energy divided by the zero-temperature Casimir asymptote.

We performed calculations [6] between two gold plates using the experimentally obtained optical constants for gold as input. We found that the results varied significantly depending on the choice of dielectric function. If we used a simple Drude model, neglecting dissipation, the way we have done at many places in this book, gave one result. Using the experimental data or using a Drude model including dissipation, the way we have done at other places in this book, gave another result. The deviations occur in the Casimir region and only for finite temperatures. The dissipation has the drastic effect that the limiting result for the Casimir force is just half the size of the value for a perfect metal or for a non-perfect metal without dissipation. This shows that there are important deviations between the idealization of a perfect metal and a real metal. Now what is interesting is that in a range around 1 μm, which is the crossover region between the van der Waals and Casimir regions, the experiments are closer to the zero-temperature results (dotted curve) and to the results without dissipation (circles). The Energy Correction Factor is the interaction energy divided by the zero-temperature Casimir asymptote. A perfect metal has no van der Waals region; the Casimir asymptote is valid all the way down to zero separation. Thus for a perfect metal the zero-temperature result would give a constant value unity in the figure. At 300 K the result for a perfect metal is given by the dashed curve. It starts out at the zero-temperature result for zero separation and increases monotonically for larger separations. The zero-temperature full result (dotted curve) starts for large separations out from the zero-temperature Casimir result; then in the crossover region between the Casimir and van der Waals regions it bends down and decreases towards zero for zero separation. For large separations both **TM** and **TE** contributions are of the same magnitude. We have shown this before neglecting dissipation; we showed this in Equation (6.72) and also for the quantum wells in Fig. 4.32. This is still valid when dissipation is taken into account.

The full result at 300 K (solid curve) agrees with the zero-temperature full result for small separations but starts to decrease in magnitude when the separation is close to the crossover region; then it starts to increase again and approaches for large separations half the value of the perfect metal result. The reason for the half is that for large separations, at finite temperature, or in the high temperature limit the **TE** modes will not give any contribution. Using a Drude approximation, neglecting dissipation, will give the erroneous result that the **TE** modes contribute for large separations, at finite temperature, or in the high temperature limit.

This is very important and interesting since it demonstrates that dissipation may have important effects on the force between objects. Let us study this in some more detail. We take a look at the integrands of the two integrals in Equation (6.80), where one is for the **TM** and one for the **TE** contributions. We perform the integration with respect to momentum and present, in the figures that follow, the integrands in the remaining integral with respect to frequency, or energy. In all Figures 6.10-6.12 we have plotted the results for energies up to $3\hbar c/d$. We present in Figure 6.10 the integrands at the separation 5 μm. The solid (dotted) curve is for the **TM** (**TE**) contribution. We see that the two integrands are almost the same and hence the two contributions are almost equal. The only difference is that the integrand for the **TE** contribution tends to zero for small frequencies. This is caused by the dissipation. So for zero temperature the two contributions are almost the same. However, for 300 K this is not so. Indicated with circles in the figure are the contributions to the summands in Equation (6.81). At this separation basically only one frequency value contributes, the $n = 0$ term, in the summation; all other points are outside the plotted range in the figure where the integrand is small. Thus at this separation, at 300 K, only the **TM** modes contribute. We have now looked at a point to the right in Figure 6.9.

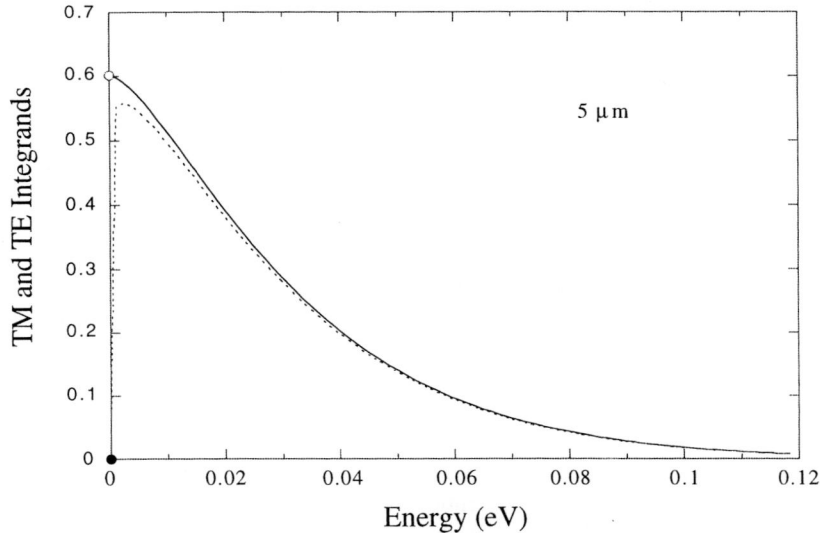

Fig 6.10: Comparison of the integrands for the **TM** (solid curve) and **TE** (dotted curve) contributions to the energy between two gold plates at separation 5 μm. The circles indicate the frequencies contributing to the free energy at 300 K.

Let us now in Figure 6.11 study one point where the experiments and theory seem to disagree, at 1 μm. Here the deviation between the two types of contribution starts to grow. The **TE** contribution starts to be drastically smaller than the **TM** contribution also at zero temperature. We further notice that we now have four frequencies contributing at 300 K and the $n = 0$ term will have a smaller relative importance. We should also remember that this $n = 0$ term should be multiplied by a factor of one half.

Our last example, Figure 6.12, is for the separation 0.1 μm. The trends continue; the **TE** contribution continues to drop in value as compared to the **TM** contribution; the number of frequencies contributing to the finite temperature result has increased to around 40. Here the discrete summation can, to a good approximation, be replaced by an integration and there are no important temperature effects.

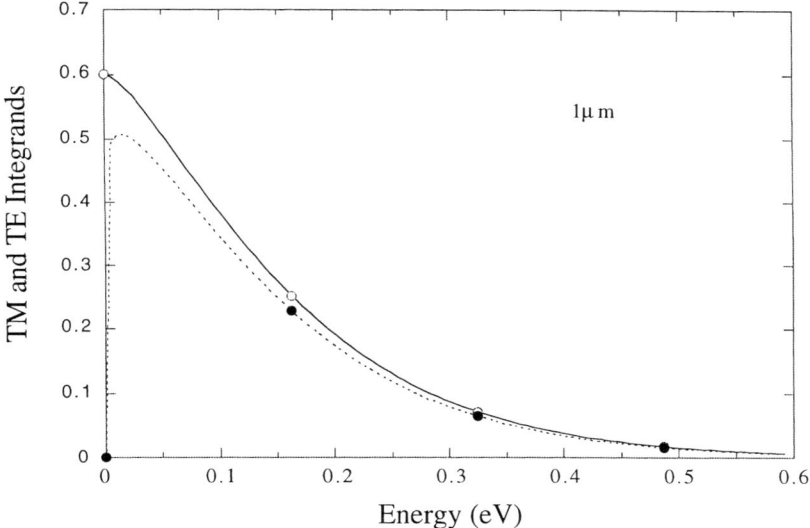

Fig 6.11: Same as Figure 6.10, but for the separation 1 μm.

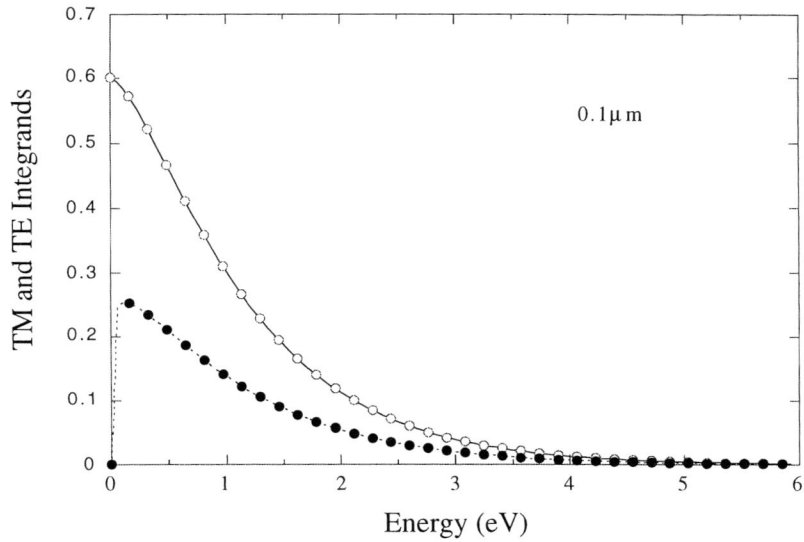

Fig 6.12: Same as Figures 6.10 and 6.12, but for the separation 0.1 μm.

6.9 Adhesion, cohesion, and wetting

In this section we will discuss phenomenon like adhesion, cohesion, wetting and related effects.

6.9.1 Work of adhesion and cohesion

The work of adhesion is the energy it takes to separate two pieces of material with an interface of unit area and bring the pieces to infinite separation.

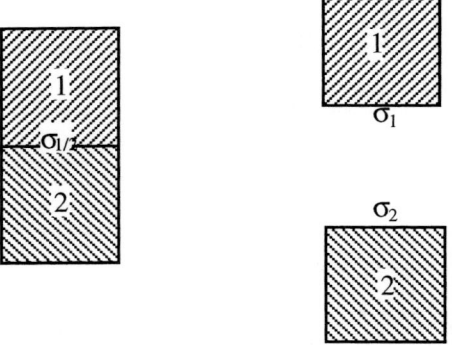

Figure 6.13:   Illustration of work of adhesion.

This energy is

$$W_a = \sigma_{1/0} + \sigma_{2/0} - \sigma_{1/2} \qquad (6.82)$$

where the indices 0 denote that the process is taking place in vacuum. The function σ is the surface energy. We could have used γ instead which usually is the character used to represent the surface tension. These two quantities are the same in the sense that they have the same values and units but they represent different physical concepts. The surface energy is an energy per unit area, the energy stored in the surface region; the surface tension is a force per unit length, and is often used in the case of liquids. One way to measure the surface tension of a liquid is to form a thin rectangular shaped film suspended by wires according to the figure, below. The U-shaped wire makes up three sides of the rectangle and is fixed in position while the last side is formed by a movable wire of the length l.

6.9 Adhesion, Cohesion, and Wetting

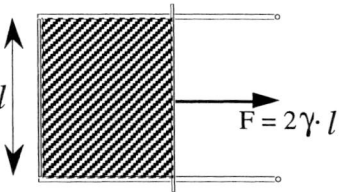

Figure 6.14: Equipment for measuring surface tension.

To note here is that the thin film has two surfaces. The force needed to pull the movable wire is then the one indicated in the figure. If the wire is pulled the distance Δx, a new amount of surface area $\Delta A = 2 \cdot \Delta x \cdot l$ is created and the work $F \cdot \Delta x$ is performed on the film. Thus the energy per unit area stored in this additional surface area is

$$\sigma = \frac{F \cdot \Delta x}{\Delta A} = \frac{2 \cdot \gamma \cdot l \cdot \Delta x}{2 \cdot \Delta x \cdot l} = \gamma \tag{6.83}$$

The unit for surface energy is erg/cm^2 and for surface tension dyne/cm. These units are the same and in SI units they correspond to 10^{-3} Nm/m^2 and 10^{-3} N/m, respectively. The room-temperature surface-energy of water is 73 erg/cm^2 and of mercury (Hg) 490 erg/cm^2, respectively.

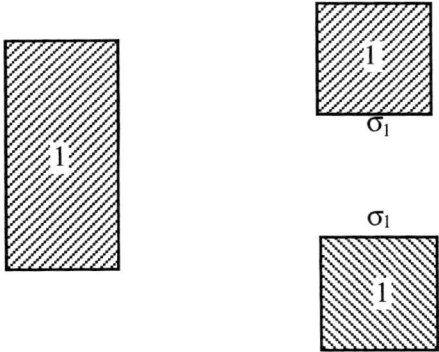

Figure 6.15: Illustration of work of cohesion.

In the special case when the two materials are the same, Figure 6.15, we do not speak about the work of adhesion but the work of cohesion:

$$W_c = 2\sigma_{1/0} \tag{6.84}$$

We may drop the second index if we let just one index mean that we work in vacuum:

$$W_c = 2\sigma_1 \tag{6.85}$$

Let us now attach indices to W to specify the two materials and the surrounding medium. If we just have two subscript indices on the W we mean vacuum, a third superscript index denotes the surrounding medium, if not vacuum. If the two subscripts are the same we have work of cohesion, otherwise work of adhesion. The following relations should make the notation clear:

$$W_{12} = \sigma_1 + \sigma_2 - \sigma_{1/2}; \tag{6.86}$$

$$\begin{aligned}W_{12}^3 &= \sigma_{1/3} + \sigma_{2/3} - \sigma_{1/2} \\ &= [\sigma_1 + \sigma_3 - W_{13}] + [\sigma_2 + \sigma_3 - W_{23}] - [\sigma_1 + \sigma_2 - W_{12}]; \\ &= \sigma_3 - W_{13} + \sigma_3 - W_{23} + W_{12} = W_{12} - W_{13} - W_{23} + W_c^3\end{aligned} \tag{6.87}$$

$$W_{11}^3 = W_{11} + W_c^3 - 2W_{13} = W_{11} + W_{33} - 2W_{13} = W_c^1 + W_c^3 - 2W_{13} \tag{6.88}$$

6.9.2 Wetting

If we put a small droplet of a liquid on the surface of a solid the resulting shape of the droplet depends on the relative size of the three surface energies or surface tensions, σ_S, σ_L and $\sigma_{S/L}$. For larger drops the shape doesn't scale with size. If we pour out one liter of water on a large flat surface the result does not resemble that when we pour out one mm^3 of water. The difference lies in that we in the last case can forget about gravitational effects while they are important in the first case. If we were to perform the experiment in outer space the situation would be different.

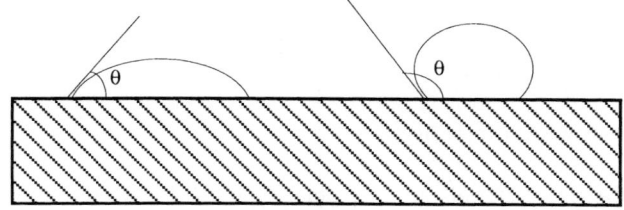

Figure 6.16: Illustration of wetting and non-wetting liquids.

Here, it is convenient to think about surface tensions; three forces are acting on the circular boundary line between the three surfaces, all three perpendicular to the line. Since the solid surface is flat, and horizontal, the horizontal component of the total force is zero:

$$\sigma_S = \sigma_{S/L} + \sigma_L \cdot \cos(\theta) \tag{6.89}$$

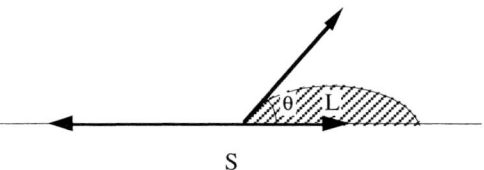

Figure 6.17: Forces acting on the borderline between the three surfaces.

The liquid is said to wet the solid if the contact angle, θ, is sharp, that is, $< 90°$. If the surrounding medium is vacuum we always have:

$$\sigma_{S/L} < \sigma_S + \sigma_L \Rightarrow \sigma_{S/L} - \sigma_L < \sigma_S \tag{6.90}$$

The work of adhesion in vacuum is always positive.

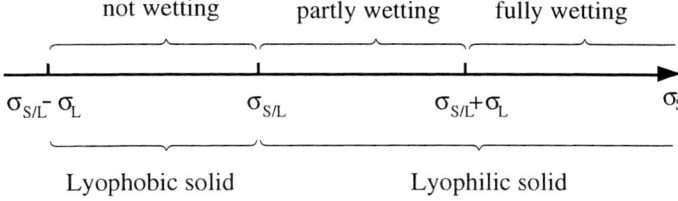

Figure 6.18: Definition of not wetting, partly wetting and fully wetting.

If $\sigma_{S/L} < \sigma_L$ the left limit is 0.

If both substances are liquids the flexibility at the interface increases, see Figure 6.19.

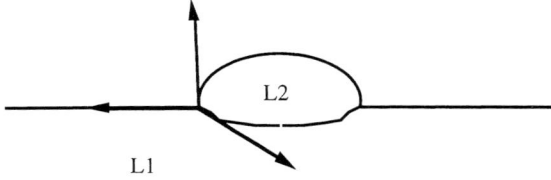

Figure 6.19: A liquid drop on a liquid surface.

Given the values of the three surface tensions the directions are such that the vector sum is zero. One immediately realizes that there is no solution if any of the tensions is larger than the sum of the other two. If this is the case another configuration results:

$$\sigma_{L1} > \sigma_{L2} + \sigma_{L1/L2} \quad ; \quad \begin{cases} \text{Complete wetting. A layer of} \\ \text{L2 covers the surface of L1} \end{cases}$$

$$\sigma_{L2} > \sigma_{L1} + \sigma_{L1/L2} \quad ; \quad \{\text{L2 will be engulfed by L1}\} \tag{6.91}$$

$$\sigma_{L1/L2} > \sigma_{L1} + \sigma_{L2} \quad ; \quad \text{Will never happen}$$

It is interesting to notice that if $\sigma_{L1/L2} < \sigma_{L2}$ it is energetically favorable if the droplet is entirely submerged as compared to fully above the surface, but it is even more favorable if the droplet penetrates the surface; not until $\sigma_{L1/L2} < \sigma_{L2} - \sigma_{L1}$ will the droplet let go of the surface and enter the interior; if $\sigma_{L2} < \sigma_{L1}$ this will never happen.

If we are not in space and the droplet is not small enough there will be effects from the gravitation. The surface of liquid *1* can be deformed upwards or downwards, depending on the relative densities of the two liquids. This is schematically illustrated in Figure 6.20.

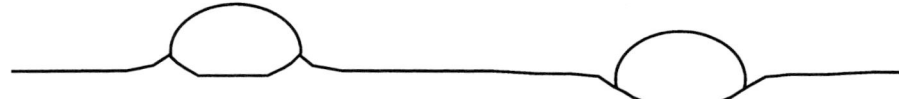

Figure 6.20: Illustration of a liquid drop on a liquid surface for different relative densities of the two liquids.

6.9.3 Model calculations

In this section we will make very simple and rough model calculations of the parameters important for the work of cohesion and adhesion and for wetting. Even though the calculations are rough they help us understand why, for example, some liquids wet a material and some do not. We will also demonstrate the effect that, given a choice, a material prefers staying close to another piece of material of the same or similar kind to staying close to a different material. We all have heard expressions like fat solves fat. We will in these model calculations represent the materials with simple model-dielectric functions.

6.9.3.1 Modelling of adhesion, cohesion and wetting

Let us now make a simple model calculation. Let both materials involved be represented by a dielectric function of the form

$$\varepsilon_i(\omega) = 1 - \frac{2a_i\omega_i}{\omega^2 - \omega_i^2} \quad ; \quad i = 1, 2 \tag{6.92}$$

Each system is characterized by a single excitation energy. We furthermore assume that

6.9 Adhesion, Cohesion, and Wetting

the number of carriers per unit volume in the two substances is the same; this gives the same numerator according to the *f-sum rule*; we put it equal to $2a$. In the case of a liquid drop on a solid surface we let the index *1* represent the liquid and *2* the solid. We keep the frequency ω_{01} fixed, and vary the excitation frequency ω_{02}. Thus, our bulk dielectric functions for the two substances are

$$\varepsilon_i(\omega) = 1 - \frac{2a}{\omega^2 - \omega_{0i}^2} \quad ; \quad i = 1, 2 \tag{6.93}$$

and the frequencies of the normal modes are

$$\omega_i^{bulk} = \sqrt{\omega_{0i}^2 + 2a} \quad ; \quad i = 1, 2 \tag{6.94}$$

The surface modes, the normal modes localized to the free surfaces of the two materials, are solutions to the equations

$$\frac{\varepsilon_i(\omega) + 1}{2} = 0 \tag{6.95}$$

The results are

$$\omega_i = \sqrt{\omega_{0i}^2 + a} \quad ; \quad i = 1, 2 \tag{6.96}$$

If we bring the two materials together so that they have a surface in common there will be normal modes localized to this surface or interface. The normal modes are obtained as solutions to the equation

$$\frac{\varepsilon_1(\omega) + \varepsilon_2(\omega)}{2} = 0 \tag{6.97}$$

The solutions are

$$\omega_\pm = \sqrt{\left(\omega_{01}^2 + \omega_{02}^2\right)/2 + a \pm \sqrt{\left(\omega_{01}^2 - \omega_{02}^2/2\right)^2 + a^2}} \tag{6.98}$$

In our example, whose results are presented in Figure 6.21, we have let both a and ω_{01} be equal to unity. For the following we should remember, from Chapter 3, that the contribution to the energy from a mode is $\hbar/2$ times the shift in frequency when the interaction is turned on.

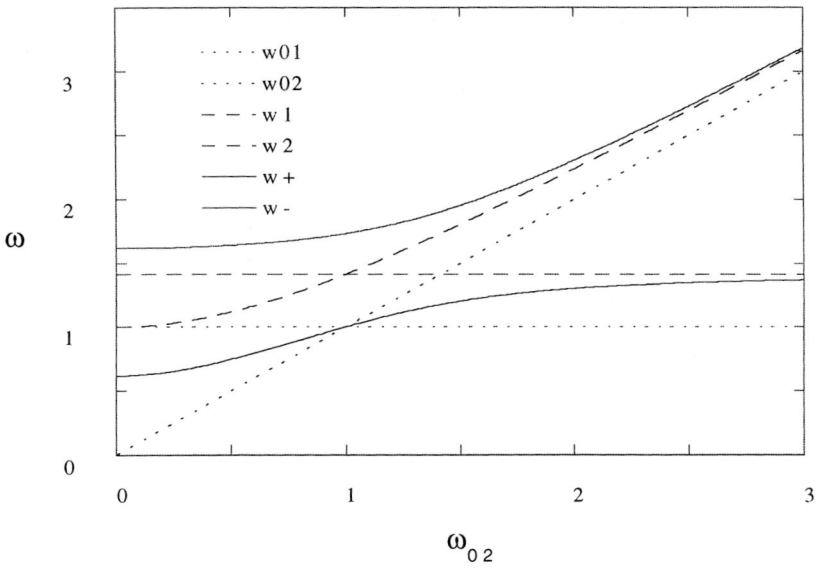

Figure 6.21: Variation of the normal modes with the excitation energy of material 2.

The surface energies are

$$\begin{cases} \sigma_1 = \omega_1 - \omega_{01} - \frac{1}{2}\left(\omega_1^{bulk} - \omega_{01}\right) = \sqrt{2} - 1 - \frac{1}{2}\left(\sqrt{3} - 1\right) \\ \sigma_2 = \omega_2 - \omega_{02} - \frac{1}{2}\left(\omega_2^{bulk} - \omega_{02}\right) = \sqrt{\omega_{02}^2 + 1} - \omega_{02} - \frac{1}{2}\left(\sqrt{\omega_{02}^2 + 2} - \omega_{02}\right) \\ \sigma_{1/2} = \omega_+ + \omega_- - \omega_{01} - \omega_{02} - \frac{1}{2}\left(\omega_1^{bulk} - \omega_{01}\right) - \frac{1}{2}\left(\omega_2^{bulk} - \omega_{02}\right) \\ \phantom{\sigma_{1/2}} = \sqrt{\frac{\left(1+\omega_{02}^2\right)}{2} + 1 + \sqrt{\left(\frac{1-\omega_{02}^2}{2}\right)^2 + 1}} + \sqrt{\frac{\left(1+\omega_{02}^2\right)}{2} + 1 - \sqrt{\left(\frac{1-\omega_{02}^2}{2}\right)^2 + 1}} \\ \phantom{\sigma_{1/2}} \quad - \frac{1}{2}\left(\sqrt{3} + 1\right) - \frac{1}{2}\left(\sqrt{\omega_{02}^2 + 2} + \omega_{02}\right) \end{cases} \quad (6.99)$$

We have here assumed that there are N modes (**q**-values) and that all modes give the same contribution. The surface energies are here expressed in units of N/A, where A is the surface area.

6.9 Adhesion, Cohesion, and Wetting

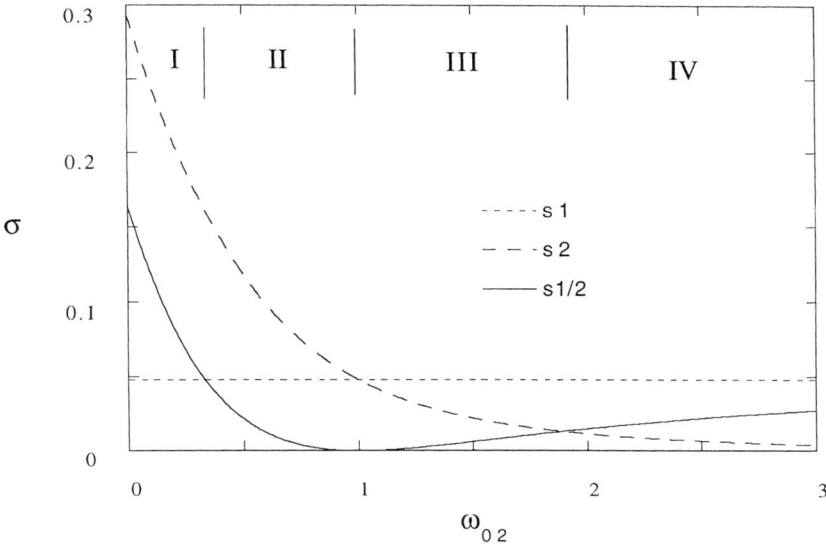

Figure 6.22: Variation of the surface energies with the excitation energy of material 2

We find four characteristic regions:

$$
\begin{aligned}
I: &\quad \sigma_2 > \sigma_{1/2} > \sigma_1 \\
II: &\quad \sigma_2 > \sigma_1 > \sigma_{1/2} \\
III: &\quad \sigma_1 > \sigma_2 > \sigma_{1/2} \\
IV: &\quad \sigma_1 > \sigma_{1/2} > \sigma_2
\end{aligned}
\tag{6.100}
$$

The surface energy of the interface is smallest when the two media have the same dielectric properties; in this case it is zero. It is also always smaller than the largest of the two surface energies of the free surfaces. We may say that the solid is metal like towards the left in Figure 6.22 and insulator like towards the right. If we return to the criterion for wetting we may indicate these regions in Figure 6.23.

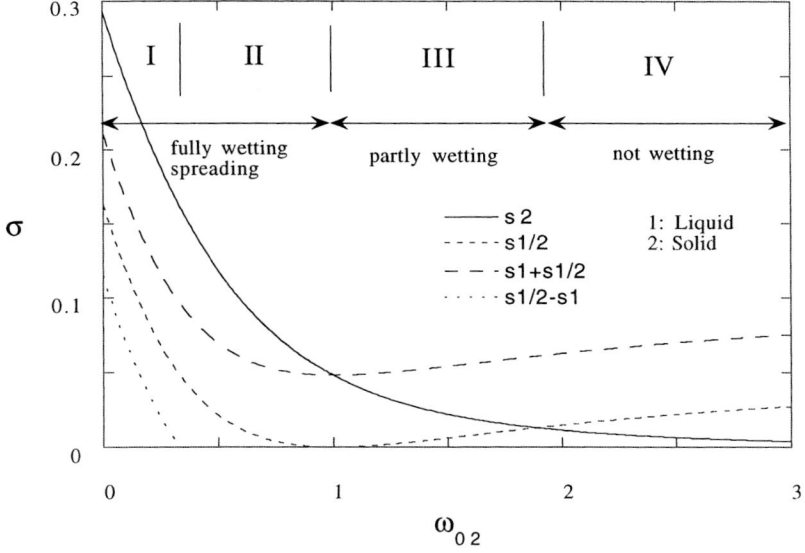

Figure 6.23: Wetting regions in excitation energy of material 2.

We may get the rough indication that metals have a tendency to be wet while wetting is less likely for insulators like, teflon. By reducing the surface energy of the liquid we may extend the region of spreading. Surfactants have this property. They also reduce the tension at the interface and extend the partly wetting region. Surfactants have a tendency to accumulate at the interface. We will return to surfactants later.

With our model we have two characteristic parameters representing the dielectric properties of our material; the frequency and the numerator or coupling strength. We have now varied the frequency. Let us just very briefly vary the numerator instead. We let a in the numerator for material 2 now be b. We then get the modes in Figure 6.24.

6.9 Adhesion, Cohesion, and Wetting

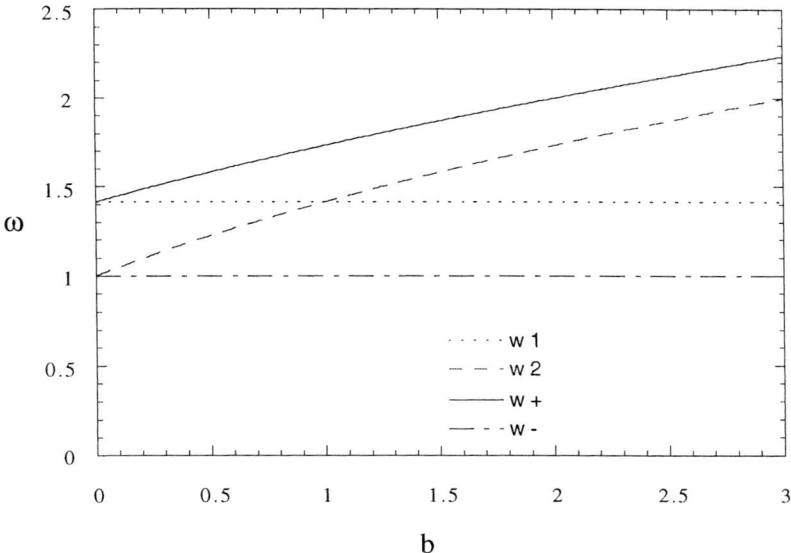

Figure 6.24: The variation of the energy of the modes with the interaction-strength parameter of material 2.

The surface energies are given in Figure 6.25,

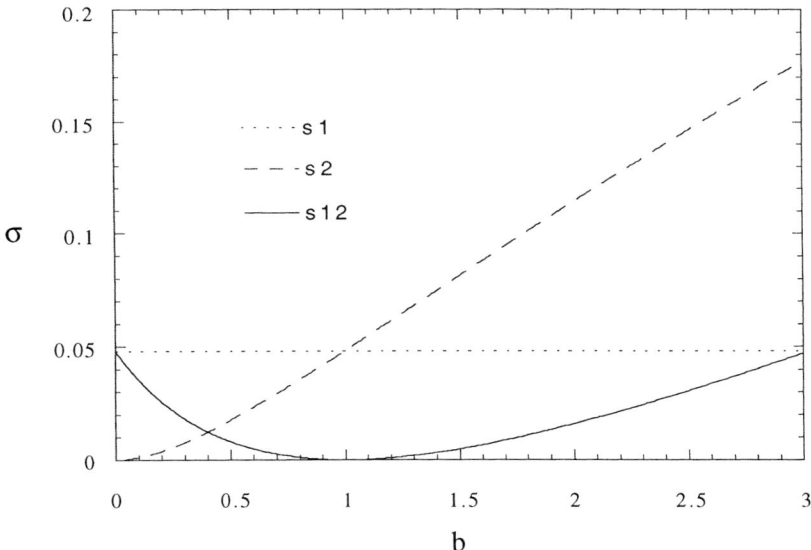

Figure 6.25: The variation of the surface energies with the interaction-strength parameter of material 2.

and the relations important for the wetting properties in Figure 6.26.

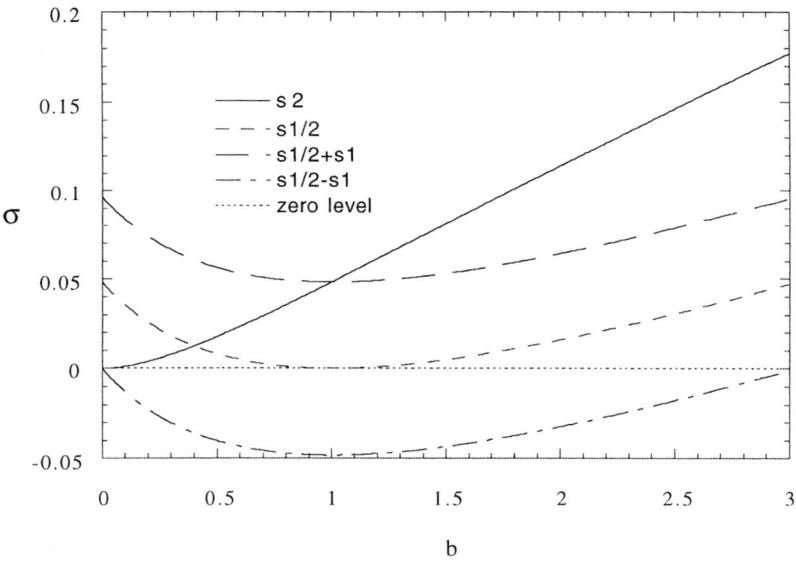

Figure 6.26: Wetting relations as functions of the interaction-strength parameter of material 2.

6.9.3.2 Birds of a feather flock together

I guess that all of us have noticed that there is a tendency for substances to stay together and not mix with others. A cloud stays together. Fog often has sharp boundaries. We will here use the simple model from the previous section to try to show why this is so. The example we will study is the following: We have two equal volumes of materials *1* and *2*. They both have the same number of atoms. The question is what happens with the energy when we combine them into one material with the same density as before. There are other contributions in this problem like the change in entropy (the entropy of mixing) that has effects on the willingness of the substances to mix which we do not consider here.

6.9 Adhesion, Cohesion, and Wetting

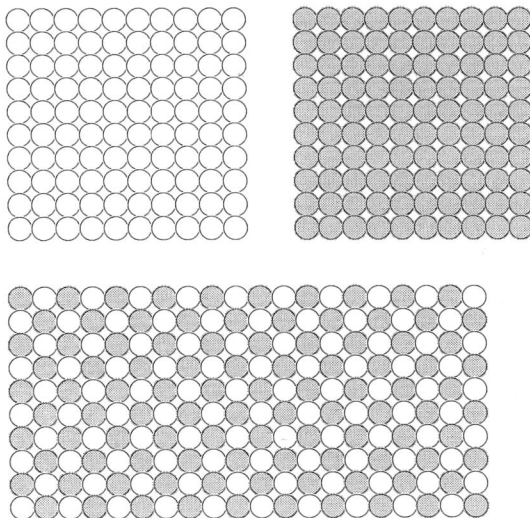

Figure 6.27: Two materials represented by white and gray atoms are solved.

We use the same notation as in Section 6.9.3.1 and find that the energy change is proportional to:

$$E = \sqrt{\frac{(\omega_{01}^2 + \omega_{02}^2)}{2} + a + \sqrt{\left(\frac{\omega_{01}^2 - \omega_{02}^2}{2}\right)^2 + a^2}} - \sqrt{\frac{(\omega_{01}^2 + \omega_{02}^2)}{2} + \sqrt{\left(\frac{\omega_{01}^2 - \omega_{02}^2}{2}\right)^2}}$$

$$+ \sqrt{\frac{(\omega_{01}^2 + \omega_{02}^2)}{2} + a - \sqrt{\left(\frac{\omega_{01}^2 - \omega_{02}^2}{2}\right)^2 + a^2}} - \sqrt{\frac{(\omega_{01}^2 + \omega_{02}^2)}{2} - \sqrt{\left(\frac{\omega_{01}^2 - \omega_{02}^2}{2}\right)^2}}$$

$$- \frac{1}{2}\left[\sqrt{\omega_{01}^2 + 2a} - \omega_{01} + \sqrt{\omega_{02}^2 + 2a} - \omega_{02}\right]$$

(6.101)

With the parameter choice that a and ω_{01} both unity we find the result in Figure 6.28.

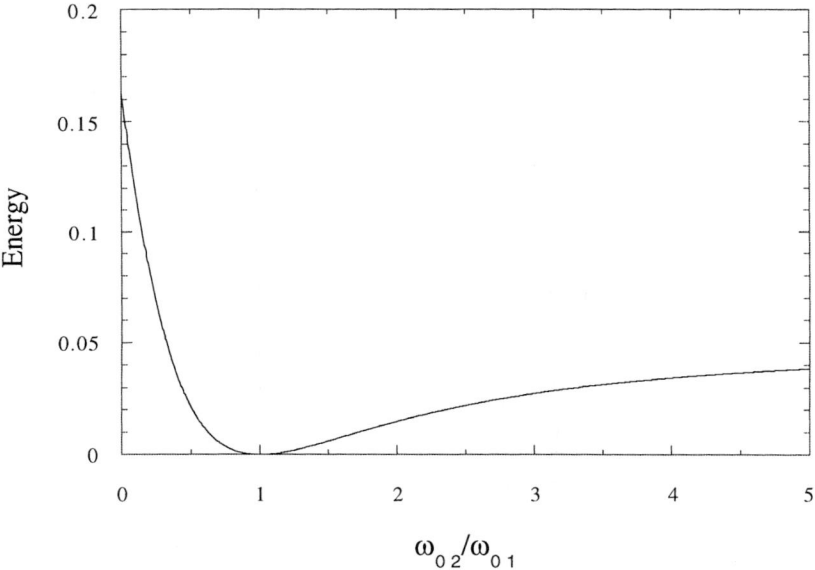

Figure 6.28: The energy as function of relative excitation frequency of the two materials.

Thus it costs energy to solve the two substances if their dielectric properties are different. Similar substances are easier to solve.

If we have two phases mixed the energy is lowered if they separate, that is, if precipitation occurs. However at the same time an interface is created between the two materials and that, we know, costs energy. There is no net energy gain until the number of atoms in the precipitate exceeds a critical value. This means that there has to be some nucleation centers or seeds present in order to get the process going. We should also notice that the energy gain per atom increases with the size of the precipitate. This means that large precipitates grow at the cost of smaller ones. This is a typical behavior. It is called particle coarsening, or sometimes Oswald ripening or Ostwald ripening.

6.9.4 Capillary rise

In this section we will discuss the capillary rise of a liquid in a thin tube and the shape of the liquid surface in the meniscus on top of the liquid pillar.

The upper surface of a liquid in a beaker will not be flat all the way out to the wall of the beaker. If the liquid wets the material of the beaker the surface will bend upwards; if not it will bend downwards.

6.9 Adhesion, Cohesion, and Wetting

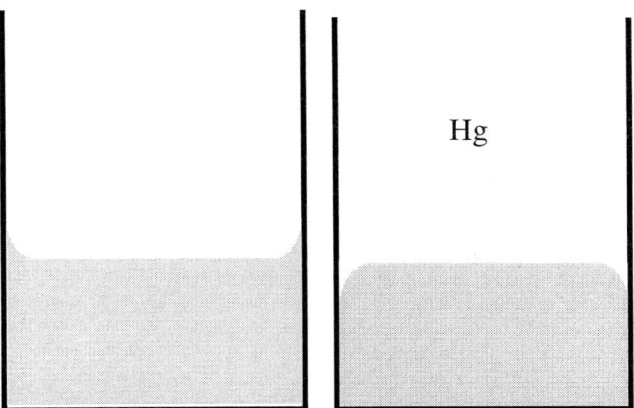

Figure 6.29: The surface curvature for a wetting and non-wetting liquid.

If a tube is placed in the liquid the surface will bend in a similar way also at the surface of the tube. If the inner diameter of the tube is thin enough the liquid will rise inside the tube.

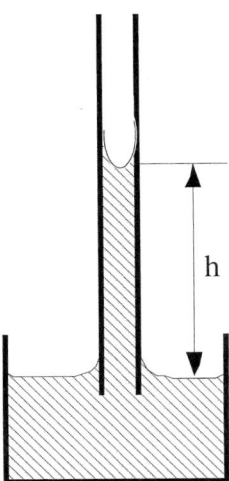

Figure 6.30: The capillary rise in a thin tube.

The capillary rise, h, depends on the surface energies of the different interfaces, the density of the liquid and on the radius of the tube, R. The rise and the shape of the liquid surface can be calculated by minimization of the energy; the energy consists of the total surface energies of the interfaces and the potential energy of the liquid. The calculation is complicated. One can come rather far by using an approximation where the weight of the meniscus, at the top of the liquid pillar inside the tube, is neglected.

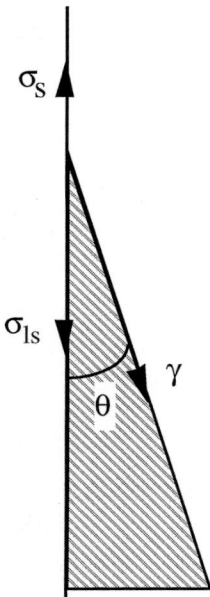

Figure 6.31: The forces per unit length acting on the line connecting the three interfaces.

The three interfaces, liquid-air, liquid-solid and solid-air, all meet along a circular curve. Since this curve does not move the total force acting on it in the vertical direction is zero. The force per unit length is

$$\sigma_s - \sigma_{ls} - \gamma \cos(\theta) = 0 \qquad (6.102)$$

The total external force in the vertical direction acting on the liquid pillar of mass m is

$$(\sigma_s - \sigma_{ls})2\pi R - mg = 0 \qquad (6.103)$$

or

$$\gamma \cos(\theta)2\pi R - mg = 0 \qquad (6.104)$$

Neglecting the weight of the meniscus we have

$$2\pi R \gamma \cos(\theta) = \pi R^2 h \rho g \qquad (6.105)$$

This gives for the height of the liquid pillar:

6.9 Adhesion, Cohesion, and Wetting

$$h = \frac{2\gamma \cos(\theta)}{R\rho g} \tag{6.106}$$

This result may seem strange in that it does not contain the surface energies of the solid and the solid-liquid interfaces. However, their effects are hidden in the angle θ. If the liquid completely wets the material of the tube, like in the case of water in a glass tube, the angle, θ, is zero.

Let us now make a more complete derivation of the shape of the meniscus. We will end up with a differential equation which can be solved numerically. We will arrive at the differential equation along two alternative routes.

Along the first it is determined by a minimization of the total energy. The energies involved are the surface energies of all three interfaces and the potential energy of the liquid inside the tube. The problem can however be solved in a simpler way. The surface energies at the tube wall determine the wetting angle, θ, at the tube surface. We need to minimize the surface energy times the total area of the liquid surface plus the potential energy and have the wetting angle as a boundary condition.

The energy can be written as

$$\begin{aligned} W &= \gamma A + E_{pot} \\ &= \gamma \int_0^R 2\pi r \, dS + \int_0^R \frac{y}{2} \rho g y 2\pi r \, dr \\ &= \int_0^R \left[\gamma 2\pi r \sqrt{1+(y')^2} + \frac{\rho g y^2}{2} 2\pi r \right] dr \end{aligned} \tag{6.107}$$

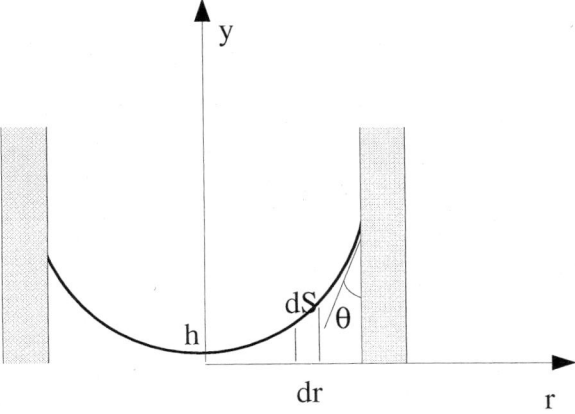

Figure 6.32: Definition of various parameters used in the text for the liquid surface in a capillary tube.

Thus, if we strip away unnecessary constants we are supposed to minimize the expression

$$\int_0^R f(r,y,y')dr \qquad (6.108)$$

where

$$f(r,y,y') = r\left[\gamma\sqrt{1+(y')^2} + \frac{\rho g y^2}{2}\right] \qquad (6.109)$$

Euler-Lagrange's equation for this problem is

$$\frac{d}{dr}\frac{\partial f}{\partial y'} = \frac{\partial f}{\partial y} \qquad (6.110)$$

and gives for the local extremum

$$\frac{d}{dr}\frac{r\gamma y'}{\sqrt{1+(y')^2}} = r\rho g y \qquad (6.111)$$

Performing the derivative on the left-hand side we obtain

$$\frac{\gamma y'}{\sqrt{1+(y')^2}} + \frac{r\gamma y''}{\left[1+(y')^2\right]^{3/2}} = r\rho g y \qquad (6.112)$$

or

$$r\gamma y'' + \gamma y'\left[1+(y')^2\right] - r\rho g y\left[1+(y')^2\right]^{3/2} = 0 \qquad (6.113)$$

We will now derive the same differential equation along a different route. We study the net vertical component of the force on a cylindrical volume element according to Figure 6.33:

6.9 Adhesion, Cohesion, and Wetting

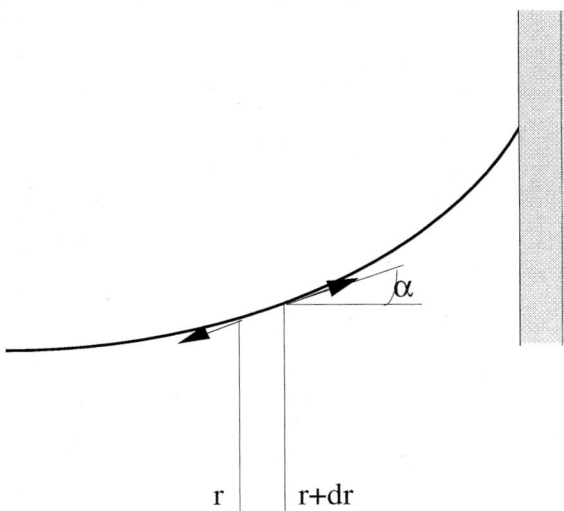

Figure 6.33: Definition of volume elements and angles used in the text for the liquid surface in a capillary tube.

The surface force acts upwards and compensates the gravitational force that points downwards. Thus we have

$$\gamma 2\pi (r+dr)\sin\alpha(r+dr) - \gamma 2\pi r \sin\alpha(r) = y 2\pi r dr \rho g \tag{6.114}$$

Expanding this, keeping terms linear in *dr* and making use of the relations

$$\sin(\alpha) = \sqrt{\sin^2(\alpha)} = \sqrt{\frac{\sin^2(\alpha)}{\sin^2(\alpha)+\cos^2(\alpha)}} = \frac{\tan(\alpha)}{\sqrt{1+\tan^2(\alpha)}} = \frac{y'}{\sqrt{1+(y')^2}} \quad ;$$

$$\frac{d\sin(\alpha)}{dr} = y'' \left[1+(y')^2\right]^{-3/2} \tag{6.115}$$

we find

$$\gamma 2\pi(r+dr)\left\{\sin\alpha(r)+dry''\left[1+(y')^2\right]^{-3/2}\right\}-\gamma 2\pi r\sin\alpha(r)$$

$$=\gamma 2\pi\left[\sin\alpha(r)+ry''\left[1+(y')^2\right]^{-3/2}\right]dr \qquad (6.116)$$

$$=\gamma 2\pi\left[y'\left[1+(y')^2\right]^{-1/2}+ry''\left[1+(y')^2\right]^{-3/2}\right]dr$$

$$=y2\pi r\rho g dr$$

and

$$\gamma y'\left[1+(y')^2\right]^{-1/2}+r\gamma y''\left[1+(y')^2\right]^{-3/2}=yr\rho g \qquad (6.117)$$

Finally we have rederived the equation

$$r\gamma y''+\gamma y'\left[1+(y')^2\right]-r\rho gy\left[1+(y')^2\right]^{3/2}=0 \qquad (6.118)$$

This differential equation has to be solved numerically. If this is done for water in a glass tube with 1 mm radius we have the parameters:

$$\gamma = 73\,erg/cm^{-2}\,(73\,mJ/m^{-2})\,;$$
$$\rho = 1\,g/cm^{-3}\,(1000\,kg/m^{-3})\,;$$
$$g = 981\,cm/s^2\,(9.81\,m/s^2)\,; \qquad (6.119)$$
$$R = 1mm\,;$$
$$\theta = 0$$

and we get the numerical value for the rise (at the center of the tube): 14.6 mm. If we instead use the simple expression obtained earlier in Equation (6.106) we find:

$$h = \frac{2\gamma\cos(\theta)}{R\rho g} = 14.9\text{ mm} \qquad (6.120)$$

Thus we get a little too high a value from Equation (6.106). The thinner the tube the smaller effect has the gravitational field on the shape of the liquid surface. The limiting shape is part of a spherical surface. If θ is zero it approaches a half-sphere. We may now return to Equation (6.104):

$$\gamma\cos(\theta)2\pi R - mg = 0 \qquad (6.121)$$

6.9 Adhesion, Cohesion, and Wetting

which was obtained on the way to the approximate expression. Now we no longer neglect the meniscus. We let m be the mass of all the liquid in the tube:

$$m = \rho \left[\pi R^2 h + \left(\pi R^3 - \frac{1}{2} \frac{4\pi R^3}{3} \right) \right] = \rho \pi R^2 \left[h + \frac{R}{3} \right] \tag{6.122}$$

and

$$h = \frac{2\gamma \cos(\theta)}{\rho g R} - \frac{R}{3} \tag{6.123}$$

Thus in the thin tube limit the simple result should be corrected by subtraction of one third of the radius. In our example with water we thus obtain 14.5, which is very close to the result from the numerical calculation.

In the above derivations we have treated the different interfaces as independent. This is a good approximation in most cases. However, in regions where the liquid film is very thin the modes on the two sides of the film interfere; the same is true for a thin air gap in the case of non-wetting liquids. We demonstrated these effects in Section 4.2.

Let us now determine the thickness, and its variation with height, of a thin liquid film adhered to a vertical solid surface. We are interested in corrections to the results neglecting interference. In the case of a capillary tube we limit ourselves to the region where the film is very thin, that is, at large heights. In this situation we may treat the film as having constant thickness at each height and we may disregard the change in surface energy with any deformation of the film. We make use of the theory developed in Section 6.5.

Let us study a system where we have two half-spaces of two different media *2* and *3* separated by a thin film of medium *1*. In our case medium *1* is the liquid, medium *2* the solid, and medium *3* air. Since we are only considering thin films we can neglect retardation effects. Then extending Equation (6.64) somewhat we may express the interaction energy between the air and the solid as

$$\begin{aligned}\Delta E(d) &= \frac{\hbar}{4\pi^2 d^2} \int_0^\infty \int_0^\infty d\omega dx\, x \ln\left\{ 1 - e^{-2x} \left[\frac{\varepsilon_2'(\omega) - \varepsilon_1'(\omega)}{\varepsilon_2'(\omega) + \varepsilon_1'(\omega)} \right] \left[\frac{\varepsilon_3'(\omega) - \varepsilon_1'(\omega)}{\varepsilon_3'(\omega) + \varepsilon_1'(\omega)} \right] \right\} \\ &= -\frac{\hbar}{4\pi^2 d^2} \int_0^\infty \int_0^\infty d\omega dx\, x \sum_n \frac{1}{n} e^{-2xn} \left\{ \left[\frac{\varepsilon_2'(\omega) - \varepsilon_1'(\omega)}{\varepsilon_2'(\omega) + \varepsilon_1'(\omega)} \right] \left[\frac{\varepsilon_3'(\omega) - \varepsilon_1'(\omega)}{\varepsilon_3'(\omega) + \varepsilon_1'(\omega)} \right] \right\}^n \\ &= -\frac{\hbar}{4\pi^2 d^2} \int_0^\infty d\omega \sum_{n=1}^\infty \frac{1}{4n^3} \left\{ \left[\frac{\varepsilon_2'(\omega) - \varepsilon_1'(\omega)}{\varepsilon_2'(\omega) + \varepsilon_1'(\omega)} \right] \left[\frac{\varepsilon_3'(\omega) - \varepsilon_1'(\omega)}{\varepsilon_3'(\omega) + \varepsilon_1'(\omega)} \right] \right\}^n \\ &\approx -\frac{\hbar}{16\pi^2 d^2} \int_0^\infty d\omega \left\{ \left[\frac{\varepsilon_2'(\omega) - \varepsilon_1'(\omega)}{\varepsilon_2'(\omega) + \varepsilon_1'(\omega)} \right] \left[\frac{\varepsilon_3'(\omega) - \varepsilon_1'(\omega)}{\varepsilon_3'(\omega) + \varepsilon_1'(\omega)} \right] \right\} \end{aligned} \tag{6.124}$$

where in the last line we have assumed that the dielectric properties of the three media are not very different from each other; if this is the case it is enough to keep just the first term of the expansion. We should also notice that if the middle medium has dielectric properties in between the other two the interaction is repulsive.

The force per unit area, or pressure, is

$$P(d) = -\frac{\partial}{\partial d}\Delta E(d)$$

$$= -\frac{\hbar}{2\pi^2 d^3}\int_0^\infty d\omega \sum_{n=1}^\infty \frac{1}{4n^3}\left\{\left[\frac{\varepsilon_2'(\omega)-\varepsilon_1'(\omega)}{\varepsilon_2'(\omega)+\varepsilon_1'(\omega)}\right]\left[\frac{\varepsilon_3'(\omega)-\varepsilon_1'(\omega)}{\varepsilon_3'(\omega)+\varepsilon_1'(\omega)}\right]\right\}^n \quad (6.125)$$

$$\approx -\frac{\hbar}{8\pi^2 d^3}\int_0^\infty d\omega\left\{\left[\frac{\varepsilon_2'(\omega)-\varepsilon_1'(\omega)}{\varepsilon_2'(\omega)+\varepsilon_1'(\omega)}\right]\left[\frac{\varepsilon_3'(\omega)-\varepsilon_1'(\omega)}{\varepsilon_3'(\omega)+\varepsilon_1'(\omega)}\right]\right\}$$

The Hamaker constant for the system considered is

$$A = \frac{3\hbar}{\pi}\int_0^\infty d\omega \sum_{n=1}^\infty \frac{1}{4n^3}\left\{\left[\frac{\varepsilon_2'(\omega)-\varepsilon_1'(\omega)}{\varepsilon_2'(\omega)+\varepsilon_1'(\omega)}\right]\left[\frac{\varepsilon_3'(\omega)-\varepsilon_1'(\omega)}{\varepsilon_3'(\omega)+\varepsilon_1'(\omega)}\right]\right\}^n$$

$$\approx \frac{3\hbar}{4\pi}\int_0^\infty d\omega\left\{\left[\frac{\varepsilon_2'(\omega)-\varepsilon_1'(\omega)}{\varepsilon_2'(\omega)+\varepsilon_1'(\omega)}\right]\left[\frac{\varepsilon_3'(\omega)-\varepsilon_1'(\omega)}{\varepsilon_3'(\omega)+\varepsilon_1'(\omega)}\right]\right\} \quad (6.126)$$

and

$$\Delta E(d) = -\frac{A}{12\pi d^2} \; ; \quad (6.127)$$

$$P(d) = -\frac{A}{6\pi d^3} \quad (6.128)$$

If the three media are diluted or if we use summation over pair interactions to calculate the Hamaker constant for the system this constant may be written as

$$A = A_{11} + A_{23} - A_{12} - A_{13} \quad (6.129)$$

With the London approximation these constants are

$$A_{ij} = \frac{3\pi^2 n^2 \alpha_i(0)\alpha_j(0)\hbar\omega_{0,i}\hbar\omega_{0,j}}{2(\hbar\omega_{0,i} + \hbar\omega_{0,j})} \quad (6.130)$$

Let us now study the free energy of a unit area of the film. It is

$$W(d) = -\frac{A}{12\pi d^2} + \rho dgh \qquad (6.131)$$

The minimum defines the equilibrium thickness:

$$\frac{\partial}{\partial d}W(d) = \frac{A}{6\pi d^3} + \rho gh \Rightarrow d_0 = \left(\frac{-A}{6\pi \rho gh}\right)^{1/3} \text{ if } A \text{ negative} \qquad (6.132)$$

If the Hamaker constant is positive there is an attractive force between the free surface of the film and the solid-liquid interface; this tends to thin the film and in this case one needs to include other energy terms as well to get a non-zero thickness. The dielectric constant of liquid helium is very close to unity. This means that its dielectric constant lies between those of most solids and air, with the effect that the Hamaker constant for a helium film on a solid is negative. This has the effect that the film thickens which explains why the liquid climbs the walls of a beaker.

6.10 Finding the pair interactions

In Section 5.12 we showed how to get a fast, order-of-magnitude, estimate of the force between objects by summing over pair interactions. We introduced the pair potentials for a pair of polarizable atoms or molecules as: $V = -Br^{-\gamma}$. We derived the pair of B and γ values for the non-retarded and retarded limits starting from the equation of motion for two coupled dipoles in Sections 6.1.1-6.1.2. Here we will derive the same results by using the expression for the interaction energy between two half-spaces made up from the polarizable atoms, taking the diluted limit and identifying the result with the expression in Equation (5.89). It is always good to have alternative ways to derive things; it is useful as a test to detect any mistakes in the derivations and also as a test of the consistency of the theory.

6.10.1 Non-retarded limit

We start from Equation (6.64) for the non-retarded interaction energy between two half-spaces:

$$\Delta E(d) = \frac{\hbar}{4\pi^2 d^2} \int_0^\infty \int_0^\infty d\omega dx\, x \ln\left\{1 - e^{-2x}\left[\frac{\varepsilon'(\omega)-1}{\varepsilon'(\omega)+1}\right]^2\right\}$$

$$= \frac{\hbar}{4\pi^2 d^2} \int_0^\infty \int_0^\infty d\omega dx\, x \ln\left\{1 - e^{-2x}\left[\frac{4\pi n \alpha'(\omega)}{4\pi n \alpha'(\omega)+2}\right]^2\right\}$$

$$\approx -\frac{\hbar}{4\pi^2 d^2} \int_0^\infty \int_0^\infty d\omega dx\, x e^{-2x}\left[\frac{4\pi n \alpha'(\omega)}{2}\right]^2 \quad (6.133)$$

$$\approx -\frac{\hbar n^2}{d^2} \int_0^\infty \int_0^\infty d\omega dx\, x e^{-2x} \alpha'(\omega)^2 \approx -\frac{\hbar n^2}{4d^2} \int_0^\infty d\omega\, \alpha'(\omega)^2$$

$$\approx -\frac{\pi n^2 \hbar \omega_0 \alpha(0)^2}{16 d^2} = -\frac{B 2\pi n^2}{(\gamma-2)(\gamma-3)(\gamma-4)d^{(\gamma-4)}}$$

In the first step we use the expression for the dielectric function in terms of the atomic polarizability given in Equation (2.8); in next step we take the diluted limit and keep only the term to the lowest order in n (the density of atoms); next step is just a rearrangement; in next we integrate over x; next step is the result from using the London approximation in the same way as we did in deriving Equation (5.40); in the last step we identify this result with what we found in the pair-interaction treatment in Equation (5.89), taking the limit of infinite thick plates. The identification gives:

$$B = \frac{3\hbar \omega_0 \alpha(0)^2}{4} \; ; \; \gamma = 6 \quad (6.134)$$

6.10.2 Retarded limit

We start from Equation (6.70) for the retarded interaction energy between two half-spaces at large separations:

$$\Delta E(d)|_{d \text{ large}} = \frac{\hbar}{32\pi^2 d^3} \int_0^\infty \int_0^\infty d\omega dk k \ln\left\{1 - e^{-\sqrt{k^2+\left(\frac{\omega}{c}\right)^2}} \left[\frac{\varepsilon'(0)\sqrt{k^2+\left(\frac{\omega}{c}\right)^2} - \sqrt{k^2+\varepsilon'(0)\left(\frac{\omega}{c}\right)^2}}{\varepsilon'(0)\sqrt{k^2+\left(\frac{\omega}{c}\right)^2} + \sqrt{k^2+\varepsilon'(0)\left(\frac{\omega}{c}\right)^2}}\right]^2\right\}$$

$$+ \frac{\hbar}{32\pi^2 d^3} \int_0^\infty \int_0^\infty d\omega dk k \ln\left\{1 - e^{-\sqrt{k^2+\left(\frac{\omega}{c}\right)^2}} \left[\frac{\sqrt{k^2+\varepsilon'(0)\left(\frac{\omega}{c}\right)^2} - \sqrt{k^2+\left(\frac{\omega}{c}\right)^2}}{\sqrt{k^2+\varepsilon'(0)\left(\frac{\omega}{c}\right)^2} + \sqrt{k^2+\left(\frac{\omega}{c}\right)^2}}\right]^2\right\}$$
(6.135)

make the substitution: $\omega \to \omega c$, introduce the expression for the dielectric function and expand for small densities

$$\Delta E(d) \approx \frac{\hbar c}{32\pi^2 d^3} \int_0^\infty \int_0^\infty d\omega dk k \ln\left\{1 - e^{-\sqrt{k^2+\omega^2}} \left[2\pi n\alpha(0) - \frac{\pi n\alpha(0)\omega^2}{k^2+\omega^2}\right]^2\right\}$$

$$+ \frac{\hbar c}{32\pi^2 d^3} \int_0^\infty \int_0^\infty d\omega dk k \ln\left\{1 - e^{-\sqrt{k^2+\omega^2}} \left[\frac{\pi n\alpha(0)\omega^2}{k^2+\omega^2}\right]^2\right\}$$
(6.136)

We may expand also the logarithm:

$$\Delta E(d) \approx -\frac{\hbar c}{32\pi^2 d^3} \int_0^\infty \int_0^\infty d\omega dk k e^{-\sqrt{k^2+\omega^2}} \left[2\pi n\alpha(0) - \frac{\pi n\alpha(0)\omega^2}{k^2+\omega^2}\right]^2$$

$$- \frac{\hbar c}{32\pi^2 d^3} \int_0^\infty \int_0^\infty d\omega dk k e^{-\sqrt{k^2+\omega^2}} \left[\frac{\pi n\alpha(0)\omega^2}{k^2+\omega^2}\right]^2$$
(6.137)

$$\approx -\frac{\hbar c n^2 \alpha(0)^2}{32 d^3} \int_0^\infty \int_0^\infty d\omega dk k e^{-\sqrt{k^2+\omega^2}} \left[4 + \frac{2\omega^4}{\left(k^2+\omega^2\right)^2} - \frac{4\omega^2}{k^2+\omega^2}\right]$$

Now, transform to polar coordinates by letting $k \to x = r\cos(\theta)$; $\omega \to y = r\sin(\theta)$:

$$\Delta E(d) \approx -\frac{\hbar c n^2 \alpha(0)^2}{32 d^3} \int_0^\infty dr \int_0^{\pi/2} d\theta r \cos(\theta) e^{-r} \left[4 + \frac{2[r\sin(\theta)]^4}{r^4} - \frac{4[r\sin(\theta)]^2}{r^2} \right]$$

$$\approx -\frac{\hbar c n^2 \alpha(0)^2}{32 d^3} \int_0^\infty dr r e^{-r} \int_0^1 d[\sin(\theta)] \left[4 + 2[\sin(\theta)]^4 - 4[\sin(\theta)]^2 \right] \quad (6.138)$$

$$\approx -\frac{\hbar c n^2 \alpha(0)^2}{32 d^3} 2\left[4 + \frac{2}{5} - \frac{2}{3} \right] = -\frac{\hbar c n^2 \alpha(0)^2 2(60 + 6 - 20)}{32 d^3 15}$$

$$\approx -\frac{\hbar c n^2 \alpha(0)^2 23}{120 d^3} = -\frac{B 2\pi n^2}{(\gamma - 2)(\gamma - 3)(\gamma - 4) d^{(\gamma - 4)}}$$

The identification in the last step gives:

$$B = \frac{23 \hbar c \alpha(0)^2}{4\pi} \quad ; \quad \gamma = 7 \quad (6.139)$$

We ended this chapter by deriving the interactions between two polarizable atoms both in the van der Waals and Casimir limits, starting from the interaction between two macroscopic objects. The results were the same as the ones we found earlier. It has a touch of magic.

References

[1] J. Israelachvili, *Intermolecular & Surface Forces,* (Academic London, 2nd Ed., 1997), pp 203-204.
[2] I. S. Grigoriev and E. Z. Meilikhov, *Handbook of Physical Quantities*, (CRC Press, New York, 1997) p. 10.3, p. 14.1.
[3] U. Mohideen and Anushree Roy, Phys. Rev. Lett. **81**, 4549 (1998).
[4] B. W. Harris, F. Chen, and U. Mohideen, Phys. Rev. **A62**, 052109 (2000).
[5] S. K. Lamoreaux, Phys. Rev. Lett. **78**, 5 (1997).
[6] M. Boström and Bo E. Sernelius, Phys. Rev. Lett. **84**, 4757 (2000).

7 Modes at non-planar interfaces

When the coffee drips into the coffee pot, in the drip coffee maker, sometimes the drops land on the surface of the coffee and stay there for some second before coalescing with the bulk liquid. Why is this?

So far we have found the electromagnetic surface modes on single planar surfaces and planar configurations containing more than one interface. In this chapter we treat objects that are not planar. We neglect retardation effects and find the modes from the scalar potential, which obeys the Laplace equation. There are eleven coordinate systems in which the Laplace equation is separable. Four of these are cylindrical: rectangular; circular; elliptical; parabolic. The standard procedure to solve the Laplace equation is the following: We make an ansatz where the function is a product of three factors, each dependent of only one of the coordinates. We put this into Laplace's equation and end up with standard differential equations, having standard functions as solutions. These functions are either available in computers, in tables or can be obtained with recurrence relations.

We start by studying spherical objects.

7.1 Modes at the surface of a sphere

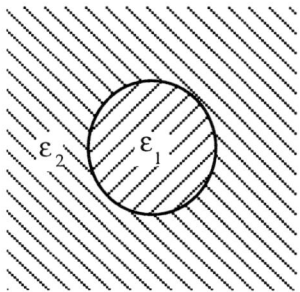

Figure 7.1: Sphere of dielectric function ε_1 imbedded in a medium of dielectric function ε_2.

We study a system consisting of a spherical dielectric imbedded in another dielectric. The full problem with retardation effects is rather complicated so we will restrict ourselves to the limit where retardation effects are negligible and only give the condition for modes, when retardation is included, at the end of the section. This is a rather weak limitation in this case since here the structure is finite in size and limited in all directions. If we assume that the fields of our modes die off rather quickly and are negligible outside of a region of diameter d, we need the restriction

$$\omega \ll \frac{c}{d\varepsilon_2(\omega)} \tag{7.1}$$

We start from the scalar potential in Coulomb gauge. It satisfies the Laplace equation:

$$\nabla^2 \phi = 0 \tag{7.2}$$

in absence of external charges, according to Equation (1.41). Since we neglect retardation effects we have

$$\mathbf{E} = -\nabla \phi \tag{7.3}$$

according to Equation (1.29). The general solution to Equation (7.2) is

$$\phi(r,\theta,\phi,t) = \sum_{l=0}^{\infty} \sum_{m=-l}^{l} \left[A_{lm} r^l + B_{lm} r^{-(l+1)} \right] Y_{lm}(\theta,\phi) e^{-i\omega t} \tag{7.4}$$

where the functions $Y_{lm}(\theta,\phi)$ are the so-called Spherical Harmonics [1]. We study one of the components (which is analogous to taking the Fourier transform in planar geometry):

$$\phi_{lm}(r,\theta,\phi) = \begin{cases} A r^l Y_{lm}(\theta,\phi) & ; \quad r < R \\ B r^{-(l+1)} Y_{lm}(\theta,\phi) & ; \quad r > R \end{cases} \tag{7.5}$$

Only one of the two terms has survived in each region as a result of our demand that the potential stays finite. The potential is continuous at the spherical surface, which gives

$$B = A R^{2l+1} \tag{7.6}$$

Laplace's equation does not contain any time derivatives which means that the general solution can have any time dependence. We have, in Equation (7.4), studied a general Fourier component with respect to time. The boundary conditions determine which Fourier components survive.

Now, the normal component of the **D** vector and the tangential components of the **E** vector are continuous, that is,

7.1 Modes at the Surface of a Sphere

$$\begin{cases} \varepsilon_1 \dfrac{\partial \phi}{\partial r}\bigg|_{r=R-0} = \varepsilon_1 A l R^{l-1} Y_{lm}(\theta,\phi) = -\varepsilon_2(l+1)AR^{l-1} Y_{lm}(\theta,\phi) = \varepsilon_2 \dfrac{\partial \phi}{\partial r}\bigg|_{r=R+0} \\ \dfrac{1}{r}\dfrac{\partial \phi}{\partial \theta}\bigg|_{r=R-0} = AR^{l-1}\dfrac{\partial}{\partial \theta}Y_{lm}(\theta,\phi) = AR^{l-1}\dfrac{\partial}{\partial \theta}Y_{lm}(\theta,\phi) = \dfrac{1}{r}\dfrac{\partial \phi}{\partial \theta}\bigg|_{r=R+0} \\ \dfrac{1}{r\sin\theta}\dfrac{\partial \phi}{\partial \varphi}\bigg|_{r=R-0} = AR^{l-1}\dfrac{1}{\sin\theta}\dfrac{\partial}{\partial \varphi}Y_{lm}(\theta,\phi) = AR^{l-1}\dfrac{1}{\sin\theta}\dfrac{\partial}{\partial \varphi}Y_{lm}(\theta,\phi) = \dfrac{1}{r\sin\theta}\dfrac{\partial \phi}{\partial \varphi}\bigg|_{r=R+0} \end{cases}$$
(7.7)

Thus we have the following condition for modes:

$$\frac{\varepsilon_1(\omega)}{\varepsilon_2(\omega)} = -\frac{l+1}{l} \tag{7.8}$$

which was a result from the first of Equations (7.7). The two others gave no new information. We should note that: when l increases towards infinity the modes are the same as for a planar interface; the result is independent of the size of the sphere; the different m components are degenerate. Another interesting feature is the behavior of the polarizability and absorption of spheres. The polarizability, according to Equation (5.68), is

$$\alpha = R^3 \frac{\varepsilon_1(\omega) - \varepsilon_2(\omega)}{\varepsilon_1(\omega) + 2\varepsilon_2(\omega)} = R^3 \frac{-(l+1)/l - 1}{-(l+1)/l + 2} = -R^3 \frac{2l+1}{l-1} \tag{7.9}$$

This is illustrated in Figure 7.2 and in Figure 7.3 we show how the charges

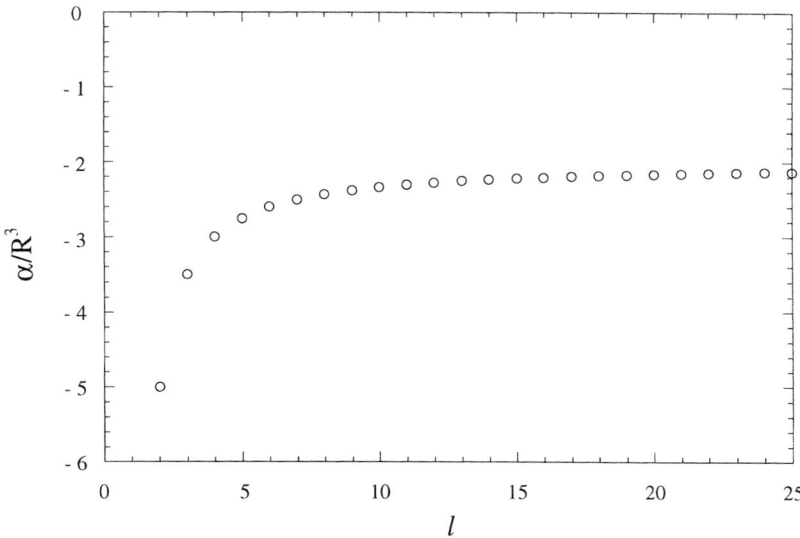

Figure 7.2: The polarizability of a sphere.

oscillate back and forth for different types of excitation. The polarizability diverges for $l = 1$. There is a giant absorption for the corresponding energy. This is further the dipolar mode which can be excited by absorption of electromagnetic radiation. The other modes can be excited in Raman light scattering experiments.

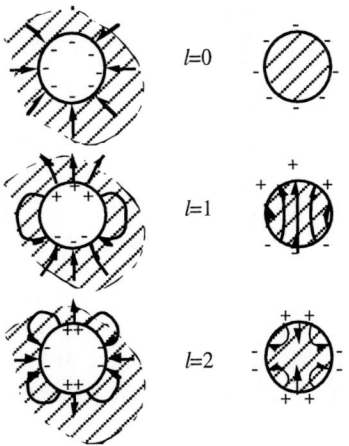

Figure 7.3: The movement of charges at the different modes for a spherical cavity and a spherical object.

If the system has a large number of small spheres distributed homogeneously throughout the material the dielectric function of the system is

$$\varepsilon(\omega) = \varepsilon_2(\omega)[1 + 4\pi n\alpha]$$
$$= \varepsilon_2(\omega)\left[1 + 4\pi nR^3 \frac{\varepsilon_1(\omega) - \varepsilon_2(\omega)}{\varepsilon_1(\omega) + 2\varepsilon_2(\omega)}\right] \quad (7.10)$$
$$= \varepsilon_2(\omega)\left[1 + 3\eta \frac{\varepsilon_1(\omega) - \varepsilon_2(\omega)}{\varepsilon_1(\omega) + 2\varepsilon_2(\omega)}\right]$$

where η is the volume fraction of spheres.

We have so far neglected retardation. Including retardation leads to the following conditions for modes in general [2]:

$$\text{TM:} \quad \left.\frac{\varepsilon_2(\omega)h_l(x)}{\partial[xh_l(x)]/\partial x}\right|_{x=\omega\sqrt{\varepsilon_2(\omega)}R/c} = \left.\frac{\varepsilon_1(\omega)j_l(x)}{\partial[xh_l(x)]/\partial x}\right|_{x=\omega\sqrt{\varepsilon_1(\omega)}R/c} \quad (7.11)$$

$$\text{TE:} \quad \left.\frac{h_l(x)}{\partial[xh_l(x)]/\partial x}\right|_{x=\omega\sqrt{\varepsilon_2(\omega)}R/c} = \left.\frac{j_l(x)}{\partial[xj_l(x)]/\partial x}\right|_{x=\omega\sqrt{\varepsilon_1(\omega)}R/c} \quad (7.12)$$

7.1 Modes at the Surface of a Sphere

where j_l and h_l are the spherical Bessel and Hankel functions [3], respectively. All these solutions have complex valued frequencies. This means that the modes have finite life time, they radiate. It furthermore means that they may be excited by light without the need for a prism or surface roughness, as is needed in the planar case; this is discussed in Chapter 8. Surface modes only exist of the **TM** type.

7.1.1 Metal sphere in vacuum

In the case of a metal sphere in vacuum we have the following dielectric functions for the two regions:

$$\varepsilon_1(\omega) = 1 - \frac{\omega_{pl}^2}{\omega^2} \quad ; \quad \varepsilon_2(\omega) = 1 \tag{7.13}$$

and the resulting modes are:

$$\omega = \omega_{pl}\sqrt{\frac{l}{2l+1}} \quad ; \quad l = 0, 1, 2\ldots \tag{7.14}$$

These results are shown in Figure 7.4.

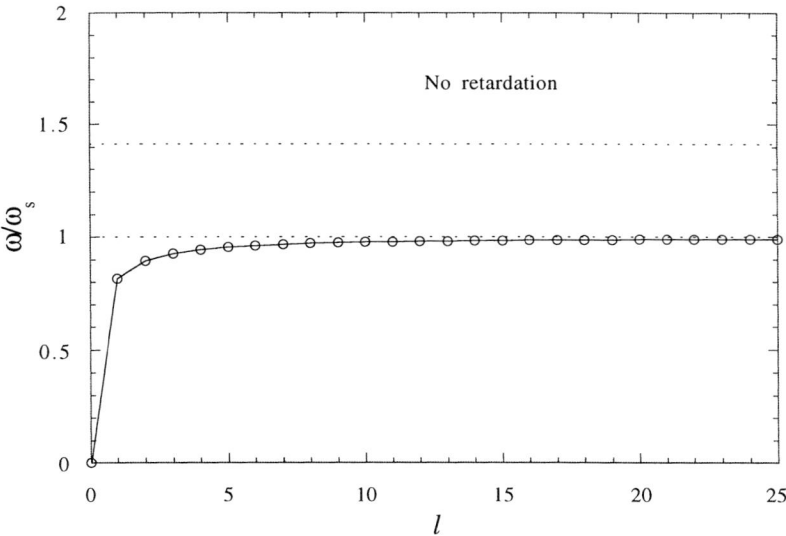

Figure 7.4: Modes for a metal sphere in vacuum. The reference energy is the surface plasmon energy.

7.1.2 Dielectric sphere in vacuum

In the case of a dielectric sphere in vacuum we have the following condition for modes:

$$\varepsilon_\infty \frac{\omega^2 - \omega_L^2}{\omega^2 - \omega_T^2} = -\frac{l+1}{l} \tag{7.15}$$

which results in the following modes:

$$\omega = \sqrt{\frac{l\varepsilon_\infty \omega_L^2 + (l+1)\omega_T^2}{l\varepsilon_\infty + (l+1)}} \quad ; \quad l = 0, 1, 2 \ldots \tag{7.16}$$

The results are shown in figure 7.5.

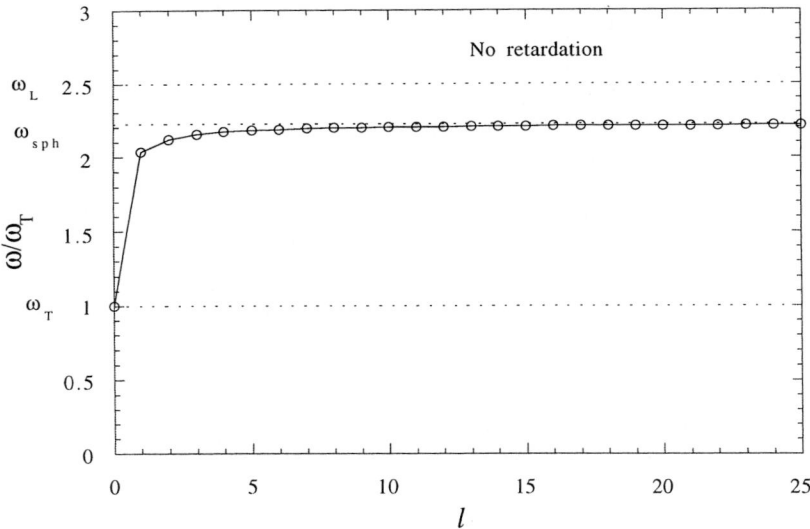

Figure 7.5: Modes for a dielectric sphere in vacuum.

7.1 Modes at the Surface of a Sphere

7.1.3 Spherical void in a metal

In the case of a spherical void in a metal the dielectric functions for the two regions are

$$\varepsilon_1(\omega) = 1 \; ;$$
$$\varepsilon_2(\omega) = 1 - \frac{\omega_{pl}^2}{\omega^2} \quad (7.17)$$

and the resulting modes are

$$\omega = \omega_{pl}\sqrt{\frac{l+1}{2l+1}} \; ; \quad l = 0,1,2\ldots \quad (7.18)$$

These results are shown in Figure 7.6.

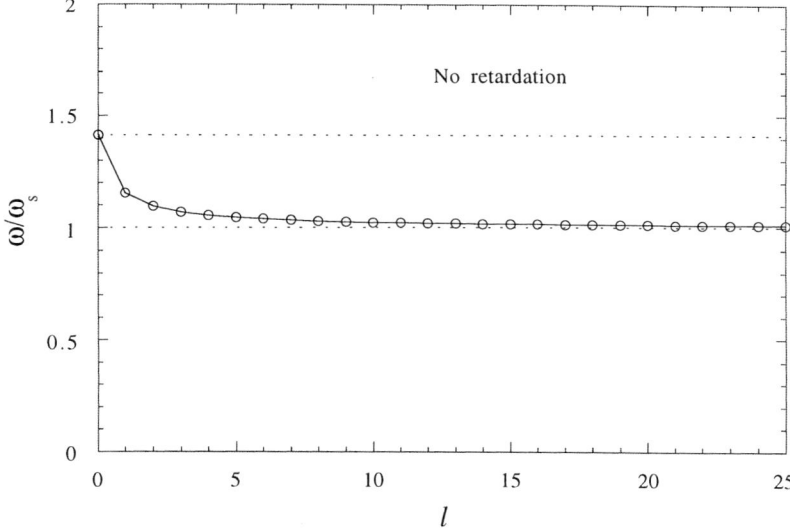

Figure 7.6: Modes for a spherical void in a metal. The reference energy is the surface plasmon energy.

7.1.4 Spherical void in a dielectric

In the case of a spherical void in a dielectric the condition for modes becomes

$$\varepsilon_\infty \frac{\omega^2 - \omega_L^2}{\omega^2 - \omega_T^2} = -\frac{l}{l+1} \tag{7.19}$$

and the resulting modes are

$$\omega = \sqrt{\frac{(l+1)\varepsilon_\infty \omega_L^2 + l\omega_T^2}{(l+1)\varepsilon_\infty + l}} \quad ; \quad l = 0,1,2\ldots \tag{7.20}$$

These results are shown in Figure 7.7.

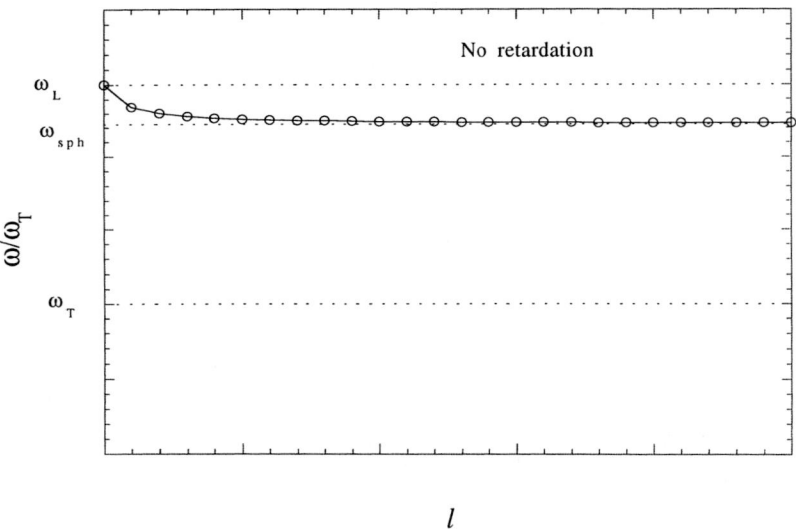

Figure 7.7: Modes for a spherical void in a dielectric.

7.1.5 Modes in a layered sphere

We study a system consisting of a layered dielectric sphere imbedded in another dielectric, as is illustrated in Figure 7.8.

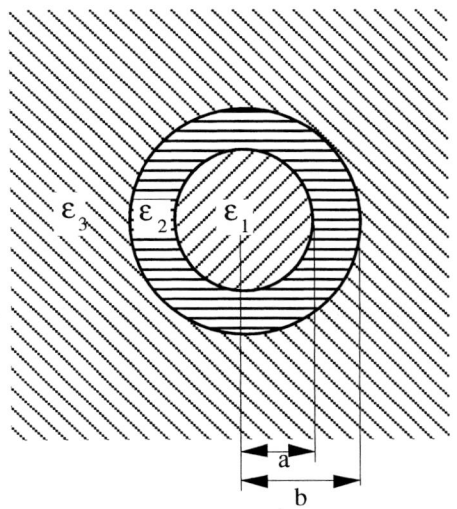

Figure 7.8: Definition of parameters for the layered sphere.

The problem is similar to that in section 7.1, but slightly more complicated. The solutions in the three regions are sought and these are matched at the two interfaces. As before, the general solution is

$$\phi(r,\theta,\phi,t) = \sum_{l=0}^{\infty} \sum_{m=-l}^{l} \left[A_{lm} r^l + B_{lm} r^{-(l+1)} \right] Y_{lm}(\theta,\phi) e^{-i\omega t} \qquad (7.21)$$

We study one of the components and demand that it is finite everywhere:

$$\phi_{ml}(r,\theta,\phi) = \begin{cases} A r^l Y_{lm}(\theta,\phi) & ; \; r < a \\ \left(C r^l + D r^{-(l+1)} \right) Y_{lm}(\theta,\phi) & ; \; a < r < b \\ B r^{-(l+1)} Y_{lm}(\theta,\phi) & ; \; b < r \end{cases} \qquad (7.22)$$

The potential is continuous at the spherical surface, which gives

$$A = C + Da^{-(2l+1)} \; ;$$
$$B = Cb^{2l+1} + D \tag{7.23}$$

Now, the normal component of the **D** vector and the tangential components of the **E** vector are continuous. The inner surface gives

$$\begin{cases} \varepsilon_1 \dfrac{\partial \phi}{\partial r}\bigg|_{r=a-0} = \varepsilon_1 \left(C + Da^{-(2l+1)}\right) la^{l-1} Y_{lm}(\theta,\phi) \\ \qquad = \varepsilon_2 \left[C - \dfrac{(l+1)}{l} Da^{-(2l+1)}\right] la^{l-1} Y_{lm}(\theta,\phi) = \varepsilon_2 \dfrac{\partial \phi}{\partial r}\bigg|_{r=a+0} \\ \dfrac{1}{r} \dfrac{\partial \phi}{\partial \theta}\bigg|_{r=a-0} = \left(C + Da^{-(2l+1)}\right) a^{l-1} \dfrac{\partial}{\partial \theta} Y_{lm}(\theta,\phi) \\ \qquad = \left(C + Da^{-(2l+1)}\right) a^{l-1} \dfrac{\partial}{\partial \theta} Y_{lm}(\theta,\phi) = \dfrac{1}{r} \dfrac{\partial \phi}{\partial \theta}\bigg|_{r=a+0} \\ \dfrac{1}{r\sin\theta} \dfrac{\partial \phi}{\partial \varphi}\bigg|_{r=R-0} = \left(C + Da^{-(2l+1)}\right) a^{l-1} \dfrac{1}{\sin\theta} \dfrac{\partial}{\partial \varphi} Y_{lm}(\theta,\phi) \\ \qquad = \left(C + Da^{-(2l+1)}\right) a^{l-1} \dfrac{1}{\sin\theta} \dfrac{\partial}{\partial \varphi} Y_{lm}(\theta,\phi) = \dfrac{1}{r\sin\theta} \dfrac{\partial \phi}{\partial \varphi}\bigg|_{r=a+0} \end{cases} \tag{7.24}$$

with the result from the first equation:

$$\left(C + Da^{-(2l+1)}\right) = \varepsilon_2 \left[C - \frac{(l+1)}{l} Da^{-(2l+1)}\right] \tag{7.25}$$

The two others gave no new information. The outer surface gives

$$\begin{cases} \varepsilon_2 \dfrac{\partial \phi}{\partial r}\bigg|_{r=b-0} = \varepsilon_2 \left[lC - (l+1)Db^{-(2l+1)}\right] b^{l-1} Y_{lm}(\theta,\phi) \\ \qquad = -\varepsilon_3 \left(C + Db^{-(2l+1)}\right)(l+1) b^{l-1} Y_{lm}(\theta,\phi) = \varepsilon_2 \dfrac{\partial \phi}{\partial r}\bigg|_{r=b+0} \\ \dfrac{1}{r} \dfrac{\partial \phi}{\partial \theta}\bigg|_{r=b-0} = \left(C + Db^{-(2l+1)}\right) b^{l-1} \dfrac{\partial}{\partial \theta} Y_{lm}(\theta,\phi) \\ \qquad = \left(C + Db^{-(2l+1)}\right) b^{l-1} \dfrac{\partial}{\partial \theta} Y_{lm}(\theta,\phi) = \dfrac{1}{r} \dfrac{\partial \phi}{\partial \theta}\bigg|_{r=b+0} \\ \dfrac{1}{r\sin\theta} \dfrac{\partial \phi}{\partial \varphi}\bigg|_{r=R-0} = \left(C + Db^{-(2l+1)}\right) b^{l-1} \dfrac{1}{\sin\theta} \dfrac{\partial}{\partial \varphi} Y_{lm}(\theta,\phi) \\ \qquad = \left(C + Db^{-(2l+1)}\right) b^{l-1} \dfrac{1}{\sin\theta} \dfrac{\partial}{\partial \varphi} Y_{lm}(\theta,\phi) = \dfrac{1}{r\sin\theta} \dfrac{\partial \phi}{\partial \varphi}\bigg|_{r=b+0} \end{cases} \tag{7.26}$$

7.1 Modes at the Surface of a Sphere

with the result from the first of Equations (7.26):

$$\varepsilon_2\left[C - \frac{(l+1)}{l} D b^{-(2l+1)}\right] = -\varepsilon_3\left(C + D b^{-(2l+1)}\right)\frac{(l+1)}{l} \tag{7.27}$$

The two others gave no new information.
Thus we have

$$\varepsilon_1\left(C + D a^{-(2l+1)}\right) = \varepsilon_2\left[C - \frac{(l+1)}{l} D a^{-(2l+1)}\right];$$
$$\varepsilon_2\left[C - \frac{(l+1)}{l} D b^{-(2l+1)}\right] = -\varepsilon_3\left(C + D b^{-(2l+1)}\right)\frac{(l+1)}{l} \tag{7.28}$$

which leads to the following condition for having modes:

$$\left(\frac{b}{a}\right)^{2l+1}\left(\frac{\varepsilon_1}{\varepsilon_2} + \frac{l+1}{l}\right)\left(\frac{\varepsilon_2}{\varepsilon_3} + \frac{l+1}{l}\right) = -\frac{(l+1)}{l}\left(\frac{\varepsilon_2}{\varepsilon_3} - 1\right)\left(\frac{\varepsilon_1}{\varepsilon_2} - 1\right) \tag{7.29}$$

7.1.5.1 Metallic spherical shell in vacuum

In the case of a metallic spherical shell in vacuum the dielectric function for the three regions are

$$\varepsilon_1 = \varepsilon_3 = 1 \,;$$
$$\varepsilon_2 = 1 - \frac{\omega_{pl}^2}{\omega^2} \qquad (7.30)$$

and the resulting modes are

$$\omega = \frac{\omega_{pl}}{\sqrt{2}} \sqrt{1 \pm \frac{1}{2l+1} \sqrt{1 + 4l(l+1)\left(\frac{a}{b}\right)^{2l+1}}} \,; \quad l = 0, 1, 2 \ldots \qquad (7.31)$$

The results in the case of an inner- to outer-radius ratio of 1/2 are shown in Figure 7.9.

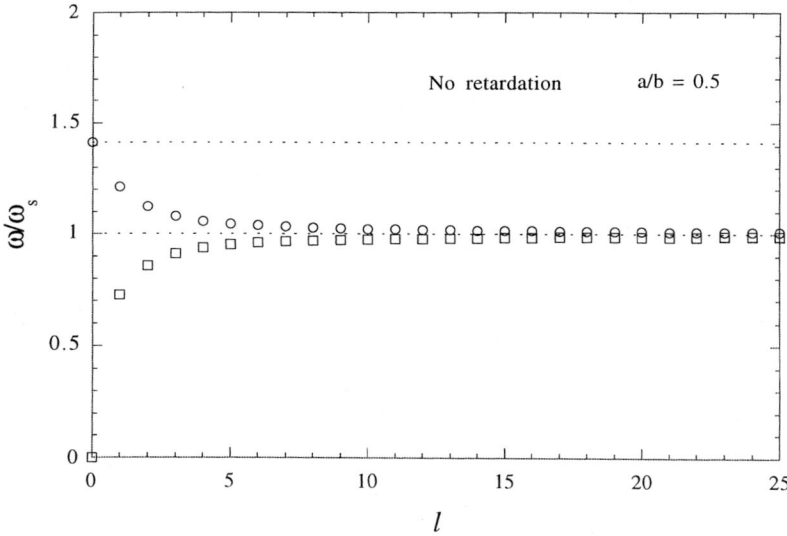

Figure 7.9: Modes for a metallic spherical shell in vacuum. The reference energy is the surface plasmon energy.

7.1.6 When liquids stay dry

(The title of this section has been borrowed from an article with the same title in Physics Today [4]. The material in this section is based on that article.)

This section is about *noncoalescence*. We have all witnessed the phenomenon when the drip coffee maker is preparing coffee: When the coffee drips into the coffee pot sometimes the drops land on the surface of the coffee and stay there for some second before coalescing with the bulk liquid. We have seen before that several factors are in favor of coalescence: The gravity is in favor; there is always an attractive van der Waals force between objects of the same material whatever the surrounding medium; the surface area decreases at coalescence and with it the surface energy. Still there seems to be some resistance to it.

Similar effects but of different origins occur in other situations: If one drops water into a hot frying pan the water drops dance around on the surface of the pan. This effect is quite different and easy to explain. The water closest to the hot surface evaporates and the drop floats around on the water vapor. It is a known fact that sometimes two drops of mercury show resistance to coalescence. In this case it is caused by surface contamination. Mercury has a very large surface tension. This means that surface contamination easily leads to a lowering of the surface energy and the impurities may cause an energy barrier to form for the process of coalescence.

The phenomenon of noncoalescence has been known for more than a century [5,6] but its detailed explanation has not been found until recently. One way to observe permanent coalescence is to heat two drops of low-viscosity silicon oil to different temperatures. Silicon oil is chosen because of its inertness to surface contamination. The experiments are performed with a setup shown in Figure 7.10.

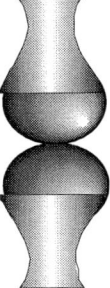

Figure 7.10: Schematic experimental setup for demonstration of noncoalescence.

Two drops are attached to the ends of two copper rods kept at different temperatures. The rods are moved together until the drops are in contact. For a proper choice of temperature difference one needs to apply a relatively large force to make the drops merge into one. The explanation is the following:

The surface energy for most materials decreases with temperature due to the thermal expansion. When the drops are brought together the hot one is cooled at the center. The surface tension increases at the center which leads to a flow of surface molecules towards the center. This is called thermal Marangoni convection.

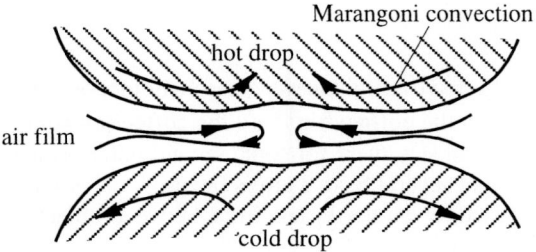

Figure 7.11: Illustration of Marangoni convection:

The cold drop is heated at the center and since the surface tension decreases there the flow of surface molecules is outwards. The moving surface molecules in the two drops drag along air molecules and there is a flow of air as sketched in figure 7.11. This flowing air film keeps the drops apart. A similar effect, wetting suppression, has also been demonstrated [7]: A silicon drop was pressed against a cooler glass surface (normally this surface is wetted by the oil) whereby it was deformed and could then be withdrawn from the surface without sticking.

Apart from the possibility to perform funny and spectacular experiments on these effects they represent a potential for important applications. They may be used for perfectly smooth, self-centering and virtually frictionless bearings for small loads; they may be utilized for measuring small loads; one may exploit the phenomena to measure attractive forces between liquid surfaces; much smaller droplets could be used in fuel droplet combustion and in the area of stability of emulsions.

7.2 Modes at the surface of a cylinder

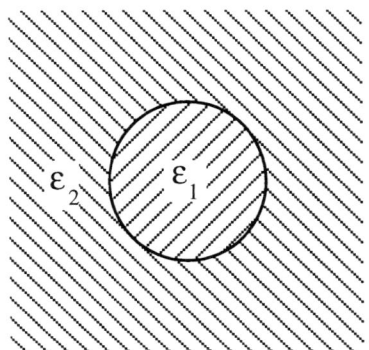

Figure 7.12: Cylinder of dielectric function ε_1 imbedded in a medium of dielectric function ε_2.

As in the treatment of an imbedded sphere we also here neglect retardation and only give the condition for modes, when retardation is included, at the end of the section. In this case the natural choice of coordinates is the cylindrical coordinates. Here, the Laplacian and gradient are

$$\nabla^2 = \frac{1}{\rho}\frac{\partial}{\partial \rho}\left(\rho \frac{\partial}{\partial \rho}\right) + \frac{1}{\rho^2}\frac{\partial^2}{\partial \phi^2} + \frac{\partial^2}{\partial z^2} ;$$

$$\nabla = \mathbf{e}_\rho \frac{\partial}{\partial \rho} + \mathbf{e}_\phi \frac{1}{\rho}\frac{\partial}{\partial \phi} + \mathbf{e}_z \frac{\partial}{\partial z}$$

(7.32)

We assume a potential solution of the form $\varphi = f(\rho)g(\phi)e^{ikz}$ and substitute this in the Laplace equation. We find:

$$\frac{1}{f\rho}\frac{\partial}{\partial \rho}\left(\rho \frac{\partial f}{\partial \rho}\right) + \frac{1}{g\rho^2}\frac{\partial^2 g}{\partial \phi^2} - k^2 = 0$$

(7.33)

We multiply by ρ^2 and obtain

$$\frac{\rho}{f}\frac{\partial}{\partial \rho}\left(\rho \frac{\partial f}{\partial \rho}\right) + \frac{1}{g}\frac{\partial^2 g}{\partial \phi^2} - \rho^2 k^2 = 0$$

(7.34)

and

$$\frac{\partial^2 f}{\partial \rho^2} + \frac{1}{\rho}\frac{\partial f}{\partial \rho} - \left(\frac{m^2}{\rho^2} + k^2\right)f = 0 \; ;$$

$$\frac{\partial^2 g}{\partial \phi^2} + m^2 g = 0 \; ; \tag{7.35}$$

$$g(\phi) = e^{im\phi} \; ; \quad m = 0, \pm 1, \pm 2\ldots$$

Now, we let $x = k\rho$:

$$x^2 \frac{\partial^2 f}{\partial x^2} + x\frac{\partial f}{\partial x} - \left(m^2 + x^2\right)f = 0 \tag{7.36}$$

This is the modified Bessel equation (9.6.1 of Reference [8]). The solutions are the modified Bessel functions $I_m(x)$ and $K_m(x)$. The first is bounded for small x and the other for large x. Thus we have

$$\varphi_m(\rho,\phi,z) = \begin{cases} A_m I_m(k\rho)e^{im\phi}e^{ikz} \; ; \; \rho < \rho_0 \\ B_m K_m(k\rho)e^{im\phi}e^{ikz} \; ; \; \rho > \rho_0 \end{cases} \tag{7.37}$$

and the matching of the potential at the cylindrical surface gives

$$B_m = A_m \frac{I_m(k\rho_0)}{K_m(k\rho_0)} \tag{7.38}$$

The matching of the normal component of the **D** field gives

$$\varepsilon_1 \frac{\partial}{\partial \rho} I_m(k\rho)\bigg|_{\rho=\rho_0} = \varepsilon_2 \frac{I_m(k\rho_0)}{K_m(k\rho_0)} \frac{\partial}{\partial \rho} K_m(k\rho)\bigg|_{\rho=\rho_0} \tag{7.39}$$

or

$$\frac{\varepsilon_1}{\varepsilon_2} = \frac{I_m(k\rho_0)K'_m(k\rho_0)}{I'_m(k\rho_0)K_m(k\rho_0)} \; ; \; m = 0, 1, 2\ldots \tag{7.40}$$

This is the condition for the modes of a cylinder. It may be of interest to note the following limits:

$$\frac{\varepsilon_1(\omega)}{\varepsilon_2(\omega)} = \frac{I'_m(k\rho_0)K'_m(k\rho_0)}{I_m(k\rho_0)K_m(k\rho_0)} \approx \begin{cases} \begin{array}{ll} -1 \; ; & m \neq 0 \\ 0 \; ; & m = 0 \end{array} \; ; \; k\rho_0 \to 0 \\ -\left(1 + \frac{1}{k\rho_0}\right) \; ; \; k\rho_0 \to \infty \end{cases} \qquad (7.41)$$

Thus, in both limits all solutions (except for $m = 0$) are degenerate and the energy equals the energy of the surface mode for a single flat surface. If we let the radius of the cylinder go to infinity we should, and we do, obtain the same modes as for a flat surface.

We have so far neglected retardation. Including retardation leads to the following condition for modes in general [9]:

TM and **TE**:

$$\left[\frac{1}{\lambda_2}\frac{H'_m(\lambda_2\rho_0)}{H_m(\lambda_2\rho_0)} - \frac{1}{\lambda_1}\frac{J'_m(\lambda_1\rho_0)}{J_m(\lambda_1\rho_0)}\right]\left[\frac{q_2^2}{\lambda_2}\frac{H'_m(\lambda_2\rho_0)}{H_m(\lambda_2\rho_0)} - \frac{q_1^2}{\lambda_1}\frac{J'_m(\lambda_1\rho_0)}{J_m(\lambda_1\rho_0)}\right] = \frac{m^2 k^2}{\rho_0^2}\left(\frac{1}{\lambda_1^2} - \frac{1}{\lambda_2^2}\right)^2 \qquad (7.42)$$

where

$$q_i = \frac{\omega\sqrt{\varepsilon_i(\omega)}}{c} \; ; \; \lambda_i = \sqrt{q_i^2 - k^2} \; ; \; i = 1, 2 \qquad (7.43)$$

and the H_m and J_m functions are cylindrical Hankel and Bessel functions, respectively. The prime on a function denotes the derivative with respect to the argument of the function.

Here the **TM** and **TE** modes mix and only $k = 0$ or $m = 0$ modes decouple into **TE** and **TM**-modes. Then letting the first (second) factor on the left-hand side of Equation (7.42) be equal to zero, defines the **TE** (**TM**) modes. The modes are either non-radiative or radiative. They radiate in the part of the ωk-plane where $\omega > ck/\sqrt{\varepsilon_2}$ and are non-radiating for $\omega < ck/\sqrt{\varepsilon_2}$.

7.2.1 Metal cylinder in vacuum

In the case of a metal cylinder of radius ρ_0 we have the following modes:

$$\omega_m(k\rho_0) = \frac{\omega_{pl}}{\sqrt{1 - \frac{I_m(k\rho_0)K'_m(k\rho_0)}{I'_m(k\rho_0)K_m(k\rho_0)}}} \quad ; \quad m = 0, 1, 2 \ldots \tag{7.44}$$

The results are shown in Figures 7.13 and 7.14.

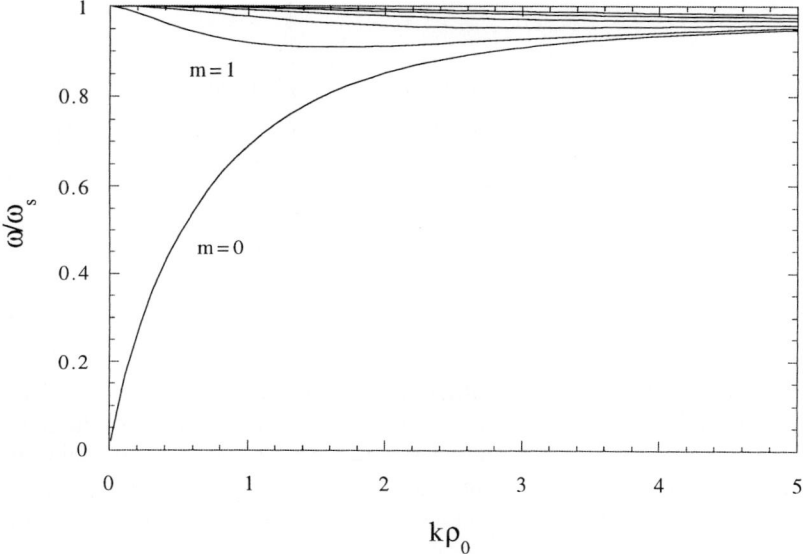

Figure 7.13: Modes for a metallic cylinder of radius ρ_0 in vacuum. The reference energy is the surface plasmon energy.

7.2 Modes at the Surface of a Cylinder

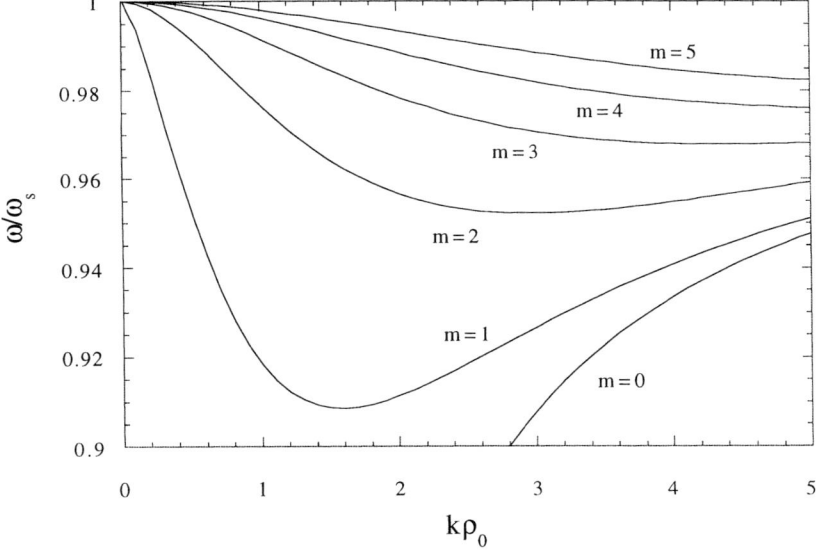

Figure 7.14: Modes for a metallic cylinder in vacuum, expanded scale. The reference energy is the surface plasmon energy.

7.2.2 Cylindrical void in a metal

In the case of a cylindrical void of radius ρ_0 in a metal we have the following modes:

$$\omega_m(k\rho_0) = \frac{\omega_{pl}}{\sqrt{1 - \frac{I'_m(k\rho_0)K_m(k\rho_0)}{I_m(k\rho_0)K'_m(k\rho_0)}}} \quad ; \quad m = 0, 1, 2 \ldots \quad (7.45)$$

The results are shown in Figures 7.15 and 7.16.

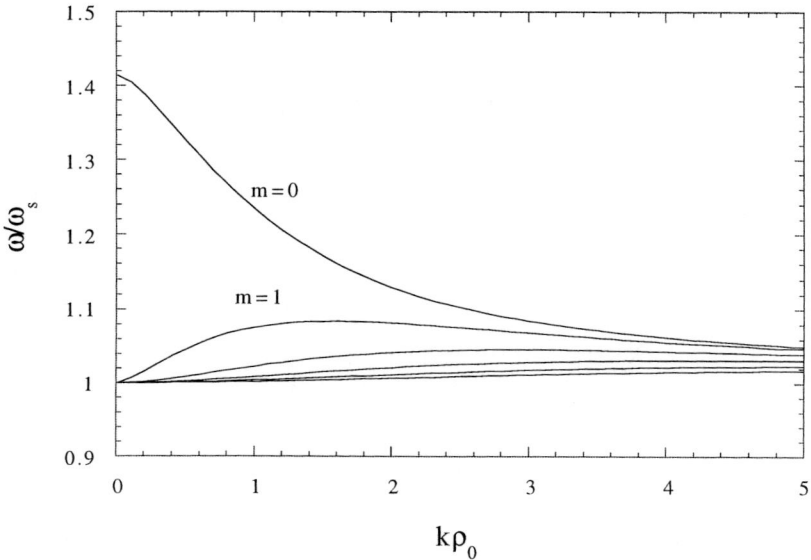

Figure 7.15: Modes for a cylindrical void of radius ρ_0 in a metal. The reference energy is the surface plasmon energy.

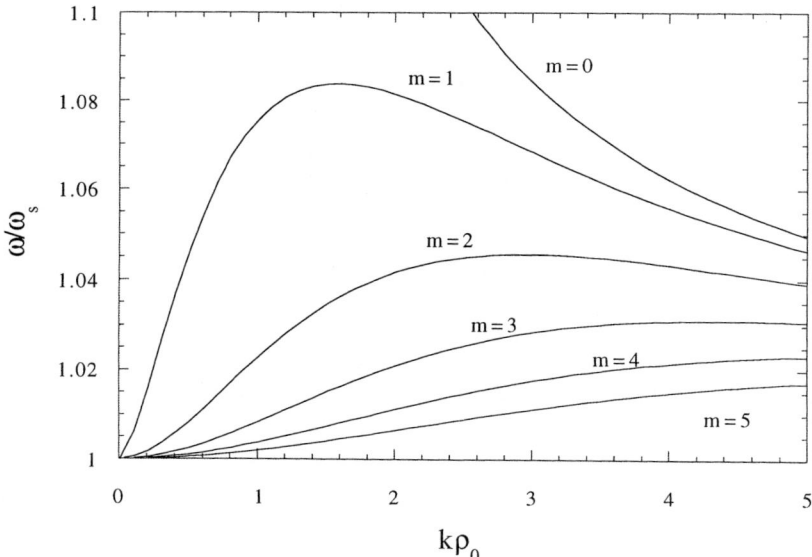

Figure 7.16: Modes for a cylindrical void in a metal, expanded scale. The reference energy is the surface plasmon energy.

7.3 Modes at an edge

The solution of the problem near sharp edges or corners often leads to spurious solutions. These can be avoided by using rounded off corners or points. We will here represent the edge by one with a parabolic profile. In this case parabolic cylindrical coordinates are the evident choice to make:

$$\begin{cases} x = \frac{1}{2}\left(u^2 - v^2\right) \\ y = uv \\ z = z \end{cases} \qquad \begin{cases} -\infty < u < \infty \\ 0 \leq v \\ -\infty < z < \infty \end{cases} \qquad (7.46)$$

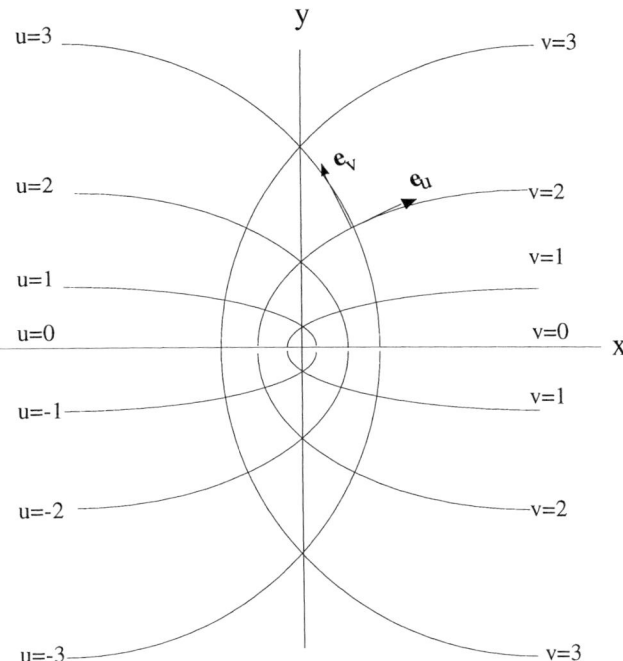

Figure 7.17: Equi-surfaces for parabolic cylindrical coordinates.

We need Laplace's operator and the gradient in these coordinates. They are

$$\nabla^2 = \frac{1}{u^2+v^2}\left(\frac{\partial^2}{\partial u^2}+\frac{\partial^2}{\partial v^2}\right)+\frac{\partial^2}{\partial z^2} \ ;$$

$$\nabla = \mathbf{e}_u \frac{1}{\sqrt{u^2+v^2}}\frac{\partial}{\partial u}+\mathbf{e}_v \frac{1}{\sqrt{u^2+v^2}}\frac{\partial}{\partial v}+\mathbf{e}_z \frac{\partial}{\partial z}$$

(7.47)

We seek a solution of the form

$$\varphi = f(u)g(v)e^{iqz} \tag{7.48}$$

This inserted in the Laplace equation gives

$$-\frac{1}{f(u)}\frac{\partial^2 f(u)}{\partial u^2}+q^2 u^2 = \frac{1}{g(v)}\frac{\partial^2 g(v)}{\partial v^2}-q^2 v^2 = E \tag{7.49}$$

where we have made use of the fact that the left-hand-side and the right-hand-side of the equation only depend on u and v, respectively, and since they are equal they have to be one and the same constant. The equation for the function $f(u)$ is identical to the Schrödinger equation for a harmonic oscillator. Its solutions are

$$f(u) = B_n e^{-qu^2/2} H_n(u\sqrt{q}) \tag{7.50}$$

where $H_n(x)$ is the Hermitian function [10] and its eigenvalues are

$$E = E_n = (2n+1)q \tag{7.51}$$

This means that the $g(v)$-function is a solution to the equation

$$\frac{\partial^2 g}{\partial(\sqrt{2q}v)^2}-\left(\frac{1}{4}(\sqrt{2q}v)^2+\left(n+\frac{1}{2}\right)\right)g = 0 \tag{7.52}$$

We have rewritten it on the form

$$\frac{\partial^2 y}{\partial x^2}-\left(\frac{1}{4}x^2+a\right)y = 0 \tag{7.53}$$

which is the standard form (19.1.2. of Reference [1]).
We define our edge to be limited by a constant-v curve, $v = v_0$. Thus, we have

7.3 Modes at an Edge

$$\varphi_n(u,v,z,t) = \begin{cases} A_n e^{-qu^2/2} H_n(u\sqrt{q}) V\left(n+\tfrac{1}{2},\sqrt{2q}v\right) e^{iqz} e^{-i\omega t} & ; v < v_0 \\ B_n e^{-qu^2/2} H_n(u\sqrt{q}) U\left(n+\tfrac{1}{2},\sqrt{2q}v\right) e^{iqz} e^{-i\omega t} & ; v > v_0 \end{cases} \quad (7.54)$$

The potential and the normal component of the **D** field are continuous at this surface. The matching of the potential gives

$$B_n = A_n \frac{V\left(n+\tfrac{1}{2},\sqrt{2q}v_0\right)}{U\left(n+\tfrac{1}{2},\sqrt{2q}v_0\right)} \quad (7.55)$$

The matching of the **D** field gives

$$\varepsilon_1 \frac{\partial V\left(n+\tfrac{1}{2},\sqrt{2q}v\right)}{\partial v}\bigg|_{v=v_0} = \varepsilon_2 \frac{V\left(n+\tfrac{1}{2},\sqrt{2q}v_0\right)}{U\left(n+\tfrac{1}{2},\sqrt{2q}v_0\right)} \frac{\partial U\left(n+\tfrac{1}{2},\sqrt{2q}v\right)}{\partial v}\bigg|_{v=v_0} \quad (7.56)$$

or

$$\frac{\varepsilon_1(\omega)}{\varepsilon_2(\omega)} = \frac{V\left(n+\tfrac{1}{2},\sqrt{2q}v_0\right) U'\left(n+\tfrac{1}{2},\sqrt{2q}v_0\right)}{U\left(n+\tfrac{1}{2},\sqrt{2q}v_0\right) V'\left(n+\tfrac{1}{2},\sqrt{2q}v_0\right)} \quad ; \quad n = 0,1,2\ldots \quad (7.57)$$

where the dielectric functions ε_1 and ε_2 are the ones inside and outside the edge, respectively. The functions U and V are the so-called parabolic-cylinder functions [11]. The prime on a function here means the derivative with respect to the second argument.

It may be of interest to note the following limits:

$$\frac{\varepsilon_1(\omega)}{\varepsilon_2(\omega)} \approx \begin{cases} -\left(1+\dfrac{1}{qv_0^2}\right) & ; \sqrt{2q}v_0 \to \infty \\[2mm] \dfrac{\sin\left[\tfrac{\pi}{2}(n-1)\right]\Gamma\left[-\tfrac{1}{2}(n)\right]\Gamma\left[\tfrac{1}{2}(n+1)\right]}{\Gamma\left[-\tfrac{1}{2}(n-1)\right]\sin\left[\tfrac{\pi}{2}(n)\right]\Gamma\left[\tfrac{1}{2}(n+2)\right]} & ; \sqrt{2q}v_0 = 0 \end{cases} \quad (7.58)$$

In the first limit the modes approach the surface-plasmon or -phonon mode. In the second limit all even solutions approach zero for a metal and the transverse phonon energy for the semiconductor. The odd solutions approach the plasmon or longitudinal phonon energy, respectively. Thus for long wavelengths all odd and all even solutions, respectively, are degenerate.

7.3.1 Metallic wedge in vacuum

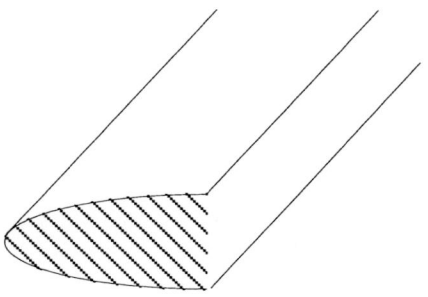

Figure 7.18: Metallic wedge.

In the case of a metallic wedge we have the following modes:

$$\omega_n = \frac{\omega_{pl}}{\sqrt{1 - \left[V\left(n+\tfrac{1}{2}, \sqrt{2q}v_0\right) U'\left(n+\tfrac{1}{2}, \sqrt{2q}v_0\right)\right] / \left[V'\left(n+\tfrac{1}{2}, \sqrt{2q}v_0\right) U\left(n+\tfrac{1}{2}, \sqrt{2q}v_0\right)\right]}} \; ;$$

$$n = 0, 1, 2 \ldots$$

(7.59)

The results are shown in Figure 7.19.

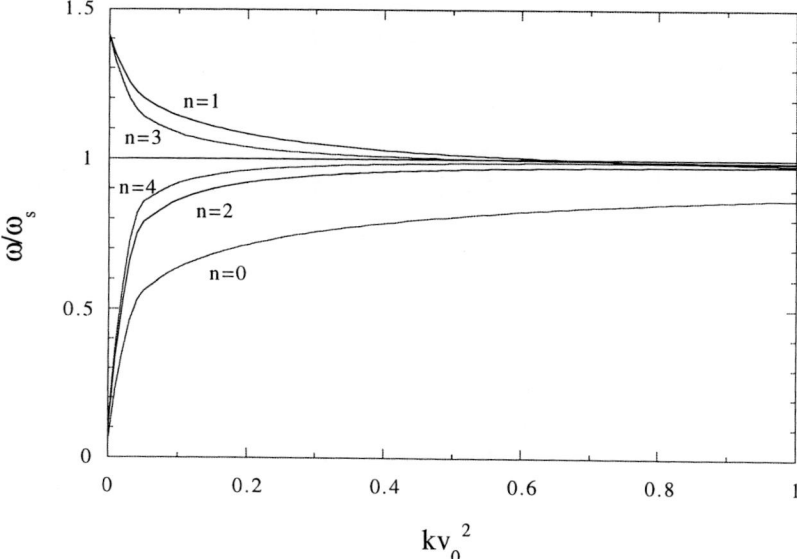

Figure 7.19: Modes for a metallic wedge in vacuum.

7.3.2 Wedge void in a metal

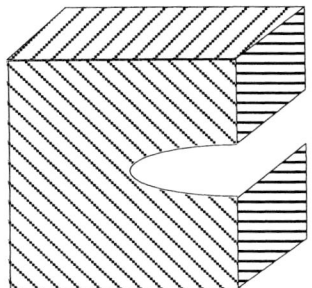

Figure 7.20: Wedge void in a metal.

In the case of a wedge void in a metal we have the following modes:

$$\omega_n = \frac{\omega_{pl}}{\sqrt{1 - \left[V'\left(n+\tfrac{1}{2},\sqrt{2qv_0}\right)U\left(n+\tfrac{1}{2},\sqrt{2qv_0}\right)\right]/\left[V\left(n+\tfrac{1}{2},\sqrt{2qv_0}\right)U'\left(n+\tfrac{1}{2},\sqrt{2qv_0}\right)\right]}} \;;$$

$$n = 0, 1, 2 \ldots$$

(7.60)

These results are shown in Figure 7.21.

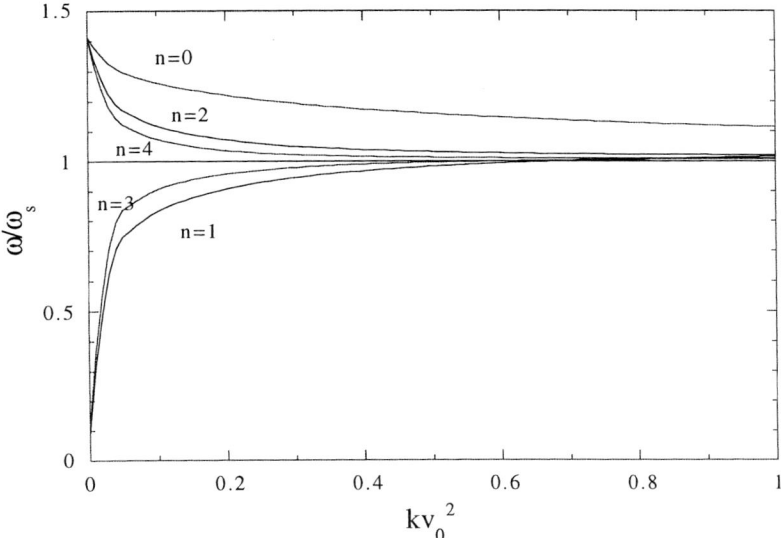

Figure 7.21: Modes for a wedge void in a metal.

7.4 Modes in a needle (a paraboloid of revolution)

In this case the natural choice of coordinates is the paraboloidal coordinates. The coordinate surfaces are obtained by revolving the parabolas in the parabolic cylinder coordinates, of Section 7.3, around the x- axis, which is then relabeled the z-axis. The third set of coordinate surfaces are planes passing through this axis:

$$\begin{cases} x = uv\cos\phi \\ y = uv\sin\phi \\ z = \tfrac{1}{2}(u^2 - v^2) \end{cases} \quad \begin{cases} 0 \le u \\ 0 \le v \\ 0 \le \phi < 2\pi \end{cases} \tag{7.61}$$

Here, the Laplacian and gradient are

$$\begin{aligned} \nabla^2 &= uv^2 \frac{\partial}{\partial u}\left(u\frac{\partial}{\partial u}\right) + u^2 v \frac{\partial}{\partial v}\left(v\frac{\partial}{\partial v}\right) + \left(u^2 + v^2\right)\frac{\partial^2}{\partial \phi^2} \ ; \\ \nabla &= \mathbf{e}_u \frac{1}{\sqrt{u^2+v^2}}\frac{\partial}{\partial u} + \mathbf{e}_v \frac{1}{\sqrt{u^2+v^2}}\frac{\partial}{\partial v} + \mathbf{e}_\phi \frac{1}{uv}\frac{\partial}{\partial \phi} \end{aligned} \tag{7.62}$$

We assume a potential solution of the form $\varphi = f(u)g(v)e^{im\phi}$ and make the substitution in the Laplace equation:

$$\left[\frac{1}{uf}\frac{\partial}{\partial u}\left(u\frac{\partial f}{\partial u}\right) - u^{-2}m^2\right] + \left[\frac{1}{vg}\frac{\partial}{\partial v}\left(v\frac{\partial g}{\partial v}\right) - v^{-2}m^2\right] = 0 \tag{7.63}$$

Since each term only depends on one of the variables the equation is separable. We have

$$\begin{aligned} u^2 \frac{\partial^2 f}{\partial u^2} + u\frac{\partial f}{\partial u} + \left(Eu^2 - m^2\right)f &= 0 \ ; \\ v^2 \frac{\partial^2 g}{\partial v^2} + v\frac{\partial g}{\partial v} - \left(Ev^2 + m^2\right)g &= 0 \end{aligned} \tag{7.64}$$

or

7.4 Modes in a Needle

$$\tilde{u}^2 \frac{\partial^2 f}{\partial \tilde{u}^2} + \tilde{u} \frac{\partial f}{\partial \tilde{u}} + \left(\tilde{u}^2 - m^2\right) f = 0 \; ;$$

$$\tilde{v}^2 \frac{\partial^2 g}{\partial \tilde{v}^2} + \tilde{v} \frac{\partial g}{\partial \tilde{v}} - \left(\tilde{v}^2 + m^2\right) g = 0 \tag{7.65}$$

where these equations are the Bessel and modified Bessel equation, respectively. The new variables are

$$\begin{aligned} \tilde{u} &= qu \; ; \\ \tilde{v} &= qv \; ; \\ q &= \sqrt{E} \end{aligned} \tag{7.66}$$

where we have assumed that E is positive. We could have chosen E negative. If so, u and v would have changed places. The choice we have made here is suitable if the surface is a constant-v surface. The Bessel functions are oscillating while the modified Bessel functions are decaying or growing.

We have

$$\varphi_m(u,v,\phi) = \begin{cases} A_m I_m(qv) H_m^{(1)}(qu) e^{im\phi} \; ; \; v < v_0 \\ B_m K_m(qv) H_m^{(1)}(qu) e^{im\phi} \; ; \; v > v_0 \end{cases} \tag{7.67}$$

where the H-functions are so-called Hankel functions.
Matching the potential at the surface gives

$$B_m = A_m \frac{I_m(qv_0)}{K_m(qv_0)} \tag{7.68}$$

The matching of the normal component of the **D** field gives

$$\varepsilon_1 \left(\frac{1}{\sqrt{u^2+v^2}} \frac{\partial}{\partial v} I_m(qv) \right)\bigg|_{v=v_0} = \varepsilon_2 \frac{I_m(qv_0)}{K_m(qv_0)} \left(\frac{1}{\sqrt{u^2+v^2}} \frac{\partial}{\partial v} K_m(qv) \right)\bigg|_{v=v_0} \tag{7.69}$$

or

$$\frac{\varepsilon_1(\omega)}{\varepsilon_2(\omega)} = \frac{I_m(qv_0) K'_m(qv_0)}{K_m(qv_0) I'_m(qv_0)} \; ; \; m = 0, 1, 2 \ldots \tag{7.70}$$

This is exactly the same relation determining the modes as in the cylinder case.

References

[1] Handbook of Mathematical Functions, edited by M. Abramowitz and I. A. Stegun, National Bureau of Standards Applied Mathematical Series, No 55 (U. S. G. P. O., Washington, D.C., 10th printing, 1972) p. 332.
[2] R. Ruppin in Electromagnetic Surface Modes, Ed. A. D. Boardman, (John Wiley, New York, 1982) p. 349.
[3] Handbook of Mathematical Functions ibid p. 437.
[4] P. Dell'Aversana, and G. P. Neitzel, Physics Today 51, (1), 38 (1998).
[5] J. W. Strutt, B. Rayleigh, Philos. Mag. 36, 321 (1899).
[6] O. Reynolds, Chem. News 44, 211 (1881).
[7] P. Dell'Aversana, V. Tontodonato, L. Carotenuto, Phys. Fluids 9, 2475 (1997).
[8] Handbook of Mathematical Functions ibid p. 374.
[9] R. Ruppin in Electromagnetic Surface Modes, Ed. A. D. Boardman, (John Wiley, New York, 1982) p. 387.
[10] Handbook of Mathematical Functions ibid p. 775.
[11] Handbook of Mathematical Functions ibid p. 686.

8 Different mode types

Why is it that the main pulse detected by the seismograph at an earthquake is preceded by a small pulse? What causes the optical glory?

In Chapter 1 we found the following equations defining the dispersions of bulk modes:

$$\omega^2 \tilde{\varepsilon}_\perp(\mathbf{q},\omega) = (cq)^2$$
$$\tilde{\varepsilon}_L(\mathbf{q},\omega) = 0$$
(8.1)

The first gives the transverse modes and the second the longitudinal ones. These relations are valid for all types of material system, metals, semiconductors, insulators, heavily doped semiconductors and so forth. As we have mentioned before the dielectric properties for different types of material are very similar in principle. There is a region between the transverse and longitudinal excitation energies where the dielectric function is negative. It is in this region that the modes we treat in this book appear. Metals are special in that the transverse frequency is zero. Thus it is enough to study a polar semiconductor and a metal in this chapter. We will now go through these two types of system in some detail. Some of the discussion will be a repetition of what we already have briefly touched upon. In each case we start by discussing the bulk modes and then the modes at a planar surface of the material. We start with a polar semiconductor or ionic insulator. We devote two subsections to very brief discussions of the effects of spatial dispersion and surface roughness. At the end of the chapter we go through one of the methods designed to detect the surface modes, the *ATR* method.

8.1 Polar semiconductors or ionic insulators

In an ionic insulator like NaCl the solid is made up by ions. In NaCl these are Na^+ and Cl^- ions. In a polar semiconductor the atoms are not ions but the charge has, to varying degree, been transferred from one atom type to the other. All semiconductors that are made up by more than one element, like GaAs or CdS, are polar. Single-element semiconductors like Si and Ge are not. There is no principle difference between ionic insulators and polar semiconductors. The difference is just in the degree of charge transfer from negative to positive atoms. We can treat the two types of solid as one.

In a material with two types of element we have apart from the acoustical phonons, found in all solids, another phonon mode, the optical phonon mode. They are called optical phonons because they couple to the light. This will be evident later. Actually all materials with more than one atom in the unit cell have optical phonon modes, but these have, for our purpose, a negligible coupling to the light if the atoms are neutral. In the optical phonons the two elements move out of phase with respect to each other. We have illustrated the two phonon-mode types in Figure 8.1. The amplitude of the oscillations is greatly exaggerated. In reality it is always much smaller than the separation between the atoms.

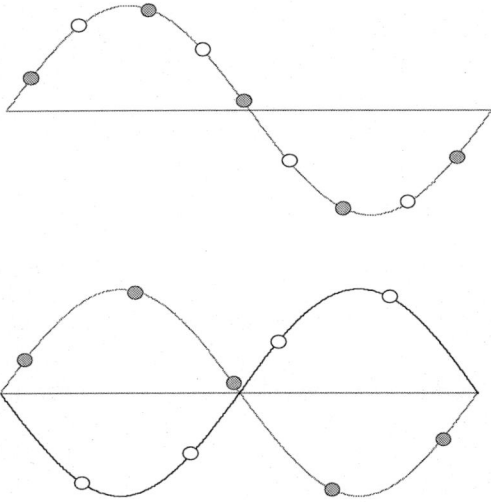

Figure 8.1: Schematic illustration of how the atoms move when phonons of different type have been excited. The upper part represents acoustical phonons and the lower optical.

The transverse optical phonon mode can be excited by electromagnetic waves. Imagine that an electromagnetic wave is travelling in the direction of the horizontal line of Figure 8.1 and that it is polarized in the plane of the paper. Then the electric field is in the plane of the paper and perpendicular to the horizontal line. The two atom types are affected by a force in opposite directions; the positive atoms are affected by a force in the direction of the electric field and the negative in the opposite direction. An excitation like the one in the lower part of the figure is stimulated. This means that the dielectric function has an imaginary part at the transverse optical phonon frequency. There is an absorption line at this energy. Let us neglect the width of this line and write the contribution to the polarizability as

$$\alpha_2^{ph}(\omega) = C\delta(\omega - \omega_T) \tag{8.2}$$

This delta function is represented by the dashed arrow in Figure 8.2. The phonon absorption band is in the infrared range of the spectrum. The system also absorbs at higher energies. There are absorption bands corresponding to excitation of electrons from the valence bands to the conduction bands, so-called interband transitions. These bands are in the visible or ultraviolet range of the spectrum.

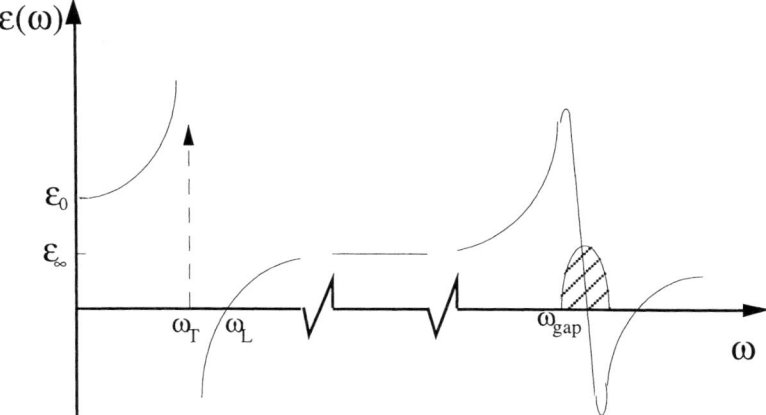

Figure 8.2: Schematic dielectric function for an ionic insulator or polar semiconductor.

Even higher up in energy, in the x-ray range, there are absorption bands corresponding to excitations from core-levels. Apart from in the phonon bands the material is transparent for photon energies smaller than the band gap. Depending on the size of the band gap the material can be opaque (Si, Ge, GaAs) or transparent (ZnO, In$_2$O$_3$, diamond) in the visible range. The phonon energies and band-gap values are typically 30 meV and 1 eV, respectively. Thus the phonon- and interband-bands are in different energy scales. This is why we have cut the frequency axis in Figure 8.2. Just below the band-gap value there are, for a real system, some structures coming from exciton excitation. We defer the discussion of these to Section 8.4.

The polarizabilities obey the *Kramers-Kronig dispersion relations:*

$$\alpha_1(\omega) = \frac{2}{\pi} P \int_0^\infty d\omega' \frac{\omega' \alpha_2(\omega')}{(\omega')^2 - \omega^2} \;;$$
$$\alpha_2(\omega) = -\frac{2}{\pi} P \int_0^\infty d\omega' \frac{\omega \alpha_1(\omega')}{(\omega')^2 - \omega^2}$$

(8.3)

All retarded correlation functions obey these relations. They are very useful both for theory and experiment. Assume that one may measure one of the components in the whole frequency range. Then it is possible to calculate the other component for arbitrary frequencies. We discussed this already in Chapter 2.

The real part of the dielectric function will in all systems have the value unity at very high frequencies. The polarizable entities like permanent dipoles, charged atoms, polarization charges and free electrons cannot, because of their finite moments of inertia or inertial mass, respond to the quickly varying fields. If we go down in energy and successively pass the absorption bands we find that, if the bands are well separated, the real part will saturate, in between the bands, at a value that each time is larger than the previous value. More and more of the entities discussed above may, via their excitation channels, follow and respond to the field. We will focus on the phonon range. The effect of the interband transitions and transitions there above is that the dielectric constant ε_∞ is "handed over" to the lower frequency range. This is the dielectric constant the system would have had, had it not been polar. Below the phonon band the dielectric function saturates at the value ε_0.

We see in Figure 8.2 that the dielectric function passes zero just above ω_T. This defines the frequency for the longitudinal optical phonon, ω_L. This illustrates a general behavior: The excitations come in pairs, one transverse and one longitudinal. If we look again at Figure 8.2 we see that above the interband absorption band the dielectric function passes zero. Here, we have another longitudinal excitation, the plasmon excitation. Even non-metals have plasmons.

Let us now determine the dielectric function in the phonon range. We need the values of ε_0, ε_∞, and ω_T as input parameters. We have the general relation

$$\varepsilon(\omega) = 1 + \alpha^e(\omega) + \alpha^{ph}(\omega) \tag{8.4}$$

when the electronic and phonon excitations are in different energy ranges. In the phonon range the first two terms saturate at ε_∞. Thus we have

$$\varepsilon(\omega) = \varepsilon_\infty + \alpha^{ph}(\omega) \tag{8.5}$$

and

$$\begin{aligned}\varepsilon_1(\omega) &= \varepsilon_\infty + \frac{2}{\pi} P \int_0^\infty d\omega' \frac{\omega' \alpha_2^{ph}(\omega')}{(\omega')^2 - \omega^2} \\ &= \varepsilon_\infty + \frac{2}{\pi} P \int_0^\infty d\omega' \frac{\omega' C\delta(\omega' - \omega_T)}{(\omega')^2 - \omega^2} \\ &= \varepsilon_\infty + \frac{2}{\pi} \frac{\omega_T C}{(\omega_T)^2 - \omega^2}\end{aligned} \tag{8.6}$$

where we have made use of Equation (8.2).
Now, using the static value of the dielectric function we find:

$$\varepsilon_0 = \varepsilon_1(0) = \varepsilon_\infty + \frac{2}{\pi} \frac{C}{\omega_T} \tag{8.7}$$

8.1 Polar Semiconductors or Ionic Insulators

and this relation determines the value of the constant C:

$$C = (\varepsilon_0 - \varepsilon_\infty)\frac{\pi\omega_T}{2} \tag{8.8}$$

Thus we have for the real and imaginary parts of the dielectric function

$$\varepsilon_1(\omega) = \varepsilon_\infty + \frac{\omega_T^2(\varepsilon_0 - \varepsilon_\infty)}{\omega_T^2 - \omega^2} = \frac{\omega_T^2\varepsilon_0 - \omega^2\varepsilon_\infty}{\omega_T^2 - \omega^2} \tag{8.9}$$

and

$$\varepsilon_2(\omega) = \frac{\pi}{2}(\varepsilon_0 - \varepsilon_\infty)\omega_T\delta(\omega - \omega_T) \tag{8.10}$$

respectively.

We may now determine the longitudinal phonon frequency as the frequency where the numerator of the expression for the real part vanishes. We find:

$$\omega_L = \omega_T\sqrt{\varepsilon_0/\varepsilon_\infty} \tag{8.11}$$

which is the so-called *Lyddane-Sachs-Teller* relation. We may rewrite the real part of the dielectric function as

$$\varepsilon_1(\omega) = \varepsilon_\infty \frac{\omega_L^2 - \omega^2}{\omega_T^2 - \omega^2} \tag{8.12}$$

to make the fact, that the real part of the dielectric function vanishes at ω_L and diverges at ω_T, more visible. The correction to the dielectric function from the phonons is more visible if we write the dielectric function as

$$\varepsilon_1(\omega) = \varepsilon_\infty\left(1 + \frac{\omega_L^2 - \omega_T^2}{\omega_T^2 - \omega^2}\right) \tag{8.13}$$

So far we have not discussed the quantum number characterizing the phonon, the momentum **q**. The optical phonons have very little dispersion. We will neglect the dispersion here. We will also disregard the fact that in anisotropic crystals there may be several phonon branches. The phonon momenta are limited to the first Brillouin zone. If we plot the phonon dispersions in a figure covering the whole Brillouin zone they look like in Figure 8.3.

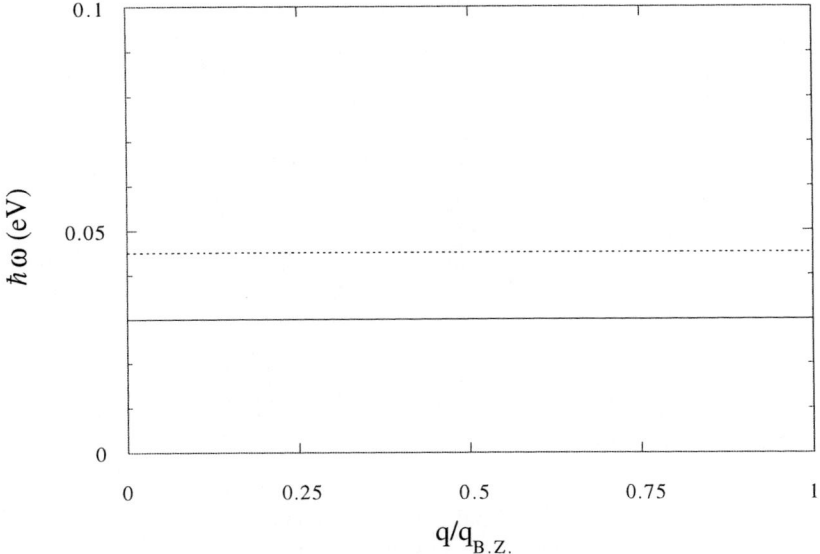

Figure 8.3: The two phonon modes throughout the Brillouin zone.

If we instead are interested in the dispersions for small momenta we find that the transverse mode has changed. The light dispersion curve and the transverse phonons curve have combined or have been mixed.

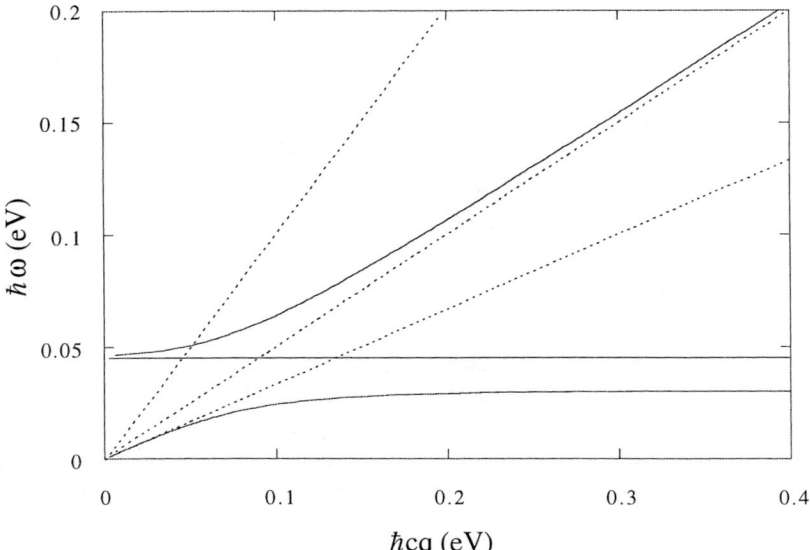

Figure 8.4: The modes for small momenta.

This occurs for very small momenta. In Figure 8.4 we have expanded our momentum scale with four orders of magnitude. The three dashed straight lines are, counted from above, the dispersion curve for light in vacuum, in a medium with dielectric constant ε_∞, and in a medium with dielectric constant ε_0, respectively. The horizontal straight line is the dispersion for the longitudinal phonon. It is unchanged. The longitudinal phonons do not couple to the light. The transverse phonon dispersion and that for the light have been mixed, hybridized. They both have light and phonon character. Hopfield [1] coined the word *polariton* to describe normal modes in solids which propagate as electromagnetic waves. The word is a combination of polarization and photon, because these modes are combinations of free photons and the polarization modes of the solid. For large momentum the two mixed modes are decoupled. We see that the lower branch starts out as a pure light wave with group velocity, $v_g = c/\sqrt{\varepsilon_0}$, which is the velocity for light in a medium with dielectric constant ε_0. For larger momentum the dispersion curve gradually bends over and ends up as the dispersion curve for transverse optical phonons. For small momentum the mode is a pure photon mode and at large momentum it is a pure phonon mode. For intermediate momentum it is a mixture of both. For the upper branch the opposite is true. For small momentum it is a pure phonon mode and for large momentum it is a pure photon mode, now for a medium with dielectric constant ε_∞. There are two modes for all momentum. However the interaction between the photons and the transverse phonons has had the effect that there is an energy gap between the transverse and longitudinal phonon energies where light no longer can travel through the system. In this region the system is totally reflecting. This region corresponds to the region where the dielectric function is negative. If we now briefly return to Figure 8.2 we see that there is a corresponding region just above the interband absorption-band where the dielectric function is also negative, and hence the system is reflecting. This is a general behavior. Between the pair of transverse and longitudinal modes there is a region of negative dielectric function (real part). These regions are very important, as we know by now, because it is in these regions those surface modes appear that we are concerned with here. While we are studying Figure 8.2 we note that the real part of the dielectric function is monotonically increasing except for in regions of absorption, that is, regions where the imaginary part of the dielectric function is finite. Because of this, these absorbing regions are called regions of anomalous dispersion. In the regions between the absorption bands the refractive index is in all systems a monotonically increasing function of energy, or frequency. This is why blue or violet light is refracted more than red in a glass prism and also why the sky is blue and the sunset is red.

Now we turn to the modes at the surface of the crystal. We will now plot our curves with the momentum parallel to the surface on the horizontal axis. We denote this momentum by **k** instead of **q** which we reserve for the total momentum. In Figure 8.5 we plot the bulk modes as a function of k or rather the projection of the bulk modes on the ωk-plane. The two horizontally hatched areas are the regions covered by bulk polaritons. The vertical hatched area is the region covered by the light outside the sample, in the vacuum. Where the two types of hatched areas overlap is where we can have a wave entering the interior of the system from the outside region. We use a diagonal hatching to indicate these areas. The frequency and the momentum parallel to the surface are conserved. This is true for an ideal surface. In presence of surface roughness this momentum conservation is relaxed. We discuss this briefly in Section 8.5. The perpendicular momentum is not conserved. The surface provides with the perpendicular momentum needed. The right-most boundary of each area looks the same as the modes in the plot with q on the abscissa. They represent waves propagating parallel to the surface. The modes at the left-most boundary, the vertical axis, represent waves propagating perpendicular to the surface. The modes in between represent waves at an angle to the surface. There are areas where the horizontally and vertically hatched regions do not overlap, which means that there are modes that can not be excited by an incoming wave. We see that the single vertically hatched area is not fully confined to the region between the longitudinal and transverse phonon energies. There is an additional small region above. The reason is the following: Where the diagonal straight line, the dispersion curve for light outside the material, crosses the upper branch of the bulk polariton dispersion the dielectric function of the medium has the value unity. For this frequency incoming light goes straight through the material unaffected, independent of the angle of incidence. For higher frequencies the material is optically denser than vacuum and an incoming wave is refracted towards the normal when impinging on the surface from outside. For smaller frequencies but still above the frequency of the longitudinal phonon the material is optically thinner and an incoming wave is refracted away from the normal when impinging on the surface from outside. In this frequency range the angle of incidence may be too large for the wave to enter the medium; the incoming wave is totally reflected.

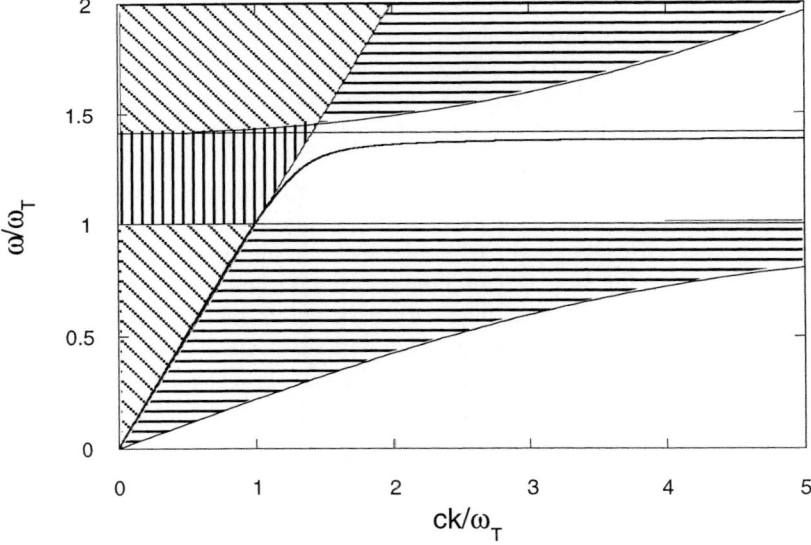

Figure 8.5: Bulk and surface modes. Bulk modes that can be excited from outside are in the diagonally hatched areas

8.2 Metallic systems

The most characteristic feature of metals is the intraband transitions. These are transverse in character and the corresponding longitudinal excitation is the plasmon. Apart from the ordinary metals also heavily doped semiconductors belong to this group of solids. Let us study an example. The plasmon mode of beryllium has an unusually strong dispersion. In most metals the deviation from a horizontal line is much smaller. Apart from this we may choose beryllium as a representative of all metals. We see that the transverse mode is not represented by a curve as in the phonon case but by a region and this region extends all the way down to the horizontal axis. This means that there is no longer a polariton mode below the transverse mode. There is only one upper polariton mode. Of course there may also be modes connected to higher lying interband transitions.

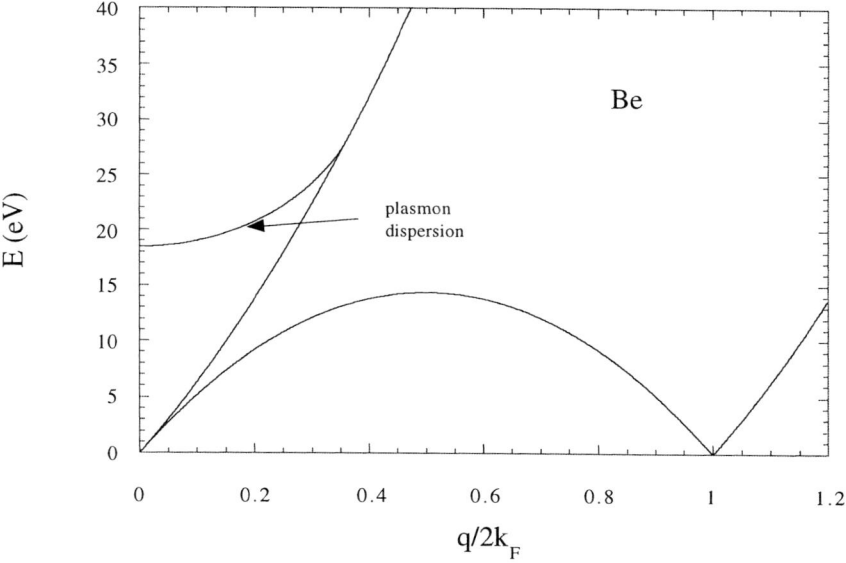

Figure 8.6: Dispersion of the collective bulk mode and boundaries of the single-particle excitation region for Be.

If we now are interested in the polariton mode we expand the horizontal axis and on the scale where the polaritons are visible the dispersions of the plasmon curve and the boundary of the single-particle continuum are suppressed.

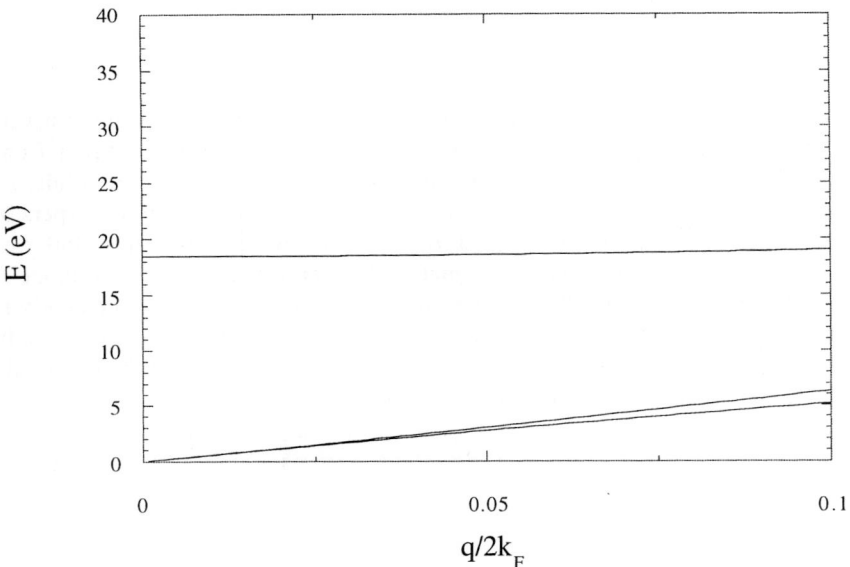

Figure 8.7: Previous figure with expanded q axis.

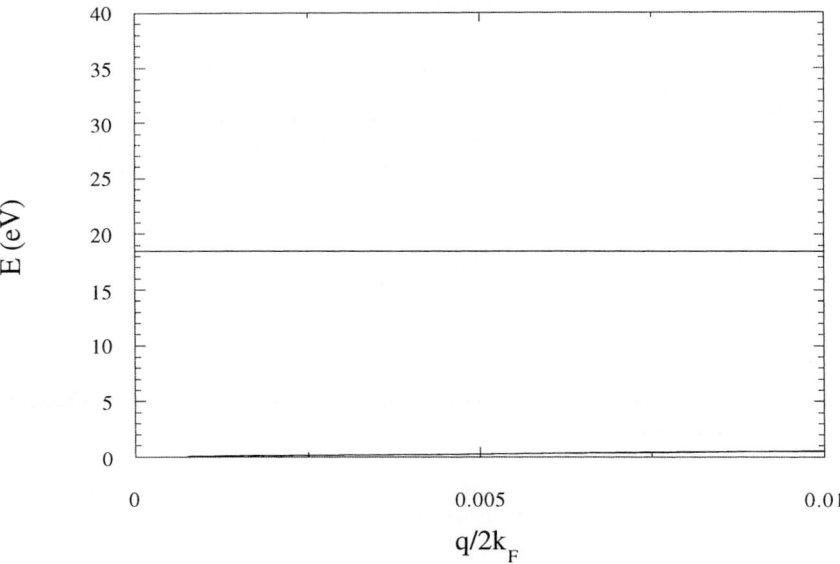

Figure 8.8: Previous figure with even more expanded q axis.

8.2 Metallic Systems

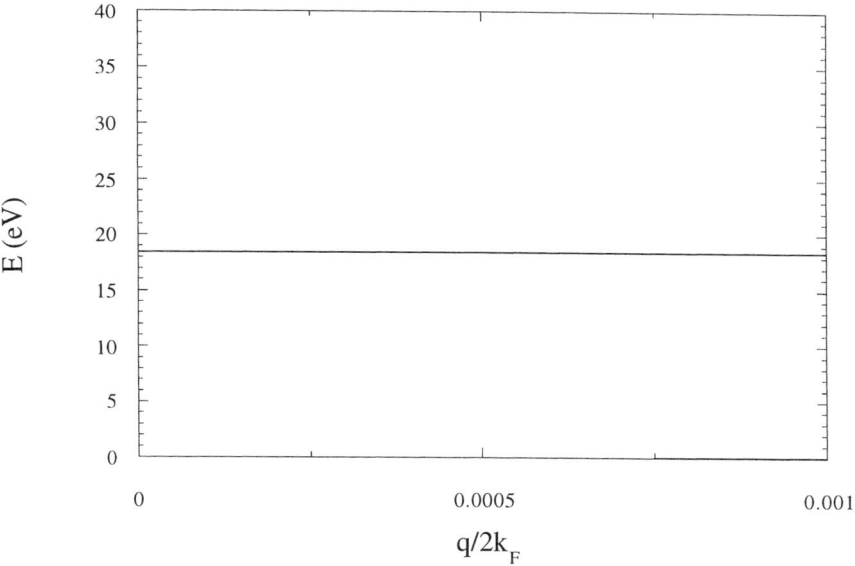

Figure 8.9: Previous figure with even more expanded q axis.

On the scale of Figure 8.9 we may replace the plasmon mode with a horizontal line at the plasmon energy and the single particle continuum with a curve at zero energy.

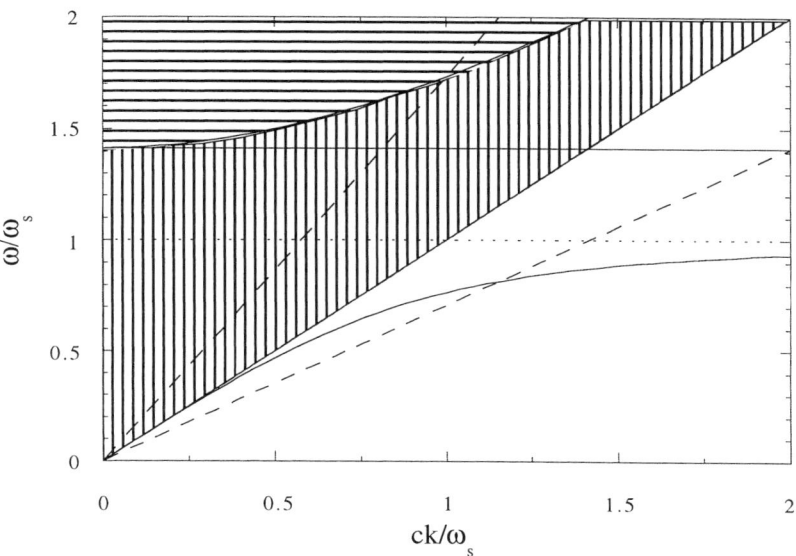

Figure 8.10: Regions and boundaries for different modes in the bulk of a metal and at a metal-vacuum interface

Thus the results from Section 8.1 can be directly used with the replacements $\omega_L \to \omega_{pl}$ and $\omega_T \to 0$. The systems treated in this section are metals or heavily doped semiconductors. In the metal case the plasmon energy is in the ultraviolet range of the spectrum, while in the semiconductor case it is in the infrared range.

The horizontally hatched region in Figure 8.10 is the region of bulk-polariton–photon coupling. The material is transparent in this region. A light wave impinging on the surface will result in a reflected wave and a refracted wave entering the bulk. A photon will either bounce off the surface or enter. In this process the energy and the momentum parallel to the surface are conserved. To be noted is that we have the momentum component parallel to the surface on our horizontal axis. The bulk polariton dispersion curve, defining the lower boundary of this region, represents the dispersion of a wave inside the material that is travelling parallel to the surface. All other waves not parallel to the surface fall inside the horizontally hatched region. The diagonal solid line represents the dispersion of a light wave outside the sample travelling parallel to the surface, $\omega = ck$. The dashed upper straight line represents the dispersion curve for a light wave outside the sample travelling at an angle to the surface: $\omega = c\sqrt{k^2 + q_z^2}$. Such a light wave can excite a bulk polariton; the dashed straight line overlaps the continuum of possible bulk waves. The slope of the curve is always larger than that for light parallel to the surface. Thus an incoming light wave cannot excite the surface polariton which is the lower solid curve in the figure since this is below the solid diagonal straight line.

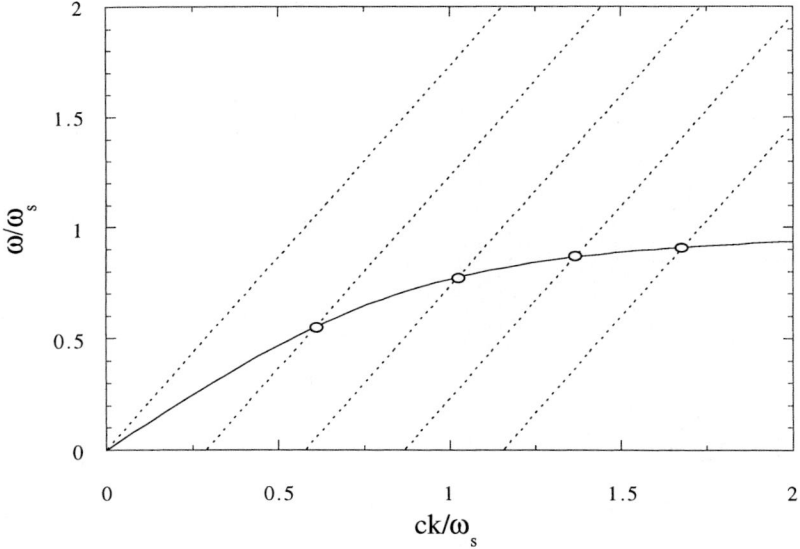

Figure 8.11: Excitation of the surface plasmon polariton with help of grating.

We can get around this problem in different ways. One way is to have a rough surface so that the parallel momentum is no longer conserved. To get a useful method to study the

8.2 Metallic Systems

surface polariton one needs to prepare the roughness in a controlled way. One can construct a grating, equidistant lines or groves, along the surface which will cause the surface to absorb the momentum in units of $2\pi/d$, where d is the period of the grating. This is illustrated in Figure 8.11. The possible surface plasmon polariton modes that can be excited for a specific grating is indicated by the small circles in the figure.

Another method is to place a prism at the surface. Light is then allowed to go through total internal reflection at the prism-air interface. An evanescent wave is then extending outside the prism and reaches inside the metal surface. This wave has an imaginary q_z. This means that the slope of the dispersion curve for this light is less than c. This is indicated by the lower dashed line in Figure 8.10. Thus this line crosses the surface polariton mode and can cause excitations. This experimental technique is called Attenuated Total Reflection (*ATR*). We will come back to this experimental technique, used to study the surface modes, in Section 8.6.

In the vertically hatched region of Figure 8.10 another type of mode can be found. This mode is termed virtual. It radiates energy (see Figure 8.12). Inside the material it is exponentially decaying away from the surface and outside it is a plane wave going outwards. It is not strictly correct to put it in a diagram like Figure 8.10, the way we have defined the axes. These modes have either complex valued frequency or complex valued momentum parallel to the surface, or both are complex valued.

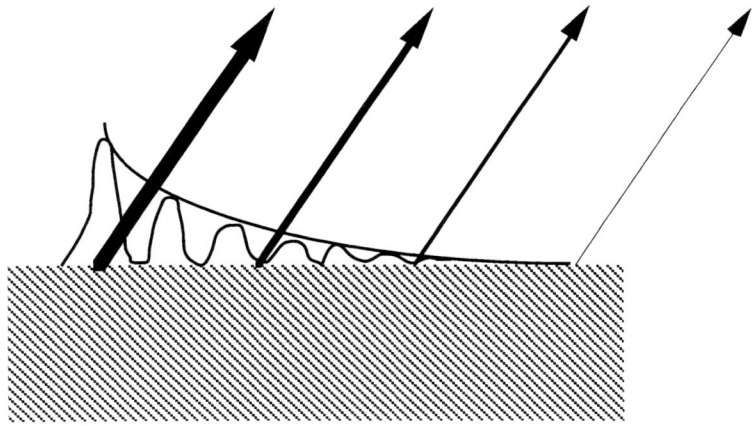

Figure 8.12: Radiative modes. The amplitudes of the fields decrease along the trajectory of the mode since the energy is radiated.

When we include these modes the momentum and frequency on the axes should be regarded as the real parts.

8.3 Characterization of different surface-mode types

The different electromagnetic surface modes can be grouped into four groups according to Table 8.1

Table 8.1: Characterization of different surface modes according to P. Halevi [2]. Reprinted by permission of John Wiley & Sons, Inc.

Mode	Dielectric function $\varepsilon = \varepsilon' + i\varepsilon''$	Wave-vector component[a] Parallel	Perpendicular	Phase velocity	Comment
Fano	$\varepsilon' \leq -1$ $\varepsilon'' \ll \|\varepsilon'\|$	$k \gg k''$	$\gamma \gg [\gamma'']$	$< c$	True surface mode
Evanescent	$-1 \leq \varepsilon' \leq 0$ $\varepsilon'' \approx \|\varepsilon'\|$	$k \approx k''$	$\gamma \approx [\gamma'']$		Rapidly attenuating
Brewster	$\varepsilon' \geq 0$ $\varepsilon'' \ll \|\varepsilon'\|$	$k \gg k''$	$\gamma \ll [\gamma'']$	$> c$	Bound to surface only in presence of damping
Zenneck	$\varepsilon' \ll -1$ $\varepsilon'' \gg \|\varepsilon'\|$	$k \gg k''$	$\gamma \approx [\gamma'']$	$\geq c$	Propagates in conducting medium

a) $\mathbf{q} = (k+ik'')\hat{x} + (-\gamma'' + i\gamma)\hat{z}$

The Fano mode is the mode we have treated here. It is the true surface mode. All other modes rely on damping in the system. To illustrate the different modes we present an example in Figure 8.13. We study a vacuum-solid interface. Let the dielectric function of the solid be

$$\varepsilon(\omega) = 1 + \varepsilon_\infty \frac{\omega_{T,1}^2 - \omega_{L,1}^2}{\omega(\omega+i\Gamma_1)^2 - \omega_{T,1}^2} + \frac{\omega_{T,2}^2 - \omega_{L,2}^2}{\omega(\omega+i\Gamma_2) - \omega_{T,2}^2} \tag{8.14}$$

where we have used the following values for the parameters:

$$\varepsilon_\infty = 6; \quad \begin{cases} \omega_{T,1} = \omega_{T,1} \\ \omega_{L,1} = 2\omega_{T,1} \end{cases} ; \quad \begin{cases} \omega_{T,2} = 5\omega_{T,1} \\ \omega_{L,2} = 6\omega_{T,1} \end{cases} ; \quad \begin{cases} \Gamma_1 = 0.05\omega_{T,1} \\ \Gamma_2 = 0.05\omega_{T,1} \end{cases} \tag{8.15}$$

8.3 Characterization of Different Surface-Mode Types

This function is not strictly correct, but if the damping is small it may be considered to fairly well represent a real dielectric function. With this choice the dielectric function looks like in Figure 8.13.

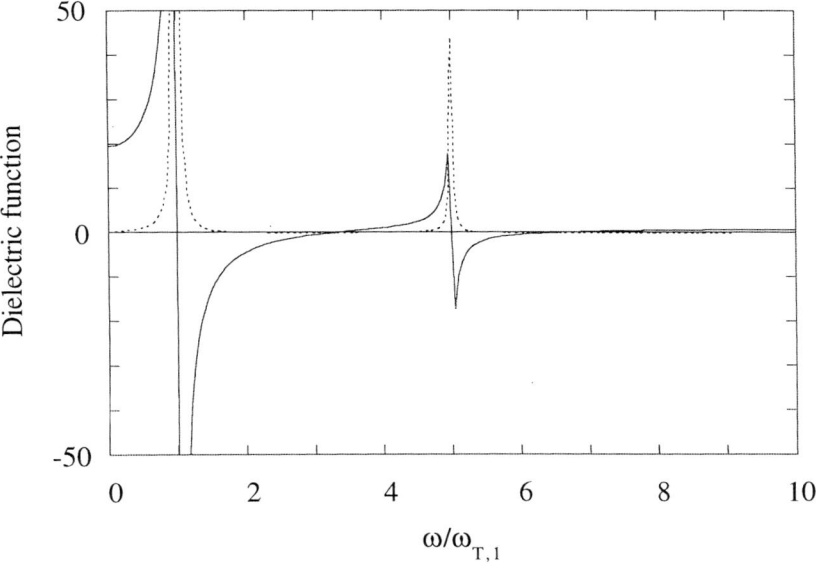

Figure 8.13: The real (solid curve) and imaginary (dotted curve) parts of the dielectric function for the example in the text.

With two absorption regions the resulting longitudinal modes are shifted in energy compared to our input values. The condition for normal surface modes leads to

$$\gamma_1 = -\gamma_2 \varepsilon_1$$
$$\Downarrow$$
$$(\gamma_1)^2 = (\gamma_2)^2 \varepsilon_1^2$$
$$\Downarrow \tag{8.16}$$
$$k^2 - \varepsilon_1(\omega/c)^2 = \left[k^2 - (\omega/c)^2\right]\varepsilon_1^2$$
$$\Downarrow$$
$$k = (\omega/c)\sqrt{\varepsilon_1(\omega)/[\varepsilon_1(\omega)+1]}$$

When the dielectric function is complex valued the momentum will also be complex valued. We chose the real part. The solution is given in Figure 8.14.

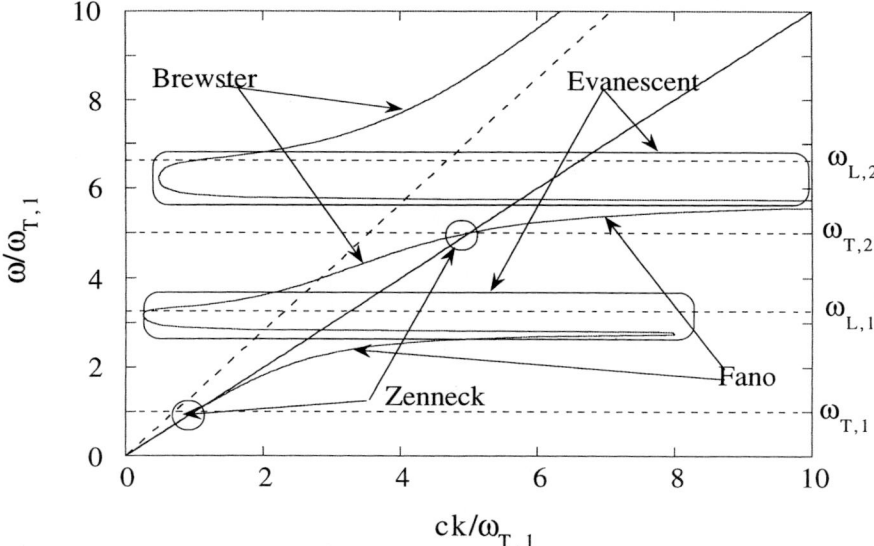

Figure 8.14: Different mode types in the example presented in the text.

The diagonal solid asymptote is the dispersion curve for light outside the surface. The dashed asymptote has a slope larger than that of the light dispersion curve. The *Brewster* mode approaches this asymptote for large frequencies. The *Fano* modes are the ones we have derived earlier. In Chapter 4 when we were solving for the modes at a semiconductor-vacuum interface we found the Brewster mode, Figure 4.6, but we never discussed it then. The Brewster mode is always above the light dispersion curve and does not qualify as a solution of the type we have been aiming at. It can be regarded as consisting of an incoming plane wave and a transmitted plane wave (no reflected wave). The so-called Brewster angle is the angle of incidence where there is no reflected power. A Brewster's angle only exists for **TM** waves. This is due to the dipoles that create the reflected wave. When a **TM** wave hits an interface dipoles are induced in the material in the direction of the **E** field. At just the right angle the dipole axis line up with the direction of the reflected wave. Since a dipole does not radiate along its axis there is no reflected wave generated. This happens when the "reflected" wave and the transmitted wave are perpendicular to each other. The *Zenneck* modes appear just below the Fano modes, in the region where the dielectric function has a large imaginary part. These modes are above but very close to the light dispersion curve. The *Evanescent* modes are the modes in the region where the Fano modes bend backwards and connect with the Brewster modes. We discussed this type of back bending behavior earlier but then for bulk modes, in Figure 2.3.

When the damping is reduced with five orders of magnitude, that is, basically neglecting damping, we have the results shown in Figure 8.15.

8.3 Characterization of Different Surface-Mode Types

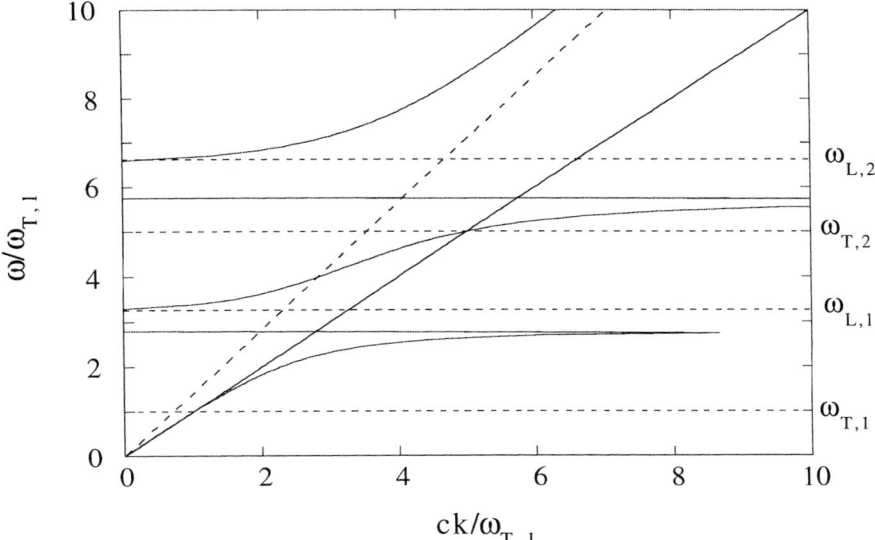

Figure 8.15: Same as Figure 8.14 but for much reduced damping

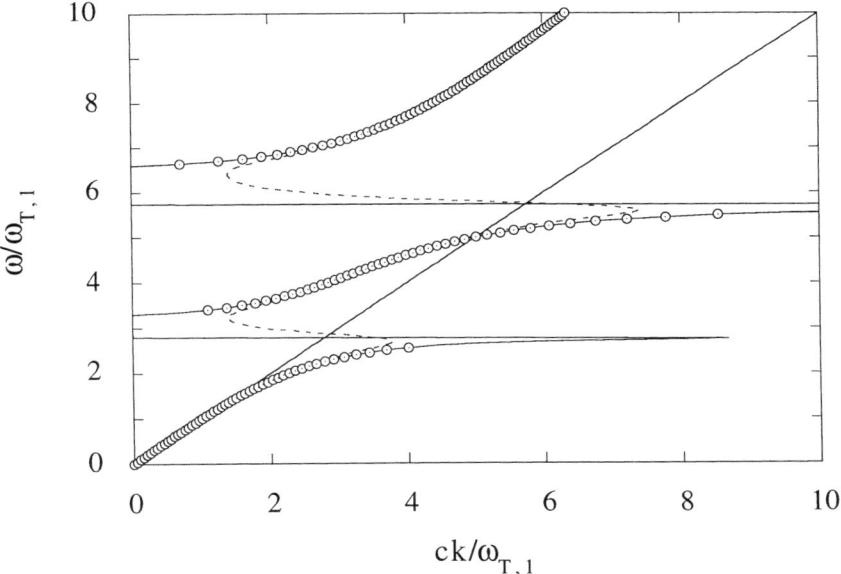

Figure 8.16: The solid and dashed curves are the dispersions of the modes without and with damping, respectively, when one demands that the frequency is real valued. If one instead demands that k/ω is real valued the real part of the complex frequency follows the circles, even in the case of damping.

In Figure 8.16 the solid curve is the result for very small damping. The dashed curve is obtained using the parameters of Equation (8.15) but with four times larger damping. In the above treatment we demanded that the frequency is real valued which then means that the momentum is complex valued. It is the real value of the momentum we have on our horizontal axis. The circles is the result from a slightly different treatment. We have let both the frequency and the momentum be complex valued and demanded that the frequency and momentum are parallel in the complex plane. Then k/ω is real valued and the square root, in Equation (8.16), has to be real valued. This gives as a solution a frequency with real and imaginary parts. The real part is plotted as circles in Figure 8.16. As can be seen this result follows more closely the undamped result and is insensitive to the damping in the system. We will in Section 8.6 demonstrate that depending on how the experiment is set up one finds a mode following more closely the dashed curve or a mode following more closely the solid curve.

8.4 Spatial dispersion

We have in the previous sections motivated the neglect of spatial dispersion. However, in some systems it cannot be neglected and there are interesting effects observed related to spatial dispersion in many different systems. Spatial dispersion is a whole research field of its own but outside the scope of this book. It should, however, be at least briefly discussed since it has implications on the modes discussed here in some systems. The interested reader is referred to a rather recent book: *Spatial Dispersion in Solids and Plasmas* [3], edited by P. Halevi, which is fully devoted to this topic. It has a broad introductory chapter covering the field and eleven specialized chapters treating different aspects of the field. There are also two chapters in the book: *Surface Polaritons* [4] edited by V. M. Agronovich and D. L. Mills which are of interest here: one by Lagois and Fischer [5] and one by Agronovich [6].

The first mention of spatial dispersion was already in 1811 when Arago [7] discovered the rotatory power of quartz. This was an example of optical activity in crystals. It can also be observed in liquids and gases. Another effect of spatial dispersion is that even cubic crystals, which in the local approximation has isotropic dielectric properties (the dielectric tensor is diagonal and all elements are equal), will have anisotropic contributions beyond the local approximation. These effects are usually very small and hard to detect but can be relatively large near absorption lines. For metals the anomalous skin effect in the microwave region is a manifestation of spatial dispersion. For metals the so-called Landau damping arise when the plasmon line enters the single-particle continuum. At finite temperatures the continuum spreads out and the damping occurs for smaller q-values. For plasmas this type of damping is much more important. The effects we have mentioned so far have only minor influences on the topic of this book. We have left the most important effect to the last—the spatial dispersion in the case of exciton polaritons.

8.4 Spatial Dispersion

When an electron is excited across the bandgap of a semiconductor, the electron in the conduction band and the hole created in the valence band are attracted to each other. They can bind and form a hydrogen-like complex in the semiconductor, an exciton. The electron and the hole in the exciton take the parts of the electron and proton in the hydrogen atom. There are two types of exciton depending on the host material: One is strongly bound to an atom of the host semiconductor; this is a *Frenkel* exciton. The other, the *Wannier* or *Mott* exciton, is delocalized and can move freely throughout the crystal. The binding energy is small and the electron and hole wave functions are extending over many lattice parameters. The spatial dispersion is most important for the Wannier/Mott exciton. At vertical transitions across the band-gap the electron in the conduction band and the hole in the valence band or the electron and the hole in the exciton obtain momentum contributions in opposite directions. The net momentum transfer to the electron-hole pair is zero; their center of mass is at rest. However, non-vertical transitions are also possible and the energy dispersions for the exciton branches are given according to:

$$E_n(\mathbf{q}) = E_g - Ry^*\left(1/n^2\right) + \hbar^2 q^2/2M \tag{8.17}$$

Where M is the mass of the exciton, the sum of the the electron and hole masses, n is the principle quantum number of the hydrogen series, and $\mathbf{q}$ is the net momentum transfer to the exciton. The *effective Rydberg constant* is denoted Ry^*.

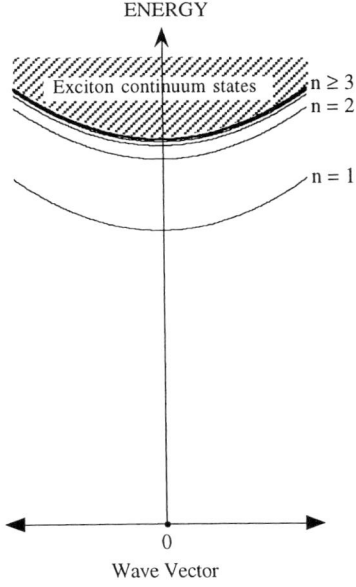

Figure 8.17: Schematic dispersion curves for excitons in a semiconductor.

The momentum dependence of the exciton dispersion curves changes the dielectric function according to:

$$\varepsilon(\omega) = \varepsilon_\infty\left(1 + \frac{\omega_L^2 - \omega_T^2}{\omega_T^2 - \omega^2}\right) \rightarrow \varepsilon(\mathbf{q},\omega) = \varepsilon_\infty\left(1 + \frac{\omega_L^2 - \omega_T^2}{\omega_T^2 - \omega^2 + (\hbar\omega_T/M)q^2}\right) \quad (8.18)$$

and the two branches of bulk polaritons, the bulk plasmon and the surface polariton change from those in Figure 8.18 to those in Figure 8.19.

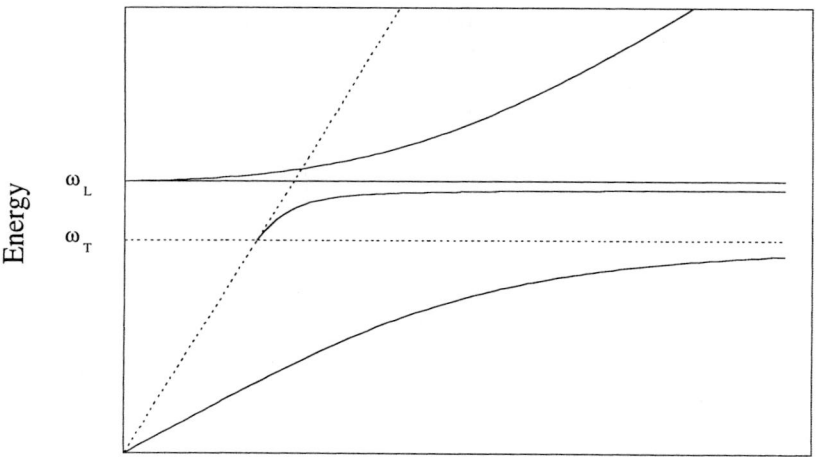

Figure 8.18: The two branches of bulk polaritons, the bulk plasmon and the surface polariton in absence of spatial dispersion.

We note in Figure 8.19 that above ω_L there are three bulk modes with the same energy, one longitudinal and two transverse. This means that if a wave is impinging on the surface of the sample three waves are appearing inside the bulk. This is illustrated in Figure 8.20. All the waves, the impinging and the refracted, have the same **k**-vector and frequency, but different **q**-vectors.

8.4 Spatial Dispersion

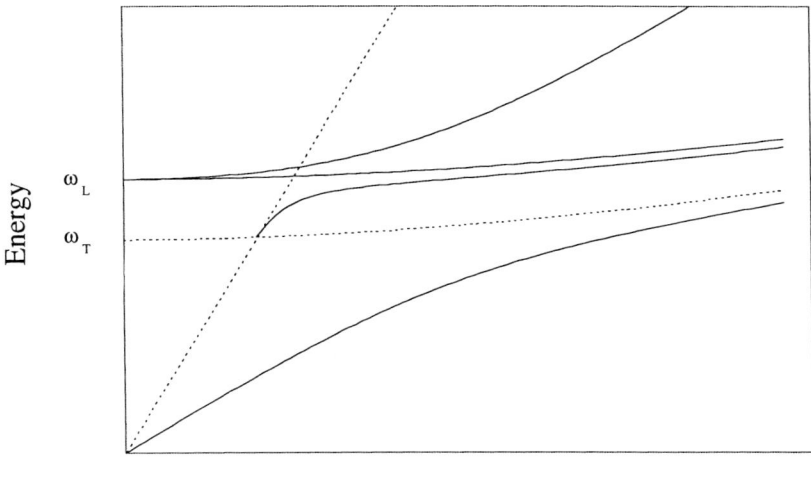

Figure 8.19: The same as Figure 8.18 but now with spatial dispersion.

We can no longer use Fresnel's equations, Equation (8.19), to determine the relative amplitudes of the three waves. We need more boundary conditions. There have been several different *Additional Boundary Conditions* (*ABC*s) proposed in the literature and they produce different results, especially for large wave vectors. Also for finding the surface modes one needs these *ABC*s and the result depends on which choice one makes.

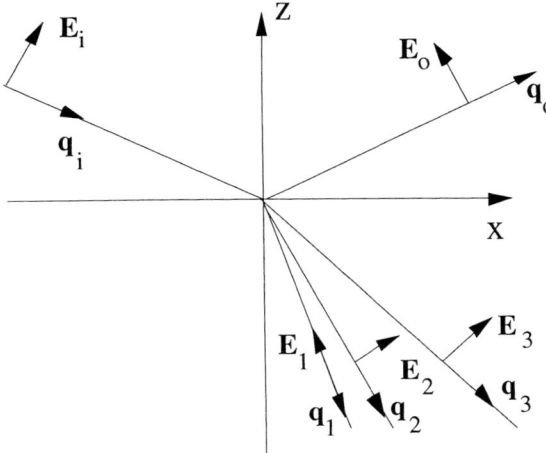

Figure 8.20: An electromagnetic wave, with energy in the region of excitonic excitations, impinging on a surface of a semiconductor is split up in three inside the material.

8.5 Surface roughness

Surface roughness is another obvious deviation from the ideal interface we have used in deriving the surface modes. The roughness is a defect that relaxes the conservation of momentum in the surface plane. The matching conditions we have used rely on this conservation. The defects can be of different type: Point defects, that is, defects that are localized to small regions; extended random defects; extended periodic defects or quasi-periodic. All these defects can be either unintentional or intentional.

The interaction of the surface modes with the surface roughness can have different consequences: The dispersion relation can be altered; the roughness may cause attenuation; the frequency may shift; new branches may appear; the modes may also become radiative.

An example of a periodic surface roughness is a diffraction grating where groves have been made in the surface. Here the surface profile is known which is a great advantage if one is trying to understand the modes. One can also produce non-stationary, moving gratings created by the passage of a Rayleigh acoustic surface wave along a dielectric medium.

The interested reader is referred to two excellent review chapters in the book: *Surface Polaritons*, edited by V. M. Agronovich and D. L. Mills that are of interest here: one by A. A. Maradudin [8] covering the experimental side and one by H. Raether [9] covering experiments. More recent work can be found in publications by J. A. Sánchez-Gil and A. A. Maradudin [10] and by A. V. Shchegrov [12] and references therein.

8.6 The ATR method

The dispersion of the surface modes are sensitive to changes at the surface. This means that one may use the modes in sensors like gas sensors and smoke detectors. One way to observe and study the modes is with optical techniques. It is however not a straight forward matter since the modes cannot be excited by an incoming electromagnetic wave from outside the surface. It is the conservation of energy and momentum that makes this impossible. There are, however, ways to get around this problem. One way is to have a rough surface so that the momentum conservation parallel to the surface is relaxed. Another way is to use a grating, for example, achieved by making parallel, equidistant groves in the surface perpendicular to the plane of incidence. A third method is to use a prism outside the surface.

We will discuss the last method in this section. There are many possible configurations, with more than one prism involved, but we will focus on the two configurations in Figure 8.21. The method relies on *Attenuated Total Reflection* (*ATR*). If the angle of incidence, α, is large enough the wave is totally reflected. At total reflection there is also, apart from the reflected wave, a wave parallel to the prism surface which decays exponentially away

8.6 The ATR Method

from the surface. This wave will travel along the interface and gradually radiate in the direction of the reflected wave. Its wave front will coincide with that of the reflected wave and is normally not seen. One situation where it is seen is when the light beam is narrow. Then the emission from the surface wave is outside that from the directly reflected light beam. There is an analogy with elastic waves. The surface wave has a velocity determined by the less dense medium of the two at the interface. At an earthquake there is a surface wave travelling faster than the bulk wave and it reaches the detector before the main wave does (see Section 8.7).

Now, if we place the prism as in Figure 8.21 (a) the surface wave will reach and extend into the sample and may excite our surface mode. This will be detected as a decrease in the reflectivity. The effect will be larger the closer to the surface we place the prism. The problem is that if we place the prism too close to the surface the presence of the prism will modify the mode. One has to be careful and optimize the experimental conditions so that one keeps the prism as far away from the sample as possible while still being able to detect the mode. The configuration in (a) is called Otto configuration or Prism-Air-Medium (PAM) configuration. Another configuration, the Kretschmann or Prism-Medium-Air (PMA) configuration is shown in Figure 8.21 (b). It can be used if the sample is in the form of a film. Here one studies the mode at the bottom side of the film. The film thickness has to be chosen with care so that one may detect the mode but the film thickness must not be too small so that the presence of the prism perturbs the result if one is interested in the mode itself. In detectors one does not need to be so careful to avoid the influence from the prism since it is only changes one is interested in. This method is well suited for gas sensors or for biological sensors where the presence of a gas or liquid modifies the mode at the bottom surface.

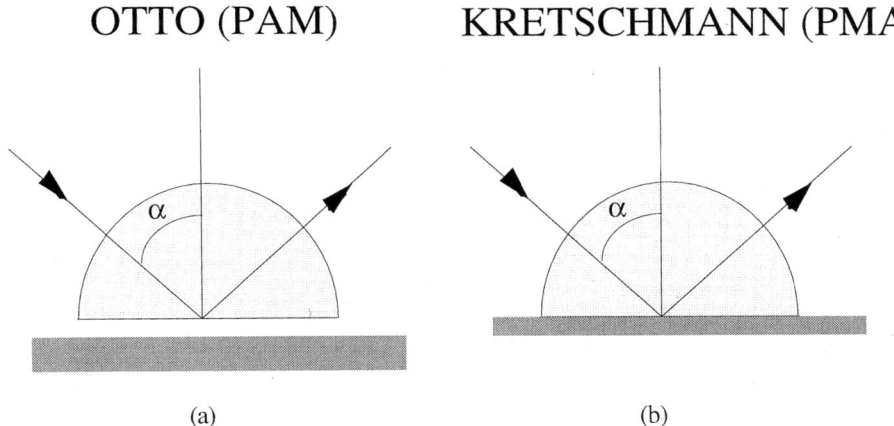

Figure 8.21: Otto (a) and Kretschmann (b) configurations in *ATR*.

Let us now derive the reflectivity in the two configurations. They both consist of three layers with different dielectric properties, the prism, the air and the medium. Let us number them *1*, *2* and *3* counted from above. Fresnel's coefficients give the relative amplitudes of the reflected and transmitted waves at an interface between two media. They are:

$$t_{ij}^s = \frac{2\sin\theta_j \cos\theta_i}{\sin(\theta_i + \theta_j)} = \frac{2n_i \cos\theta_i}{n_i \cos\theta_i + n_j \cos\theta_j}$$

$$r_{ij}^s = -\frac{\sin(\theta_i - \theta_j)}{\sin(\theta_i + \theta_j)} = \frac{n_i \cos\theta_i - n_j \cos\theta_j}{n_i \cos\theta_i + n_j \cos\theta_j}$$

$$t_{ij}^p = \frac{2\sin\theta_j \cos\theta_i}{\sin(\theta_i + \theta_j)\cos(\theta_i - \theta_j)} = \frac{2n_i \cos\theta_i}{n_j \cos\theta_i + n_i \cos\theta_j}$$

$$r_{ij}^p = \frac{\tan(\theta_i - \theta_j)}{\tan(\theta_i + \theta_j)} = \frac{n_j \cos\theta_i - n_i \cos\theta_j}{n_j \cos\theta_i + n_i \cos\theta_j}$$

(8.19)

where the superscripts s and p represent s-polarized and p-polarized waves, respectively and the angles θ_i and θ_j are the angle of incidence and angle of transmission, respectively. The optical properties of each material enter in the form of the refractive index, n. For s-polarized waves the electric field vector is perpendicular to the plane of incidence and for p-polarized waves it is in the plane of incidence. The plane of incidence is the plane defined by the incoming wave and the normal to the surface.

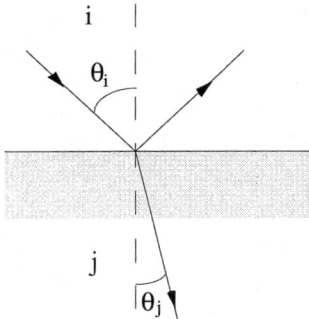

Figure 8.22: The interface between two media i and j, discussed in the text.

The Fresnel coefficients are also valid for materials with complex-valued refractive index. For our problem we need to treat two interfaces as shown in Figure 8.23. We need to take multiple reflections in the middle layer into account.

8.6 The ATR Method

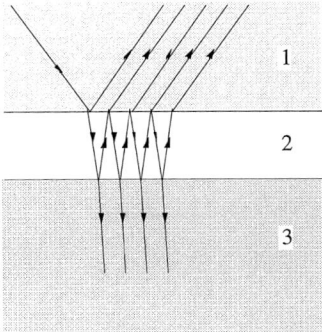

Figure 8.23: Waves contributing to the total reflected and transmitted waves in a three layer structure as discussed in the text.

The total reflected amplitude, r, is obtained from an infinite summation of waves due to the multiple reflections in the middle layer. Each time a wave impinges on an interface the Fresnel equations are used and the phases of the waves are taken into account. All coefficients are complex-valued. The phase $2\delta_2$ is the phase difference for two waves: one wave that is transmitted through the first interface, passing through layer 2, is reflected at the second interface, passing through layer 2 again and is finally transmitted through the first interface; the second wave is one that is reflected at the first interface. This second wave has furthermore traveled a longer distance in layer *1* before it impinges on the interface.

The results are obtained as follows:

$$r = r_{12} + t_{12}xt_{21} \; ; \quad x = \left\{ e^{i\delta_2} r_{23} e^{i\delta_2} [1 + r_{21}x] \right\} \; ; \quad \delta_2 = \frac{2\pi d_2}{\lambda} n_2 \cos(\theta_2)$$

$$x = \frac{e^{i2\delta_2} r_{23}}{1 - e^{i2\delta_2} r_{23} r_{21}}$$

$$r = r_{12} + \frac{e^{i2\delta_2} r_{23} t_{12} t_{21}}{1 - e^{i2\delta_2} r_{23} r_{21}} = \frac{r_{12} - r_{12} e^{i2\delta_2} r_{23} r_{21} + e^{i2\delta_2} r_{23} t_{12} t_{21}}{1 - e^{i2\delta_2} r_{23} r_{21}} \qquad (8.20)$$

$$= \frac{r_{12} + e^{i2\delta_2} r_{23}(t_{12}t_{21} - r_{12}r_{21})}{1 - e^{i2\delta_2} r_{23} r_{21}} = \frac{r_{12} + e^{i2\delta_2} r_{23}}{1 - e^{i2\delta_2} r_{23} r_{21}}$$

$$= \frac{r_{12} + e^{i2\delta_2} r_{23}}{1 + e^{i2\delta_2} r_{12} r_{23}}$$

We have made use of the following useful relations:

$$\begin{aligned} r_{21} &= -r_{12} \\ t_{12}t_{21} - r_{12}r_{21} &= 1 \end{aligned} \qquad (8.21)$$

These results are valid for both polarization directions. We are interested in the *p*-polarized case since the surface modes are of that type.

The reflectivity is

$$R_p = |r_p|^2 \tag{8.22}$$

and all we need for the calculation of the reflectivity can be summarized as:

$$\begin{aligned} R_p &= |r_p|^2 \\ r_p &= \frac{r_{12}^p + e^{i2\delta_2} r_{23}^p}{1 + e^{i2\delta_2} r_{12}^p r_{23}^p} \\ r_{12}^p &= \frac{n_2 \cos\theta_1 - n_1 \cos\theta_2}{n_2 \cos\theta_1 + n_1 \cos\theta_2} \\ r_{23}^p &= \frac{n_3 \cos\theta_2 - n_2 \cos\theta_3}{n_3 \cos\theta_2 + n_2 \cos\theta_3} \\ \cos\theta_2 &= \left[1 - (n_1/n_2)^2 \sin^2\theta_1\right]^{1/2} \\ \cos\theta_3 &= \left[1 - (n_1/n_3)^2 \sin^2\theta_1\right]^{1/2} \\ \delta_2 &= \frac{\omega d_2}{c} n_2 \cos(\theta_2) \end{aligned} \tag{8.23}$$

We have two experimental variables to vary, namely, the frequency and the angle of incidence. We will perform two calculations for the Otto configuration where we let the dielectric constant of the prism have the value 3 and use the model dielectric function used in Section 8.3 to represent the medium we are investigating:

$$\varepsilon(\omega) = 1 + \varepsilon_\infty \frac{\omega_{T,1}^2 - \omega_{L,1}^2}{\omega(\omega + i\Gamma_1)^2 - \omega_{T,1}^2} + \frac{\omega_{T,2}^2 - \omega_{L,2}^2}{\omega(\omega + i\Gamma_2) - \omega_{T,2}^2} \tag{8.24}$$

We have used the following values for the parameters:

$$\varepsilon_\infty = 6; \quad \begin{cases} \omega_{T,1} = \omega_{T,1} \\ \omega_{L,1} = 2\omega_{T,1} \end{cases}; \quad \begin{cases} \omega_{T,2} = 5\omega_{T,1} \\ \omega_{L,2} = 6\omega_{T,1} \end{cases}; \quad \begin{cases} \Gamma_1 = 0.2\omega_{T,1} \\ \Gamma_2 = 0.2\omega_{T,1} \end{cases} \tag{8.25}$$

In the first calculation we keep the angle of incidence fixed and scan the frequency. In the second we keep the frequency fixed and scan the angle of incidence. In both cases we let the air gap be 20 μm. The results of the first calculation are presented in Figure 8.24.

8.6 The ATR Method

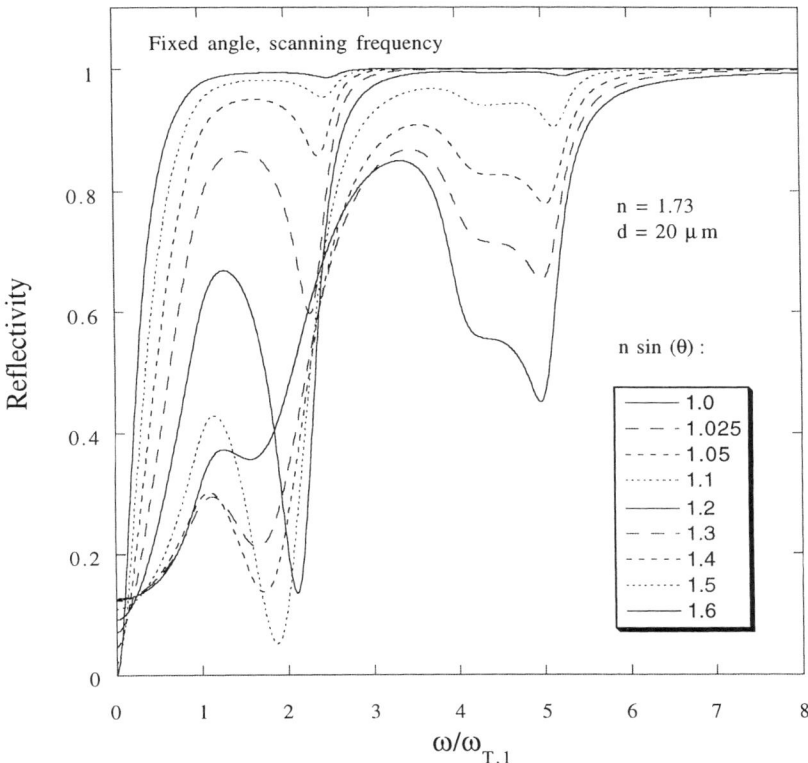

Figure 8.24: The reflectivity as function of frequency at fixed angle of incidence for a set of angle values in the range of total reflection.

We find distinct minima in the reflectivity. The left-most minimum corresponds to the lower surface mode and the right-most to the upper mode. We also find that the right-most minimum does not survive for the highest angles of incidence while the left-most does. If we had decreased the size of the air gap the minima had been deeper and one could have traced the upper mode in a larger energy range. The results are presented on an ωk plot in Figure 8.26. The momentum parallel to the surface, k, is conserved at all interfaces and it is obtained as $ck = \omega n_1 \sin(\theta_1)$, where the factor multiplying the frequency on the right hand side is just the angle parameter used to characterize the curves in Figure 8.24. We will return to these results after we have presented the second calculation.

In the second calculation we keep the frequency fixed and scan the angle. The results from this calculation are displayed in Figure 8.25.

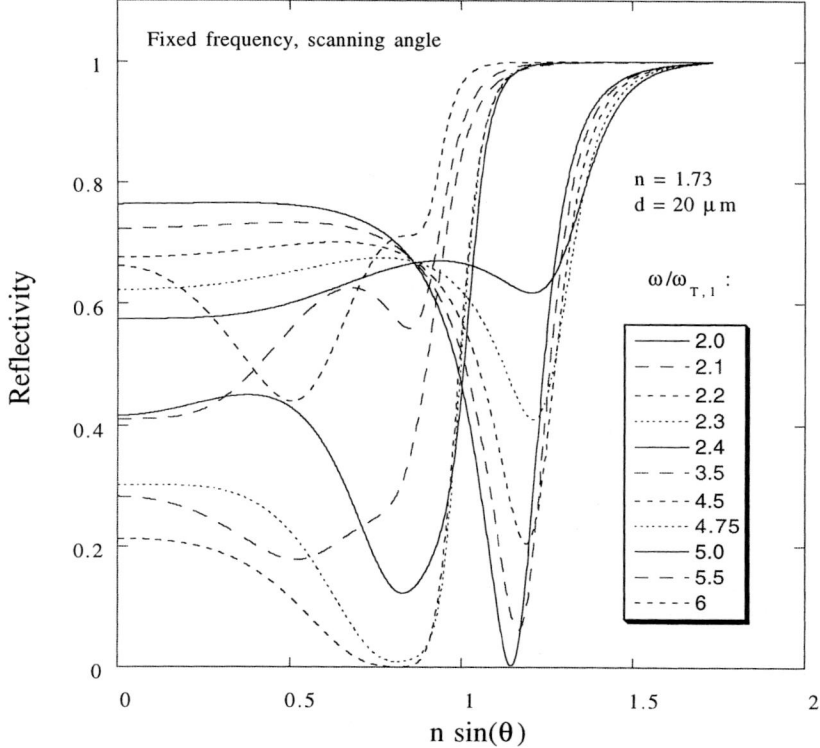

Figure 8.25: The reflectivity as function of angle of incidence for a set of frequency values. We have included all angles from 0 to $\pi/2$, but it is only angles with values on the horizontal axis larger than unity that correspond to total reflection.

Only results with $n\sin(\theta)$ larger than unity is of real interest here since these values correspond to total reflection. We find that the upper mode does not produce a minimum in this range, only the lower mode does.

The results from the two calculations are shown in Figure 8.26 together with some other results. The solid curve is the modes for the system if the damping is zero. The real modes are to the right of the solid, diagonal, straight line. The modes to the left of this line are Brewster modes and are no real surface modes. These modes approach the vertical dotted line for large momentum. When the system has damping or dissipation the frequency and momentum of the modes are no longer both real valued. One has to decide which of the variables to keep real. The dashed curve are the modes for the actual damping. They are solutions with real frequencies but complex valued momentum. Here they are plotted as function of the real value of the momentum. The open circles are solutions with the actual damping when ω/k is real valued. The frequency (and also momentum) is then complex valued, and the real part of the frequency is plotted in the figure. As can be seen this solution follows more closely the undamped modes.

8.6 The ATR Method

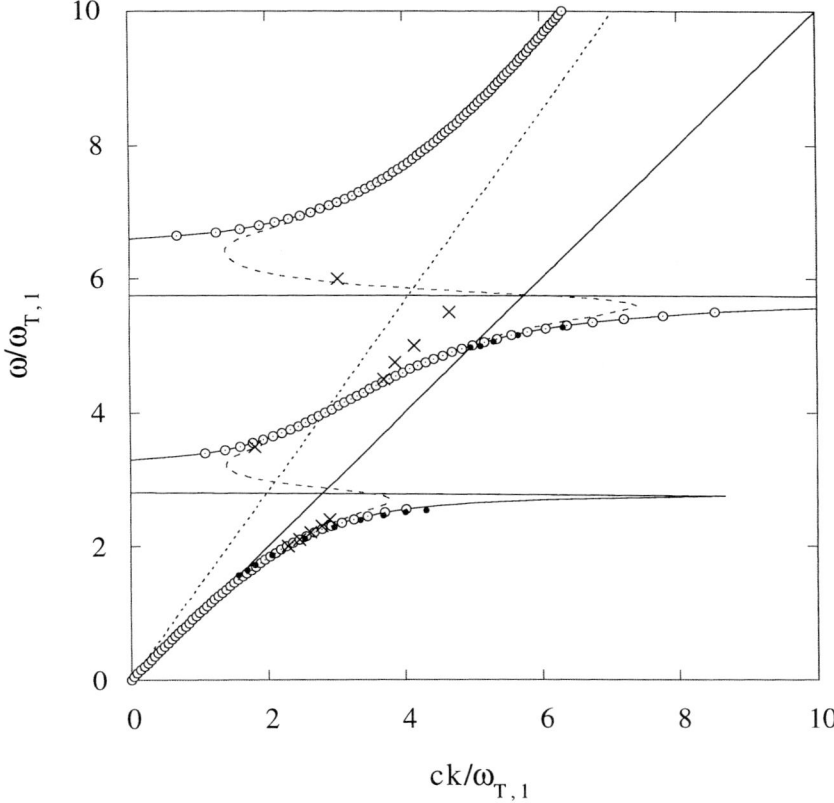

Figure 8.26: The modes as obtained for different conditions and from different experiments. See the text for details.

The solid circles are the results from the calculation of the reflectivity for the *ATR* experiment when the angle of incidence is kept constant and the frequency is scanned. It follows rather closely the undamped modes. The other calculation, where the frequency is kept fixed and the angles are scanned, gives the results represented with crosses in the figure. It has the typical back bending as has the damped mode and represents in that sense closer the damped mode.

Thus in summary we have said that for a dissipative system the modes do not have both ω and k real valued. If ω/k is real valued the solutions (real part of ω as function of real part of k) are very much alike the modes in the non dissipative system and the constant-angle frequency-scan experiment traces out a dispersion curve that is close to this. If we on the other hand keep the frequency real valued the mode (ω as function of real part of k) has a back bending and there are no gaps in the dispersion. The constant-frequency angle-scan experiment traces out a dispersion curve that also has a back bending and is in that sense similar to the real-frequency-value mode.

8.7 Earthquakes, rainbow and optical glory

In this section we briefly discuss two spectacular effects caused by surface modes.

In 1735 a French expedition led by Bouguer had gathered on top of a mountain in the Peruvian Andes. The members of the group saw their shadows projected on the clouds below. What was most remarkable was that they all saw a halo or glory around their heads, consisting of three or more concentric circles, very brightly colored. What was even more remarkable was that each one only saw the glory around his own head and not around the heads of the others. Bouguer seems to be the first to document [12] the observation of the glory phenomenon. The phenomenon involves surface modes but also rainbow modes so before we can explain the glory effect we have to discuss the rainbow modes.

In the visible range of the spectrum the water drops forming the clouds are much larger than the wave length of the light. To a good approximation one can therefore treat the light-drop interaction with geometrical optics. A ray that enters the drop is reflected once in the drop and leaves in the backward direction. An axial ray passes through the center of the drop and is reflected back the same way it came from, that is, it is reflected 180 degrees. A ray slightly off center is reflected slightly less than 180 degrees. When we go further off center the reflection angle decreases further but at a reflection angle of around 137 degrees, for water drops, the angle starts to increase again. This has the effect that the light intensity is strongly enhanced in this particular direction. The angle where this happens is slightly different for different wave lengths, or colors. This explains the rainbow. There are also higher order rainbows where the ray is reflected twice, three times and so on. The second order rainbow is quite often observed. It has the colors in opposite order as compared to the main rainbow. The rainbow phenomenon is quite well described by geometrical optics.

Light that hits the outer edge of the drop may excite surface modes. We said in Section 7.1 that when taking retardation into account the surface modes of a sphere become radiative. This means that the energies of the modes are complex valued; the modes have finite life time. It also means that the modes may be excited by incoming electromagnetic radiation. Interference effects between surface modes and higher order rainbow waves cause the glory effect. The effect is only strong in the immediate vicinity of the backward direction. This is why the people in the French expedition only saw the glory around the shadows of their own heads. One has been able to produce the glory also in laboratory experiments. The interested reader is referred to Reference [13] and references therein for a detailed explanation of these effects.

Surface modes further play an important role in connection with earthquakes. The main pulse detected by the seismograph has taken the shortest path through earth from the place of the earthquake to the detector. Another wave goes from the point of the earthquake to the surface of earth and then follows the surface of earth in the form of a surface mode. This mode has a higher group velocity than the bulk mode and arrives ahead of the main pulse.

References

[1] J. J. Hopfield, Phys. Rev. **112**, 1555 (1958).
[2] P. Halevi, in *Electromagnetic Surface Modes*, Ed. A. D. Boardman, (John Wiley, New York, 1982) Chapter 7.
[3] *Spatial Dispersion in Solids and Plasmas*, in the series: Electromagnetic Waves: Recent developments in research, Vol 1. Editor P. Halevi, North-Holland Amsterdam 1992.
[4] *Surface Polaritons*: Electromagnetic Waves at Surfaces and Interfaces, Editors V. M. Agronovich and D. L. Mills, North-Holland (Amsterdam- New York- Oxford), 1982. (Vol 1 in the series: Modern Problems in Condensed Matter Sciences, Series Editors V.M. Agranovich and A. A. Maradudin.)
[5] *Surface Polariton,* ibid.: *Surface Exciton Polariton from an Experimental Viewpoint,* pp 69-92
[6] *Surface Polariton,* ibid.: *Effects of the Transition Layer and Spatial Dispersion in the Spectra of Surface Polaritons,* pp 187-238.
[7] D. F. Arago, Mém. Cl. Sci. Math. Phys. Inst. France 1, 115 (1811).
[8] *Surface Polariton,* ibid., *Interaction of Surface Polaritons and Plasmons with Surface Roughness,* pp 405-531
[9] *Surface Polariton,* ibid., *Surface Plasmons and Roughness,* pp 331-403
[10] J. A. Sánchez-Gil and A. A. Maradudin, Phys. Rev. **B56**, 1103 (1997); Waves in Random Media **4**, 499-510 (1994)
[11] A. V. Shchegrov, Phys. Rev. **B57**, 4132 (1998)
[12] J. M. Pertner and F. M. Exner, *Meteorologissche Optik* (W. Braumüller, Vienna, 1910).
[13] V. Khare, in *Electromagnetic Surface Modes*, ibid., Chapter 11.

9 Colloids

What is a colloid? What determines the stability of colloids? What is a double layer?

The term colloid was coined in 1861 by Thomas Graham. It means "glue" in Greek. A colloid usually consists of two phases (there are also multi phase colloids); one continuous, or homogeneous, phase in which the other phase is dispersed. The size of the entities, or particles, of the dispersed phase is larger than the size of molecules and small enough for the dispersed phase to stay suspended for a longer period of time. There are no strict boundaries for the size limits. In 1903 Wolfgang Ostwald formulated the official definition of a colloid: *a system containing entities having at least one length scale in between 1nm and 1μm*. For smaller particles there are no distinct boundaries between the phases and the system is considered a solution. For larger entities the particles will fall to the bottom, or float to the top, due to the gravitational force, and the phases are separated. The particle size is in the so-called mesoscopic range in between the macroscopic and microscopic limits. One very important quality of the colloids is the large interfacial area between the dispersed and the continuous phases. This means that interface effects and hence the electromagnetic surface modes, are very important for the properties of the colloids. It costs energy to create this much surface and the particles would clump together if this were not prevented. Usually the particles are charged and hence repel each other.

There are *four states of matter*: solid, liquid, gas and plasma. The plasma colloids are not that common yet so we will not discuss them here. Thus we are in principle left with nine possible colloid types. However there are no gas-in-gas colloids, so we have eight types of common colloid, which we have summarized in Table 9.1.

Table 9.1: The eight different types of common colloid with examples.

Homogenous Phase	Dispersed Phase	Name	Examples
Solid	Solid	Solid Suspension	ruby glass, composites, ceramics, bone
Solid	Liquid	Solid Emulsion	bitumen (asphalt), opal, pearl
Solid	Gas	Solid Foam	expanded polystyrene, isolation foam, pumice
Liquid	Solid	Sol	ink, paint, blood, tooth paste, mud
Liquid	Liquid	Emulsion	milk, mayonnaise, cream
Liquid	Gas	Foam	beer foam, fire extinguisher foam, soap foam
Gas	Solid	Aerosol	smoke, dust
Gas	Liquid	Aerosol	mist, fog, clouds, household sprays

Colloidal properties are utilized in many branches of industry, for example the food industry, the cosmetic industry, the pharmaceutical industry, and the mining industry. The minerals our bodies absorb from food are in the form of colloids, and much of our body-fluids and -tissues themselves can be regarded as colloids. The large surface- or interface-area in colloids makes these systems ideal for *catalytic applications*. There is an huge potential for chemists, biologists and physicists to do cross-disciplinary research in this area.

Solid foam can have very interesting properties. One may make foam out of metals, like aluminum, that is much lighter than the pure metal but still keeps the same strength as the pure metal. This means a large potential for applications.

Aerogels are similar to solid foams. A gel is a sol, that is, a solid phase dispersed in a liquid phase, where the solid phase is lyophilic (defined in Section 9.2). A continuous solid network may form. As an example we may choose gelatin. It behaves as an elastic solid or semi-solid rather than a liquid. The particles have lost their mobility. It is possible, although tricky, to replace the liquid phase with air without damaging the solid component and keep the open structure. The result is an aerogel. These materials are very light. Other names for dried-out gels are *zerogels* or *xerogels*.

Foams have both liquid and solid properties. They can support small static shear like a solid, but can also flow and deform arbitrarily like a liquid. They can be considered complex fluids or soft matter. They are characterized by *soft order*, that is, by microstructures which are both flexible and stable. There are discussions going on about classifying these substances as a fifth state of matter.

9.1 Milk

Let us briefly discuss a colloid that we all have experienced, namely milk. This colloid consists of more than two components. The homogeneous phase is a liquid and the dispersed phase consists of fat droplets and protein particles. The fat dominates the general properties so we can consider the colloid to be an emulsion. Fat is hydrophobic (defined in Section 9.2). Those of you who have lived on a farm know that the raw milk, that is, the milk that is produced by the cow is not very stable. The dispersed phase floats towards the surface and collects there as cream and can be skimmed off. This phase separation continuous and the milk left at the bottom becomes more and more fat free. The milk we buy in the stores has, after the cream has been skimmed off, been treated to become more stable. It has been homogenized which means that the fat droplets have been split up into smaller droplets.

Whipping cream can be viewed as concentrated milk. It is still an emulsion but there is less of the liquid phase than in milk. It contains approximately 40 % fat. When we whip the cream we introduce yet another dispersed component, namely, small air bubbles. The fat droplets are collected at the surface of the air bubbles; the energy is minimized in this way. We get a more or less stable foam.

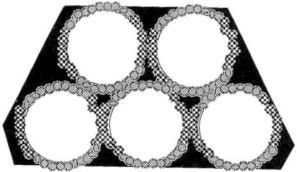

Figure 9.1: The bubbles in whipping cream.

If we continue whipping we break the bubbles and the fat now becomes the homogeneous phase with the water as the dispersed phase. We have a new emulsion where the phases have been inverted. The result is butter. Some of the water has left the material and the fat content is now around 80 %. Butter is also a colloid where the homogeneous phase is fat and the dispersed phase water. If we want to use this butter in preparing a juicy steak for dinner we put it in the frying pan. The butter, or rather the fat, melts and the water evaporates. When the butter has silenced, that is, when all water has evaporated it is time to put in the steak. At this stage the phase separation is complete; the colloid is gone and the fat phase remains.

We have so far completely neglected the proteins. The milk colloid is rather complicated and the proteins probably play an important role in the stability of the colloid. If we boil the milk the skin at the surface is protein that has been separated out. Furthermore, one can really "feel" a difference between the boiled milk after it has cooled again and unboiled milk. The taste is also different.

9.2 Stability of colloids

As we mentioned in Section 4.3, the discovery of the spectacular and very fascinating Casimir effect sprung out of the study of colloid stability. Let us spend some time on this subject in general here. We will return to the subject in more detail in later sections. Colloids are divided into two classes called *lyophobic* and *lyophilic*. In the first class the dispersed medium avoids or dislikes the solvent while in the second it prefers or likes the solvent. If the solvent is water the classes are called *hydrophobic* and *hydrophilic*, respectively. If the solvent is an organic liquid (usually fatty, for example oil) the classes are called *lipophobic* or *oleophobic* and *lipophilic* or *oleophilic*, respectively. A lyophilic colloid can be dispersed by just adding the proper solvent while in the case of lyophobic colloids one has to mechanically force the particles to disperse. A lyophobic colloid is never really thermodynamically stable. Given enough time it will eventually form aggregates. A highly stable lyophobic colloid may look homogeneous for days, weeks and even months; the formation of aggregates goes on all the time but at a slow rate.

Gelatin is an example of a hydrophilic substance. Milk is a hydrophobic colloid where a homogenization process is used to break the dispersed particles into smaller pieces and thereby making the colloid more stable.

The particles in a colloid undergo Brownian motion and collide with each other. If they come close enough to each other they get stuck due to the van der Waals attraction. As this process goes on the particles or aggregates of particles get larger and larger until gravitation leads to phase separation. The aggregates fall to the bottom of the container or float to the top. The formation of aggregates can be prevented or reduced by different mechanisms. If the solvent is properly chosen the particles become charged and are surrounded by a cloud of opposite charge – a so-called *double layer* is formed. This leads to a repulsive part of the potential between two particles. We will return to this later. Another effect is responsible for the hindrance of formation of aggregates in hydrophilic colloids. Water molecules are bound to the surface of the particles and it costs too much energy to remove the water molecules from the polar groups of the particles to let the particles come close enough to bind via the van der Waals forces. This repulsive force is called a *hydration* force. One method to increase the colloid stability is to let surfactants or polymers bind to the surface of the particles and thereby prevent particles to come close together. This leads to a repulsive so-called *steric force.*

The particles in a colloid, where the homogeneous phase is a liquid or gas, have to have the right size or rather be in a specific range of sizes to keep suspended for a reasonable long time. If the particles are too big they fall to the bottom due to the gravitational forces. This is different in the outer space, where we can neglect gravitational effects. Thus the behavior of colloids is different in a gravitational-free environment. If the particles clump together they will sooner or later reach a size where they no longer stay suspended and the colloid breaks down. If the particles bind weakly to each other the process may be reversible; this process is called *flocculation*. If the process is irreversible it is called *coagulation*. In the case of liquid droplets the two droplets usually transform into one, bigger droplet after the coagulation. This is called *coalescence*. We will see below that the interaction potential between two particles often has two minima, one weak at large separation and one deep at small separation. If the temperature is low the particles can bind weakly and reversibly. Increase of the temperature may lead to strong and irreversible binding. Thus flocculation and coagulation can occur in the same system under different circumstances. We may also have another effect if the particles are slightly soluble in the liquid and are in equilibrium with the solution. In that case larger particles grow at the cost of smaller. The net effect is the same. The particle size increases and destabilizes the colloid. This is termed *Ostwald ripening*.

We will now discuss the stability of lyophobic colloids. Examples of lyophobic colloids are: a gold sol, a silver iodide sol, a quartz suspension (in water or in an organic liquid), and the emulsions. The particles are either rigid particles, amorphous or crystalline, or small droplets. The stability of these colloids is very sensitive to electrolytes. As an example of this we may choose milk, where a small amount of citrus acid is enough to destroy the stability and cause coagulation. In general the lyophilic colloids do not show up this sensitivity, although there are exceptions. They are also affected by the electrolyte, but not in the same way.

The dispersed particles in the lyophobic colloids carry an electric charge typically of the size of hundreds of elementary charges. Since the colloid is neutral the corresponding

9.2 Stability of Colloids

opposite charge is in the form of ions in the liquid. Typically a fraction of the surface atoms of the particle is ionized. The *counter ions* in the liquid, that is, the ions with opposite charge to that of the particles, are moving around in thermal motion so at least not all are bound to the particle. They are however not homogeneously distributed throughout the solvent. They are affected by the fields from the charged particles and preferably stay close to the particles and make up an effective screening of them. If we consider one particle separately it is surrounded by an *electric double layer*. One layer of this double layer is made up by the charges in the surface layer of the particle and the second by the neutralizing ions in the liquid. One may to a fairly good approximation treat the surface charge as homogeneously distributed over the surface of the particle. The concentration of charges in the outer layer decreases with distance from the particle and constitutes a so-called *diffuse layer*. The thickness of the double layer is in general of the same order as the size of the particles; it is sometimes larger and sometimes smaller than the size of the particles. As we will see, this formation of a double layer is very important for the stability of the colloid. The double layer is sensitive to electrolytes and also to temperature. This means that the stability of the colloid may be manipulated by adding electrolytes or by changing the temperature. For some colloids the formation and stabilization occur spontaneously when adding the solvent, for others one needs to add small amounts of specific electrolytes. The process in which the dispersed phase is broken up in small particles with the formation of a double layer around each is called *peptization*. The inverse process is called *depeptization*.

When adding a proper electrolyte the ions in the diffuse layer are sometimes neutralized and the stability is reduced, but this effect is not so common. The stability is in most situations changed without affecting the ions bound to the particles or the ions of opposite charge in the solution. We will come back to the explanation later. For each combination of colloid and electrolyte one may determine how much electrolyte is needed to cause flocculation. It turns out that this amount is strongly dependent on the valency of the counter ion. Which the *anions* and *cations* are is much less important. The size of the counter ions has importance, although small; the larger the counter ion the less amount of electrolyte is needed. In the tables below, experimental results clearly demonstrate these effects. Table 9.2 is for negatively charged sols and Table 9.3 for positively charged sols. These results are very old, and many of the values are probably out of date, but the tables can still serve as illustrations of the general trends. The tables are from the classical book by Verwey and Overbeek [1].

Table 9.2: Flocculation values in millimole/liter for negatively charged sols. From Reference [1] p. 8.

Electrolyte	As_2S_3 – sol	Au – sol	AgI – sol
$LiCl$	58	–	–
$LiNO_3$	–	–	165
$NaCl$	51	24	–
$NaNO_3$	–	–	165
KCl	49.5	–	–
KNO_3	–	25	136
$1/2 K_2SO_4$	65.5	23	–
$RbNO_3$	–	–	126
$MgCl_2$	0.72	–	–
$Mg(NO_3)_2$	–	–	2.53
$MgSO_4$	0.81	–	–
$CaCl_2$	0.65	0.41	–
$Ca(NO_3)_2$	–	–	2.38
$SrCl_2$	0.635	–	–
$Sr(NO_3)_2$	–	–	2.33
$BaCl_2$	0.69	0.35	–
$Ba(NO_3)_2$	–	–	2.20
$ZnCl_2$	0.685	–	–
$Zn(NO_3)_2$	–	–	2.50
$UO_2(NO_3)_2$	0.64	2.8	3.15
$AlCl_3$	0.093	–	–
$Al(NO_3)_3$	0.095	–	0.067
$1/2 Al_2(SO_4)_3$	0.096	0.009	–
$La(NO_3)_3$	–	–	0.069
$Ce(NO_3)_3$	0.080	0.003	0.069
$Th(NO_3)_4$	–	–	0.013

Table 9.3: Flocculation values in millimole/liter for positively charged sols. From Reference [1] p. 9.

Electrolyte	Fe_2O_3 – sol	Al_2O_3 – sol
$NaCl$	9.25	43.5
KCl	9.0	46
$1/2 BaCl_2$	9.65	–
KNO_3	12	60
$1/2 Ba(NO_3)_2$	14	–
K_2SO_4	0.205	0.3
$MgSO_4$	0.22	–
$K_2Cr_2O_7$	0.195	0.63
$K_3Fe(CN)_6$	–	0.080
$K_4Fe(CN)_6$	–	0.053

9.3 Formation of the double layer

There are two types of object that are most important and will be treated here. They are objects with flat surfaces and spherical particles. Thus we will consider flat and spherical double layers. The charges bound to the surface of the particle is usually treated as a homogeneous surface charge. This is usually a good approximation especially for metal particles. We will here consider solid particles. Liquid particles have to be treated separately. We will as a first approximation neglect the fact that the charges in the diffuse layer are carried by ions with finite extension; we will treat the charges as point charges. We will furthermore assume that we are in the linear-response regime, that is, where the potential is weak enough for a linear relation between the induced charges and the potential to hold. This is not always the case in reality. If it wasn't for the Brownian motion counter ions would accumulate at the particle surface and completely neutralize the charges of the particle. In the linear-response approximation the charges are distributed according to the *Debye-Hückel screening*. One way to obtain this classical result is to use the *Thomas-Fermi approximation* and take the high-temperature limit of the result. In the high-temperature limit quantum effects go away and the system behaves classically. The dielectric function takes on the form:

$$\varepsilon(q,0) = 1 + q_{DH}^2/q^2 \qquad (9.1)$$

where

$$q_{DH}^2 = \frac{4\pi\beta n_+(Z_+e)^2}{\kappa} + \frac{4\pi\beta n_-(Z_-e)^2}{\kappa} = \frac{4\pi\beta n(Z_+ - Z_-)e^2}{\kappa} \qquad (9.2)$$

is the square of the inverse Debye-Hückel screening length. The densities n, n_+ and n_- are the average density of elementary charges of each sign, the average concentration of positive ions and average concentration of negative ions, respectively. We have assumed that a very small fraction of all charges in the system are bound to the particle so that the solution, to a good approximation, is overall charge neutral. We immediately notice that an ion of higher valency has a larger effect on the screening and hence on the stability of the colloids, than one with lower valency. This is in accordance with the experiments. However, we find that, with the used approximation, there is no difference between the effects on the screening from anions and cations. This is not in accordance with experiments where one finds that the counter ions have much stronger effect on the stability.

9.3.1 Flat double layer

The charge density per unit volume of the surface charge is in the case of a flat double layer

$$\rho(\mathbf{r}) = \sigma\delta(z) \qquad (9.3)$$

where σ is the surface-charge density, the charge density per unit area. Now, if we just place this charge density in a solution there will be screening charges on both sides of the layer. Each side contributes the same amount of charge, and the total amount compensates the charge contained in σ. This is not what we have in the real situation we are trying to model, though. The particle occupies the region to the left, say, of the surface-charge layer and there will be no screening charges there. We may simulate the real system by using the double surface-charge density and neglect what goes on to the left of the surface-charge layer. Alternatively we may introduce another charged layer to the left of the real layer. Then the right-most layer induced charges to the right and the left-most charges to the left. Let us follow the first line of approach. The Fourier transform of the charge density is

9.3 Formation of the Double Layer

$$\rho(\mathbf{q}) = \int d^3r \, e^{-i\mathbf{q}\cdot\mathbf{r}} \rho(\mathbf{r})$$
$$= \int d^3r \, e^{-i\mathbf{q}\cdot\mathbf{r}} 2\sigma\delta(z) \tag{9.4}$$
$$= \int d^3r \, e^{-i\mathbf{q}\cdot\mathbf{r}} 2\sigma\delta(z) = 2\sigma(2\pi)^2 \delta(\mathbf{q}_{//})$$

and the Fourier transform of the resulting potential:

$$\Phi(\mathbf{q}) = \frac{2\sigma(2\pi)^2 \delta(\mathbf{q}_{//}) v_\mathbf{q}/e^2}{\kappa(1+q_{DH}^2/q^2)} = \frac{2\sigma(2\pi)^2 \delta(\mathbf{q}_{//}) 4\pi}{\kappa(1+q_{DH}^2/q^2)q^2} = \frac{4\sigma(2\pi)^3}{\kappa(q_z^2+q_{DH}^2)} \delta(\mathbf{q}_{//}) \tag{9.5}$$

Taking the inverse Fourier transform of this potential results in:

$$\Phi(\mathbf{r}) = \int \frac{d^3q}{(2\pi)^3} e^{i\mathbf{q}\cdot\mathbf{r}} \Phi(\mathbf{q}) = \int_{-\infty}^{\infty} dq_z e^{iq_z z} \frac{4\sigma}{\kappa(q_z^2+q_{DH}^2)}$$
$$= \frac{8\sigma}{\kappa} \int_0^\infty dq_z \cos(q_z z) \frac{1}{(q_z^2+q_{DH}^2)} = \frac{8\sigma}{\kappa q_{DH}} \frac{\pi}{2} e^{-q_{DH} z} = \frac{4\pi\sigma}{\kappa q_{DH}} e^{-q_{DH} z} \tag{9.6}$$

From this follows that the electric field is

$$\mathbf{E} = \frac{4\pi\sigma}{\kappa} e^{-z q_{DH}} \hat{z} \tag{9.7}$$

The induced charge density is

$$\Delta\rho(\mathbf{q}) = \rho(\mathbf{q}) \left[\frac{1}{\varepsilon(\mathbf{q})} - 1\right] = -2\sigma(2\pi)^2 \delta(\mathbf{q}_{//}) \frac{q_{DH}^2}{q^2 + q_{DH}^2} \tag{9.8}$$

From this we find the induced charge density in real space:

$$\Delta\rho(\mathbf{r}) = \int \frac{d^3q}{(2\pi)^3} e^{i\mathbf{q}\cdot\mathbf{r}} \Delta\rho(\mathbf{q}) = -\int_{-\infty}^{\infty} dq_z e^{iq_z z} \frac{\sigma q_{DH}^2}{\pi(q_z^2+q_{DH}^2)}$$
$$= -\frac{\Phi(\mathbf{r}) \kappa q_{DH}^2}{4\pi} = -\sigma q_{DH} e^{-z q_{DH}} \tag{9.9}$$

With the help from Equation (9.2) we can separate the induced charge into contributions from positive and negative charges:

$$\Delta\rho_{\pm}(\mathbf{r}) = -\frac{4\pi\beta n_{\pm}(Z_{\pm}e)^2}{\kappa}\frac{\sigma}{q_{DH}}e^{-zq_{DH}} \qquad (9.10)$$

The approach we have used so far is only valid if the induced charge density of each component is small relative to the average value, that is, if

$$\frac{4\pi\beta|Z_{\pm}e\sigma|}{\kappa q_{DH}}e^{-zq_{DH}} \ll 1 \qquad (9.11)$$

If this is not fulfilled we have to go beyond the linear response approximation.

9.3.2 Spherical double layer

Let us now repeat the procedure in Section 9.3.1 for a spherical particle of radius R and with the surface-charge density:

$$\sigma = \frac{Q}{4\pi R^2} \qquad (9.12)$$

The Fourier transform of the charge density is

$$\begin{aligned}\rho(\mathbf{q}) &= \int d^3 r e^{-i\mathbf{q}\cdot\mathbf{r}}\rho(\mathbf{r}) = \int d^3 r e^{-i\mathbf{q}\cdot\mathbf{r}}\sigma\delta(r-R) \\ &= \int_0^\infty dr 2\pi r^2 \int_{-1}^1 dx e^{-iqrx}\sigma\delta(r-R) \\ &= 2\pi R^2 \sigma \int_{-1}^1 dx e^{-iqRx} = \sigma 2\pi R^2\, 2\sin(qR)/(qR) \\ &= Q\sin(qR)/(qR)\end{aligned} \qquad (9.13)$$

and the Fourier transform of the resulting potential becomes

$$\Phi(\mathbf{q}) = \frac{Q[\sin(qR)/(qR)]v_\mathbf{q}/e^2}{\kappa(1+q_{DH}^2/q^2)} = \frac{Q\sin(qR)4\pi}{\kappa(1+q_{DH}^2/q^2)Rq^3} = \frac{Q4\pi\sin(qR)}{\kappa(q^2+q_{DH}^2)qR} \qquad (9.14)$$

Taking the inverse Fourier transform results in

9.3 Formation of the Double Layer

$$\Phi(\mathbf{r}) = \int \frac{d^3q}{(2\pi)^3} e^{i\mathbf{q}\cdot\mathbf{r}} \Phi(\mathbf{q}) = \int_0^\infty dq \frac{1}{(2\pi)^3} 2\pi q^2 \frac{Q 4\pi \sin(qR)}{\kappa(q^2 + q_{DH}^2)qR} \int_{-1}^1 dx\, e^{iqrx}$$

$$= \int_0^\infty dq \frac{1}{(2\pi)^3} 2\pi q^2 \frac{Q 4\pi \sin(qR)}{\kappa(q^2 + q_{DH}^2)qR} \frac{2\sin(qr)}{qr} \tag{9.15}$$

$$= \frac{2Q}{\pi \kappa R r} \int_0^\infty dq \frac{\sin(qR)\sin(qr)}{(q^2 + q_{DH}^2)}$$

There should be no induced charges inside the particle, which we actually get with our treatment here. If we limit ourselves to small particles, that is, let qR be small the potential should still be realistic, though. Then we have

$$\Phi(\mathbf{r}) \approx \frac{2Q}{\pi \kappa r} \int_0^\infty dq \frac{q \sin(qr)}{(q^2 + q_{DH}^2)} = \frac{2Q}{\pi \kappa r} \int_0^\infty dq \frac{q \sin(qrq_{DH})}{(q^2 + 1)} = \frac{Q}{\kappa r} \exp(-q_{DH} r) \tag{9.16}$$

From this follows that the electric field is

$$\mathbf{E} = \frac{4\pi e^2 \sigma}{\kappa} e^{-r q_{DH}} \hat{\mathbf{r}} \tag{9.17}$$

The induced charge density is

$$\Delta \rho(\mathbf{q}) = \rho(\mathbf{q}) \left[\frac{1}{\varepsilon(\mathbf{q})} - 1 \right] = -Q \frac{q_{DH}^2}{q^2 + q_{DH}^2} \tag{9.18}$$

and in in real space:

$$\Delta \rho(\mathbf{r}) = \int \frac{d^3q}{(2\pi)^3} e^{i\mathbf{q}\cdot\mathbf{r}} \Delta \rho(\mathbf{q}) = -\frac{Q q_{DH}^2}{4\pi r} e^{-r q_{DH}} \tag{9.19}$$

Separated into contributions from positive and negative charges we have

$$\Delta \rho_\pm(\mathbf{r}) = -\frac{4\pi \beta n_\pm (Z_\pm e)^2}{\kappa} \frac{Q}{4\pi r} e^{-r q_{DH}} \tag{9.20}$$

The approach we have used so far is only valid if the induced charge density, for each component, is small relative to the average value, that is, if

$$\frac{4\pi\beta|Z_\pm eQ|}{\kappa 4\pi r} e^{-rq_{DH}} \ll 1 \tag{9.21}$$

Let us now go beyond the linear response regime but still treat the atoms and ions in the solvent as point particles. This means using the simple theory by Gouy [2] and Chapman [3].

9.4 Gouy and Chapman theory

The potential in the diffuse layer is given by Poisson's equation:

$$\nabla^2 \Phi = -4\pi\rho/\kappa \; ; \; \rho = \rho_+ + \rho_- = Z_+ e n_+ + Z_- e n_- \tag{9.22}$$

where $Z_\pm$ are the valencies of the cations and anions taken with signs. The charge distributions depend on the potential so Equations (9.22) and (9.23) must be solved simultaneously. At zero temperature the counter ions will be collected at the surface of the particle. This is prevented for finite temperatures due to the thermal motion of the atoms and ions. The theory assumes that the average concentration of the ions at a point in the diffuse layer is determined by the average potential at that point. The concentration is determined by a Boltzmann factor:

$$n_\pm = \tilde{n}_\pm e^{-Z_\pm e \beta \Phi} \tag{9.23}$$

where $\tilde{n}_\pm$ are the equilibrium concentrations of the ions far away from the particle, where the potential is zero.

We will now limit the treatment to the case of ions of equal valency type. We let

$$Z_+ = -Z_- = Z \tag{9.24}$$

Thus we have

$$\nabla^2 \Phi = \frac{8\pi Z e \tilde{n}}{\kappa} \sinh(Ze\beta\Phi) \tag{9.25}$$

This is a nonlinear equation for Φ.
Let us introduce new variables:

9.4 Gouy and Chapman Theory

$$y = Ze\beta\Phi \; ;$$
$$y_0 = Ze\beta\Phi_0 \; ;$$
$$q_0^2 = \frac{8\pi\tilde{n}Z^2e^2\beta}{\kappa} \; ; \qquad (9.26)$$
$$\gamma = \frac{e^{y_0/2} - 1}{e^{y_0/2} + 1}$$

This simplifies the differential equation and the result is

$$\nabla^2 y = q_0^2 \sinh(y) \qquad (9.27)$$

If y is small this may be linearized with the result

$$\nabla^2 y = q_0^2 y \qquad (9.28)$$

We are back to the linear response regime treated earlier, and the screening is the Debye-Hückel screening, just as before. Near the colloid particles y can be large and we have to keep the full non-linear differential equation. We will first treat flat double layers and then spherical double layers.

9.4.1 Gouy and Chapman theory of a flat double layer

In the case of a flat double layer the potential depends on a single variable, z, the distance from the surface and we introduce the new dimensionless variable

$$\xi = q_0 z \qquad (9.29)$$

The fundamental differential equation, Equation (9.27) turns into

$$\frac{d^2 y}{d\xi^2} = \sinh(y) \qquad (9.30)$$

This equation can be integrated:

$$2\frac{dy}{d\xi}\frac{d^2y}{d\xi^2} = \frac{dy}{d\xi}e^y - \frac{dy}{d\xi}e^{-y}$$
$$\Downarrow$$
$$\frac{d}{d\xi}\left(\frac{dy}{d\xi}\right)^2 = \frac{d}{d\xi}\left(e^y + e^{-y}\right)$$
(9.31)

Now,

$$\left(\frac{dy}{d\xi}\right)^2 = \left(e^y + e^{-y}\right) - 2 \tag{9.32}$$

where we have used that $y = 0$ and $dy/d\xi = 0$ for $\xi = \infty$ to determine the integration constant. Now,

$$\frac{dy}{d\xi} = \sqrt{\left(e^y + e^{-y}\right) - 2} = -\left(e^{y/2} - e^{-y/2}\right) \tag{9.33}$$

and

$$\frac{dy}{e^{y/2} - e^{-y/2}} = -d\xi \tag{9.34}$$

Rearrangements give

$$-d\xi = dy\left[\frac{\frac{1}{2}e^{y/2}}{e^{y/2} - 1} - \frac{\frac{1}{2}e^{y/2}}{e^{y/2} + 1}\right] = d\ln\left[\frac{e^{y/2} - 1}{e^{y/2} + 1}\right] \tag{9.35}$$

Integration leads to

$$-\xi = \ln\left[\frac{e^{y/2} - 1}{e^{y/2} + 1}\right] - \ln\left[\frac{e^{y_0/2} - 1}{e^{y_0/2} + 1}\right] \tag{9.36}$$

and

$$e^{-\xi} = \frac{\left(e^{y/2} - 1\right)\left(e^{y_0/2} + 1\right)}{\left(e^{y/2} + 1\right)\left(e^{y_0/2} - 1\right)} \tag{9.37}$$

This gives

9.4 Gouy and Chapman Theory

$$e^{y/2} = \frac{e^{y_0/2} + 1 + (e^{y_0/2} - 1)e^{-\xi}}{e^{y_0/2} + 1 - (e^{y_0/2} - 1)e^{-\xi}} \tag{9.38}$$

and y can be extracted:

$$y = 2\ln\left[\frac{e^{y_0/2} + 1 + (e^{y_0/2} - 1)e^{-\xi}}{e^{y_0/2} + 1 - (e^{y_0/2} - 1)e^{-\xi}}\right] = 2\ln\left[\frac{1 + \gamma e^{-\xi}}{1 - \gamma e^{-\xi}}\right] \tag{9.39}$$

In Figure 9.2 we show the results for different y_0-values.

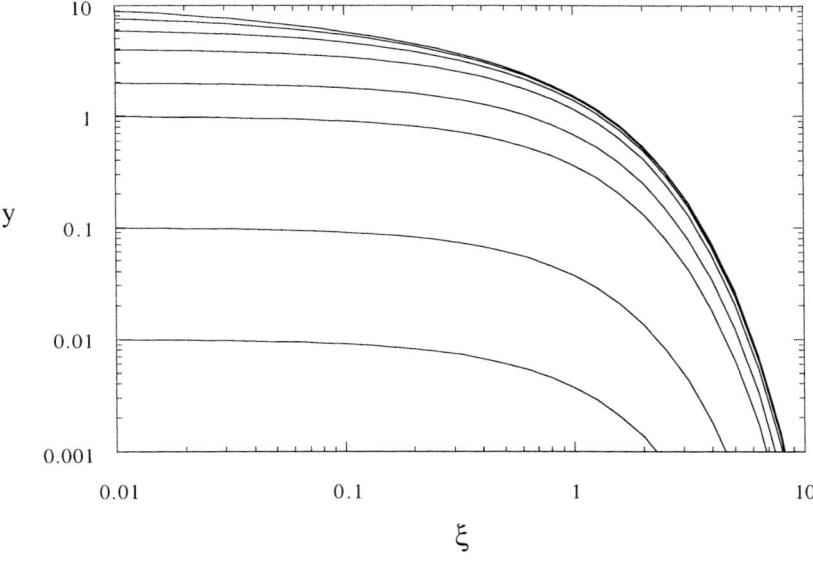

Figure 9.2: Normalized potential y as function of ξ for different y_0 values.

In the limit $y_0 \ll 1$ we may use the approximation: $y = y_0 e^{-\xi}$. This is illustrated in Figure 9.3 for y_0-values 0.01, 0.1, 1, 2, and 4. The dashed curves are the approximation. As can be seen it works well even up to $y_0 = 2$. For smaller y_0-values the approximated results cannot be distinguished from the exact results.

For $y_0 \gg 1$ we may for large distances ($\xi \gg 1$) use the approximation: $y = 4\exp(-\xi)$. This is demonstrated in Figure 9.4 for the y_0-values 6, 8 and 10.

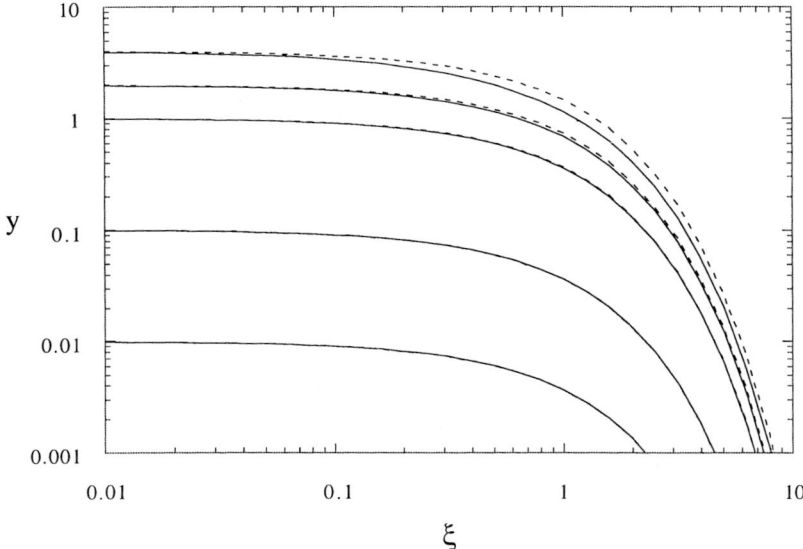

Figure 9.3: Normalized potential y in weak potential approximation as function of ξ for different y_0 values. The dashed curves are the approximation, defined in the text.

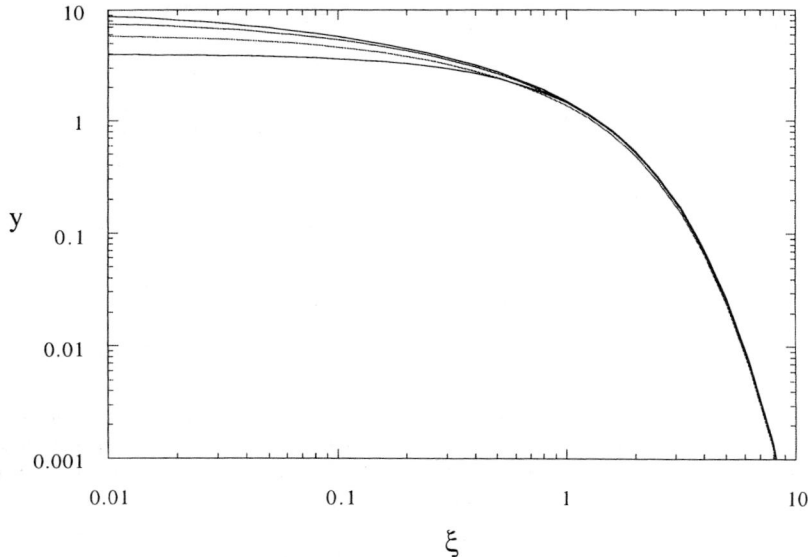

Figure 9.4: Normalized potential y in strong potential approximation as function of ξ for different y_0 values. The upper curve is the large ξ approximation.

9.4 Gouy and Chapman Theory

In the other limit ($\xi \ll 1$) the approximation: $y = y_0 - \xi \exp(y_0/2)$ is valid and is useful for finding the initial slope. This is demonstrated in Figure 9.5.

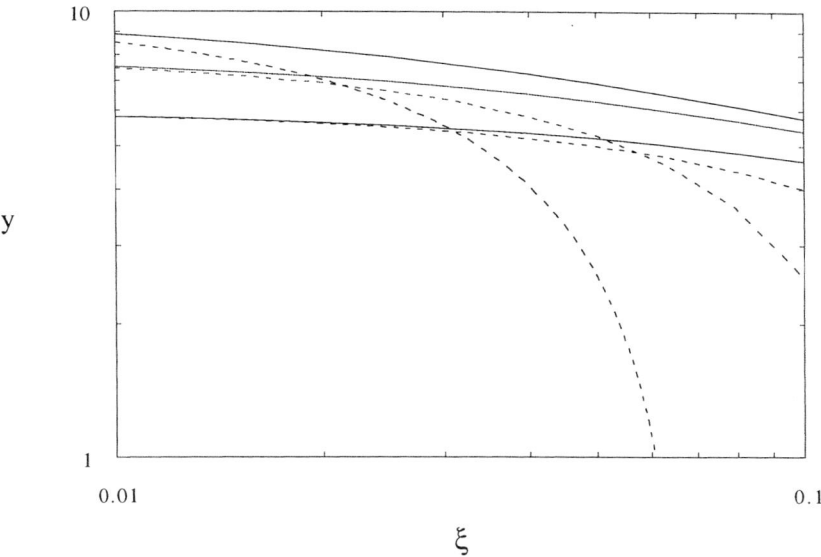

Figure 9.5: Normalized potential y in strong potential approximation as function of ξ for different y_0 values. The dashed curves are the small ξ approximation.

From this slope we may find the capacitance of the double layer, which is an important quantity. It is often used for experimental characterization of the double layer.

For arbitrary y_0 but still with ($\xi \gg 1$) we may use the approximation: $y = 4\gamma \exp(-\xi)$. This is illustrated in Figure 9.6 for the y_0-values 0.01, 0.1, 1, 2, 4, 6, 8, and 10. The dashed curves are the approximations. We find that for small y_0-values the approximation is good for all distances and for large values the approximation approaches the previous approximation and is hence good for large ξ values. We could reformulate this and in a more compact way say that this approximation is good for small y-values.

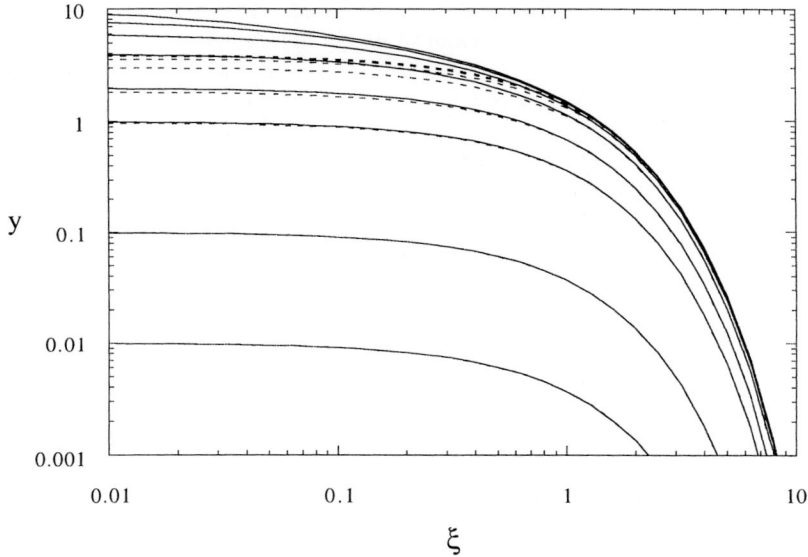

Figure 9.6: Normalized potential y as function of ξ for different y_0 values. The dashed curves are the small-y approximation defined in the text.

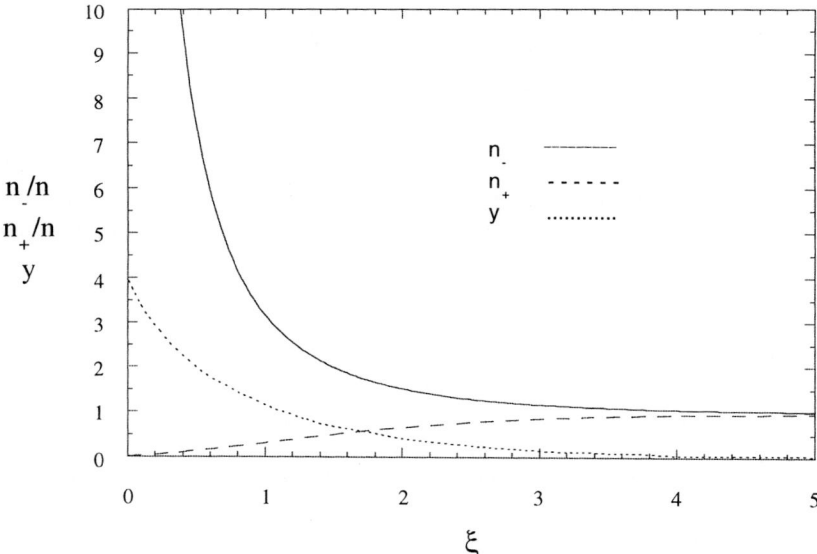

Figure 9.7: Charge distribution in the diffuse layer of a positively charged surface as function of ξ.

9.4 Gouy and Chapman Theory

Figure 9.7 shows the charge distribution in the diffuse layer of a positively charged surface. As is clearly seen the charge density in the layer is dominated by the counter ions. The total surface-charge density (the surface charge per unit area, or minus the charge of the diffuse layer per unit area) is

$$\sigma = -\int_0^\infty dz \rho(z) = \frac{\kappa}{4\pi} \int_0^\infty dz \frac{d^2\Phi}{dz^2} = -\frac{\kappa}{4\pi} \frac{d\Phi}{dz}\bigg|_{z=0}$$

$$= -\frac{\kappa}{4\pi} \frac{d\Phi}{dy} \frac{dy}{dz}\bigg|_{z=0} = -\frac{\kappa}{4\pi} \frac{d\Phi}{dy} \frac{d\xi}{dz} \frac{dy}{d\xi}\bigg|_{\xi=0}$$

$$= -\frac{\kappa}{4\pi} \frac{1}{Ze\beta} \sqrt{\frac{8\pi\beta\tilde{n}Z^2 e^2}{\kappa}} \frac{dy}{d\xi}\bigg|_{\xi=0} \qquad (9.40)$$

$$= -\sqrt{\frac{8\tilde{n}\kappa}{2\beta\pi}} \frac{dy}{d\xi}\bigg|_{\xi=0} = -\sqrt{\frac{\tilde{n}\kappa}{2\beta\pi}} [-2\sinh(y/2)]\bigg|_{\xi=0}$$

$$= \sqrt{\frac{2\tilde{n}\kappa}{\beta\pi}} \sinh(y_0/2)$$

For small y_0-values there is a linear relation between the charge in the double layer and the potential across the layer. This means that the double layer behaves as an electric condenser:

$$\sigma \approx \sqrt{\frac{2\tilde{n}\kappa}{\beta\pi}} y_0/2 = \frac{\kappa}{4\pi} \sqrt{\frac{8\pi\tilde{n}Z^2 e^2 \beta}{\kappa}} \Phi_0 = \frac{\kappa q_0}{4\pi} \Phi_0 \qquad (9.41)$$

with a constant (independent of the potential) capacitance:

$$C = \frac{\kappa q_0}{4\pi} A = \frac{\kappa}{4\pi} \frac{A}{d} \qquad (9.42)$$

The double layer behaves as a parallel plate capacitor with a plate separation of $1/q_0$. For higher potentials the capacitance increases with the potential. Capacitance measurements are very useful as diagnostic tools; to measure the valency and concentration of the electrolyte and other characteristics of the double layer. The planar double layer treated here is also of importance in applications outside the field of colloids: sensors submerged in water or other liquids; biological implants; electrodes in electrolyte batteries.

We have studied solid-liquid interfaces. In emulsions and also in soft tissues in biological systems one has liquid-liquid interfaces. In these cases the double layer consists of two diffuse layers, one on each side of the interface. The double layer potential will be divided between both liquid phases. It is fairly easy to generalize the results we have obtained to these cases. We leave this to the reader.

9.4.2 Gouy and Chapman theory of a spherical double layer

In the case of a spherical double layer the potential will be a function of a single variable, r. The fundamental differential equation, Equation (9.27), is

$$\nabla^2 y = \frac{1}{r^2}\frac{d}{dr}\left(r^2 \frac{dy}{dr}\right) = q_0^2 \sinh(y) \tag{9.43}$$

It can no longer be solved analytically. The linearized version, valid for small potentials,

$$\frac{1}{r^2}\frac{d}{dr}\left(r^2 \frac{dy}{dr}\right) = q_0^2 y \tag{9.44}$$

can be solved. We do not benefit from studying y instead of Φ in this case, so we have

$$\frac{1}{r^2}\frac{d}{dr}\left(r^2 \frac{d\Phi}{dr}\right) = q_0^2 \Phi \tag{9.45}$$

For a sphere of radius a the solution is

$$\Phi = \Phi_a \frac{a}{r} e^{-q_0(r-a)} \tag{9.46}$$

or expressed in terms of the charge of the particle as

$$\Phi = \frac{Q}{\kappa r}\frac{e^{-q_0(r-a)}}{1+q_0 a} \tag{9.47}$$

This result works well apart from close to the particle surface. However, this region is much less important for spherical particles than it was for flat ones since the volume of the region close to the particle is here relatively much smaller. This also means that the valency of the ions in the solution is less important for spherical particles than for flat ones. Now, if $q_0 a > 1$, that is, if the size of the particle is larger than the width of the double layer, the Debye-Hückel results are no longer of much use; one has to solve the fundamental differential equation numerically or resort to the results for flat double layers. In the study of flocculation in colloids the electrolyte concentration is usually so large that one is in this range.

9.5 Stern's theory of a flat double layer

So far we have neglected the finite extension of the ions. This has the effect that the concentration can become unrealistically high close to the surface. The theory cannot be realistic if a major part of the charges in the diffuse layer is found within atomic distances from the surface. Another effect, not taken into account, is the van der Waals attraction of the ions to the surface, that should come into play at very short distances. This means that ions should more or less be adsorbed to the surface. We could imagine that we have three regions: the first is the surface or wall with its surface charge; the second is a region where the counter ions are more or less stuck to a fixed position; the third is a diffuse layer as before.

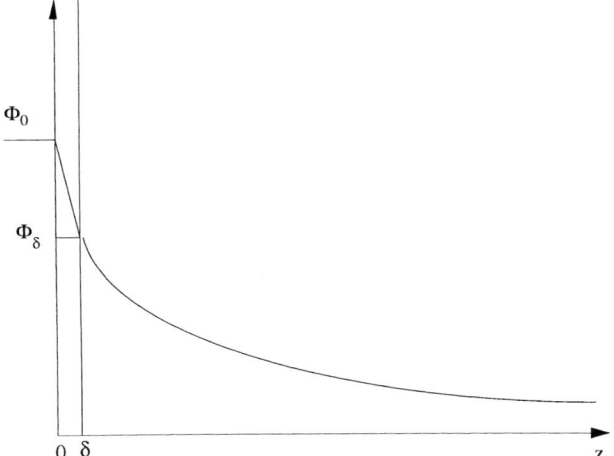

Figure 9.8: Schematic potential in Stern's theory.

Stern [4] tried to model the situation by introducing a second surface-charge layer of opposite sign a distance δ from the wall; the distance is often used as a parameter in the model but it should be of the order of the ionic radius. He derived a theory where the ions adsorbed to this layer are in equilibrium with the ions far from the surface. We will not go into any details here. The potential in Stern's theory is schematically shown in Figure 9.8.

Between the wall and the so-called *Stern-layer* the potential varies linearly; further out there is a diffuse layer for which the Gouy-Chapman theory can be applied; this layer is often called the *Gouy-layer*. The Stern-layer has the effect that the effective particle charge is reduced so that the Gouy-Chapman theory can be applied. The higher the electrolyte concentration the larger is the potential drop in the Stern layer and the potential drop in the Gouy layer is still small enough for the theory to work. Stern's theory can also be considered an improvement upon the Gouy-Chapman theory in that it explains that the capacitance does not increase indefinitely with increasing electrolyte concentration.

Experimentally it saturates at around $10\mu F/cm^2$ while in the Gouy-Chapman theory it would rise indefinitely to values more than ten times the experimental ones.

9.6 The ζ-potential

Apart from the potential at the particle surface, Φ_0, and the Stern potential, Φ_δ, at the Stern plane one frequently talks about another potential, the zeta-potential, ζ, which is the potential at the so-called shear plane.

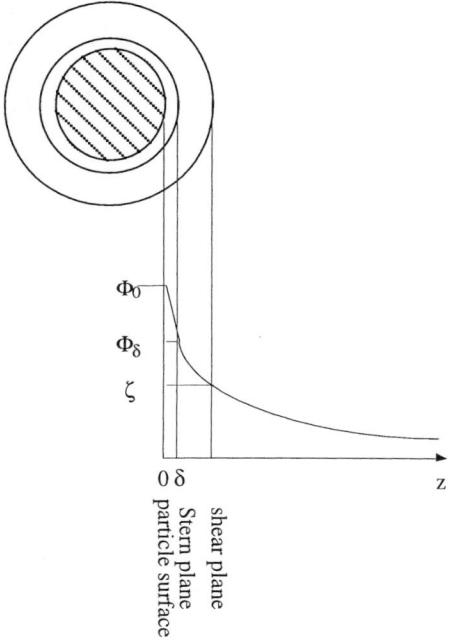

Figure 9.9: Shear plane and ζ-potential.

The size of this potential is often close to, but smaller than, the Stern-potential and depending on the system it can be much smaller. When an electric field, in a so-called electrophoresis experiment, is applied to the system the charged particles are made to move relative the liquid. Some of the liquid atoms or molecules closest to the particle are dragged along and there is a slip-plane separating these from the rest of the liquid. This is the so-called shear plane, defining the ζ-potential.

9.7 Interaction energy and force between objects with double layers

In this section we will consider the free energy of the double layer and the change in the free energy when two objects are brought together. This change is equal to the potential between the objects and results in a force. The separation dependence of this potential is very crucial for the stability of the colloid. We will study flat surfaces and spherical particles. We will follow Verwey and Overbeek (VO) [1] closely. First we present a general discussion of methods to calculate the free energy, then we treat the simplest case of flat surfaces and finally we treat spherical particles.

The formation of the double layer occurs spontaneously. In equilibrium there is an excess of one of the ionic species in the surface. There is obviously a chemical preference of these ions for the surface. If we first imagine the system before the double layer has been formed, there is a chemical free energy difference, $\Delta\mu$, for that ion type in the solution versus in the surface. This is illustrated in the first part of Figure 9.10. The hatched region indicates states below the chemical potential for the ionic species in question.

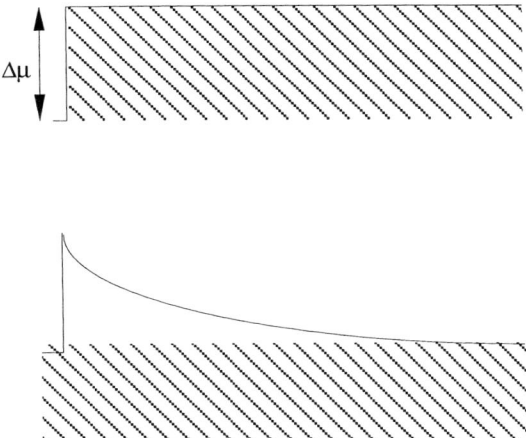

Figure 9.10: System before and after the double layer has been formed.

In the second part of the figure we show what has happened when equilibrium has been reached. Some ions have been adsorbed by the surface and there has been a potential built up so that the chemical potential for the ions is constant throughout the sample. At equilibrium we have

$$Ze\Phi_0 = \Delta\mu \tag{9.48}$$

The free energy, $\mathfrak{F}$, can be calculated in different ways. We will use two different methods here. We make the basic assumption that the double layers make up a small part of the system. We keep the temperature constant, that is, we assume that the system is in contact with a heat bath. The free energy is the work we perform on the system when we gradually build up the surface charge, σ, by taking the ions from a long distance to the surface and putting them at the surface. All other ions of same type or with opposite charge adjust themselves to the change in potential when we make this manipulation. They gain energy in doing so and this energy gain results in kinetic energy that leaves the system in the form of heat. When dealing with the free energy we do not have to bother with this energy. The free energy has in this treatment two parts; one is chemical, $\mathfrak{F}_\mu$, a negative contribution when the ion is placed in the wall; the other is electrical, $\mathfrak{F}_e$, work needed to overcome the potential:

$$\mathfrak{F} = \mathfrak{F}_\mu + \mathfrak{F}_e = -\sigma\Phi_0 + \int_0^\sigma \Phi_0 d\sigma' \tag{9.49}$$

These energies are per unit area. Now, $\mathfrak{F}_e = \mathfrak{F}_e(\sigma)$ which means that $d\mathfrak{F}_e = \Phi_0 d\sigma$. From this follows that $\mathfrak{F} = \mathfrak{F}_e - \sigma\Phi_0 \Rightarrow d\mathfrak{F} = -\sigma d\Phi_0$ and

$$\mathfrak{F} = -\int_0^{\Phi_0} \sigma' d\Phi_0' \tag{9.50}$$

In the special case of weak potential, when there is a linear relation between the surface charge and the potential, we have

$$\mathfrak{F} = -\tfrac{1}{2}\sigma\Phi_0 \tag{9.51}$$

and the electric work costs half the chemical energy gain. In other cases the net gain is less. There is a gain though, otherwise the double layer would not form spontaneously.

With this method we have derived the free energy by just considering the formation of the surface charge. In our simple expressions we have assumed a constant surface-charge density. In more complicated situations we do not have that, but this causes no big problems. We just generalize the results in the following way:

$$\mathfrak{F} = -\int_S dS \int_0^{\Phi_0} q' d\Phi_0' \tag{9.52}$$

where $q'dS$ is the charge of a surface element dS.

The problem with this method is that sometimes we do not have a simple relation between the potential and the surface charge. It can be useful to have a complementary method. We present one here. In this method we study the interaction energy due to all charges in the double layer and introduce a coupling constant integration, simulating that the

9.7 Interaction Energy and Force Between Objects with Double Layers

interactions are turned on adiabatically. We have discussed how this coupling constant comes about earlier in Chapter 3. Also the chemical shift at the surface, which in turn is the result of the interaction with some charged particles in the wall, is scaled simultaneously. In this approach we start with the surface charges in place at the outset but the interaction is turned off. This means that these charges will not have any effect on the results. We get:

$$\widetilde{\mathfrak{F}} = \int_0^1 \frac{d\lambda}{\lambda} \int\int\int dxdydz \Phi'(x,y,z,\lambda) \rho'(x,y,z,\lambda) \tag{9.53}$$

This coupling constant is defined in a slightly different way than we are used to. Usually we make the replacement $e^2 \Rightarrow \lambda e^2$ everywhere. Here, we make the replacement $e \Rightarrow \lambda e$ instead. The result is equivalent, except that the final result for the free energy is a factor of two smaller. This has been compensated above by a multiplication by a factor of two. We only need to include those ions which are present in excess in the solution and carry the liquid charge of the double layer. To be noted is that here the integration is over the whole system, while in the other method the integration was over the surface only. The two methods seem to calculate completely different things but one can show that they give the same result and are equivalent.

When two particles, each surrounded by a double layer, are brought together the free energy changes. We will in most cases assume that the potential at the surfaces are unchanged, that the surface charges change correspondingly and that there is thermodynamic equilibrium. The Brownian motion bringing the particles into contact is usually slow enough for the much quicker ions to have time to adjust to the potential changes. Sometimes in rare occasions, however, the surface charges have to overcome activation energies before they can leave the particle surface. In these cases it is more realistic to assume that the charges stay fix and the potentials change. In this case the system is not in equilibrium. It turns out that the force between the particles caused by the overlap of the double layers is the same whichever case we have.

In the case of constant charge we have a change in the electric part of the free energy, only, and the pressure or force per unit area is

$$p = \frac{Force}{unit\ area} = -\frac{\partial \widetilde{\mathfrak{F}}_e}{\partial d} = -\left(\frac{\partial}{\partial d}\right)_\sigma \int_0^\sigma \Phi_0' d\sigma' = -\int_0^\sigma \frac{\partial \Phi_0'}{\partial d} d\sigma' \tag{9.54}$$

In the other case, of constant potential we have

$$p = \frac{Force}{unit\ area} = -\frac{\partial(\mathfrak{F}_\mu + \mathfrak{F}_e)}{\partial d} = -\left(\frac{\partial}{\partial d}\right)_{\Phi_0}\left\{-\sigma\Phi_0 + \int_0^\sigma \Phi_0' d\sigma'\right\}$$

$$= \Phi_0\left(\frac{\partial\sigma}{\partial d}\right)_{\Phi_0} - \left\{\frac{\partial}{\partial\sigma}\int_0^\sigma \Phi_0' d\sigma'\right\}\left(\frac{\partial\sigma}{\partial d}\right)_{\Phi_0} - \int_0^\sigma \frac{\partial\Phi_0'}{\partial d} d\sigma' \qquad (9.55)$$

$$= \Phi_0\left(\frac{\partial\sigma}{\partial d}\right)_{\Phi_0} - \Phi_0\left(\frac{\partial\sigma}{\partial d}\right)_{\Phi_0} - \int_0^\sigma \frac{\partial\Phi_0'}{\partial d} d\sigma' = -\int_0^\sigma \frac{\partial\Phi_0'}{\partial d} d\sigma'$$

We find that the expressions for the force are equal, independent of what happens at a small variation in the separation between the particles.

9.7.1 Interaction between two flat double layers

We will use the Gouy-Chapman theory. The derivations of the quantities of importance for the interaction between two double layers are cumbersome. We will not perform the calculations in detail; we will give the starting expressions and the results. For the interested reader we refer to VO [1] and references therein.

When we bring the plates in Figure 9.11 together so that their double layers overlap the potential at the plates, Φ_0, stays constant and is not increasing. In order for this to be possible charges must leave the surfaces.

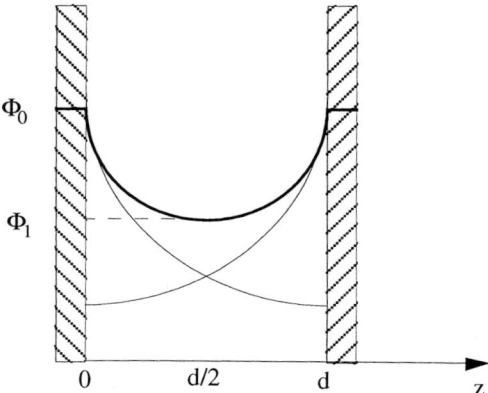

Figure 9.11: Overlapping flat double layers.

9.7 Interaction Energy and Force Between Objects with Double Layers

The basic differential equation in the Gouy-Chapman theory is summarized here:

$$\nabla^2 \Phi = -4\pi \rho / \kappa \; ;$$
$$\nabla^2 \Phi = \frac{8\pi Ze\tilde{n}}{\kappa} \sinh(Ze\beta\Phi) \; ; \tag{9.56}$$
$$\nabla^2 y = q_0^2 \sinh(y)$$

The surface-charge density is proportional to the slope of the potential at the surface:

$$\sigma = -\int_0^{d/2} \rho \, dz = -\frac{\kappa}{4\pi}\left(\frac{d\Phi}{dz}\right)_0 = -\frac{\kappa}{4\pi}\left(\frac{d\Phi}{dz}\right)_0 = -\frac{\kappa}{4\pi}\frac{q_0}{Ze\beta}\left(\frac{dy}{d\xi}\right)_0 \tag{9.57}$$

We have earlier, in Equation (9.31,) obtained:

$$\frac{d}{d\xi}\left(\frac{dy}{d\xi}\right)^2 = \frac{d}{d\xi}\left(e^y + e^{-y}\right) \tag{9.58}$$

For a single surface we solved this with the boundary condition that the slope vanishes at infinite separation. Now, we have a different boundary condition. The slope vanishes at $z = d/2$ where the potential equals $\Phi = \Phi_1$ or $y = y_1$. Thus we have

$$\left(\frac{dy}{d\xi}\right)^2 = \left(e^y + e^{-y}\right) - \left(e^{y_1} + e^{-y_1}\right) \tag{9.59}$$

and

$$\begin{aligned}\sigma &= -\frac{\kappa}{4\pi}\frac{q_0}{Ze\beta}\left(\frac{dy}{d\xi}\right)_0 = -\frac{\kappa}{4\pi}\frac{q_0}{Ze\beta}\sqrt{\left(e^{y_0} + e^{-y_0}\right) - \left(e^{y_1} + e^{-y_1}\right)} \\ &= -\frac{\kappa}{4\pi}\frac{q_0}{Ze\beta}\sqrt{2\cosh(y_0) - 2\cosh(y_1)}\end{aligned} \tag{9.60}$$

The charge and hence the slope of the potential at the surface stay unchanged until the surfaces are very close to each other. To illustrate this we study a high potential system with $y_0 = 10$. When the double layers start to overlap, that is when $d=2/q_0$, as much as 99.99 % of the full charge is still in the surface. The potential in the middle between the layers, for this system, is given in Figure 9.12. It was obtained by numerically integrating Equation (9.59). The corresponding variation of surface-charge density is depicted in Figure 9.13.

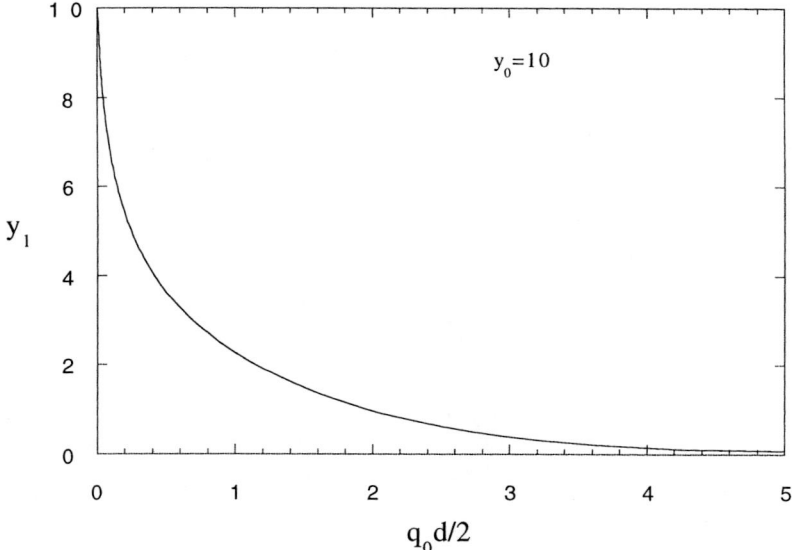

Figure 9.12: Potential in the middle between the layers for a high potential system, $y_0=10$.

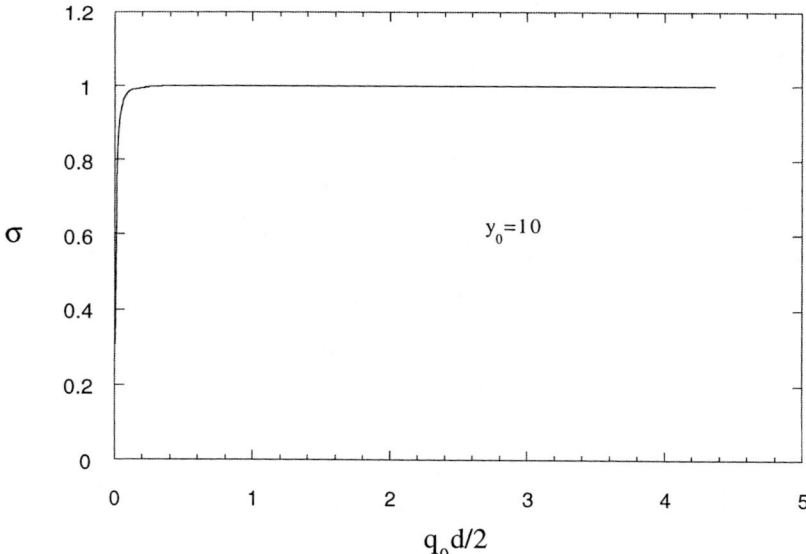

Figure 9.13: Variation of surface-charge density for a high potential system, $y_0=10$.

9.7 Interaction Energy and Force Between Objects with Double Layers

The free energy is obtained with the second method discussed earlier, that is,

$$\mathfrak{F} = \int_0^1 \frac{d\lambda}{\lambda} \int\int\int dx\,dy\,dz\, \Phi'(x,y,z,\lambda)\rho'(x,y,z,\lambda) \tag{9.61}$$

We should remember that all electron charges in the potential and in the charge density are multiplied by the coupling constant. The result is the following:

$$\mathfrak{F} = -\frac{2n}{q_0\beta}\left\{ \frac{q_0 d}{4}\left(3e^{y_1} - 2 - e^{-y_1}\right) + 2\sqrt{2\cosh(y_0) - 2\cosh(y_1)} \right.$$

$$\left. + 2\int_{y_1}^{y_0} dy \frac{e^{-y} - e^{y_1}}{\sqrt{2\cosh(y) - 2\cosh(y_1)}} \right\} \tag{9.62}$$

This energy is the free energy per unit surface of one of the layers. One finds for the energy at infinite separation the result:

$$\mathfrak{F}_\infty = -\frac{2n}{q_0\beta}\left[4\cosh\left(\frac{y_0}{2}\right) - 4\right] \tag{9.63}$$

The potential between the plates is

$$V_R = 2(\mathfrak{F} - \mathfrak{F}_\infty) \tag{9.64}$$

It is shown in Figure 9.14. The used data are taken from Table XI of VO [1]. The potential is repulsive for all distances. If we plot the potential on a logarithmic plot, Figure 9.15, we find that it decreases exponentially with separation, to a good approximation. For large potentials it decays faster than exponentially at small separations while the converse is true for weak potentials. An approximation that seems to work well even outside its range of validity, which is for weak interactions, is the following:

$$V_R \approx \frac{32n}{\beta q_0}\gamma^2[1 - \tanh(q_0 d/2)]; \quad \gamma = \frac{e^{y_0/2} - 1}{e^{y_0/2} + 1} \tag{9.65}$$

We will use this approximation extensively in connection with stability calculations later.

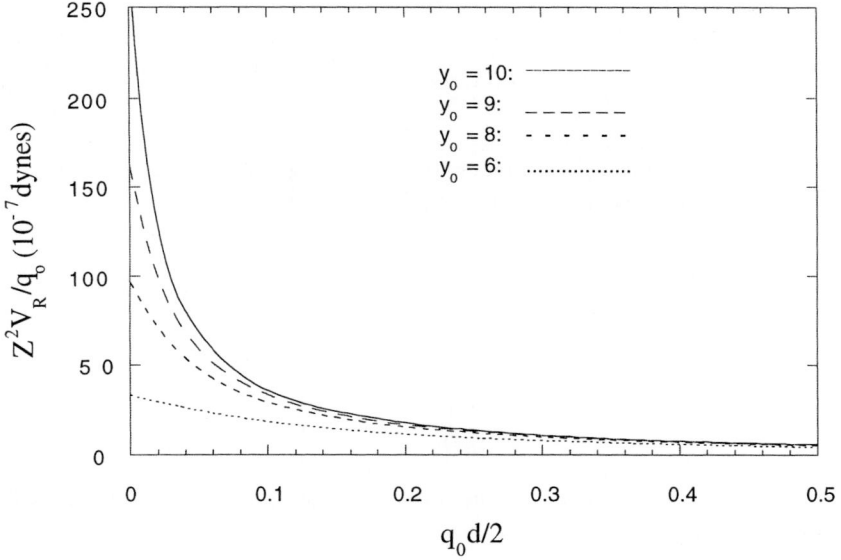

Figure 9.14: Repulsive potential between the layers.

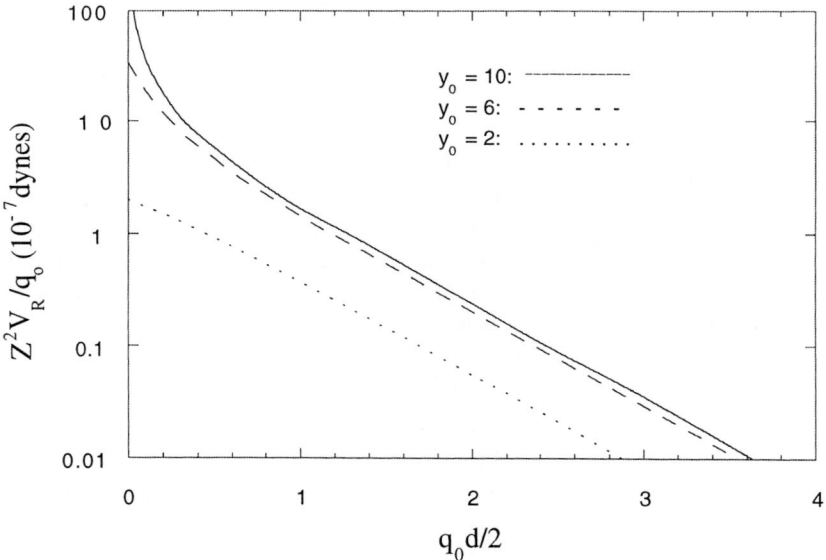

Figure 9.15: Same as Figure 9.14 but on a logarithmic plot and for another choice of y_0-values.

9.7.1.1 Total potential between two layers

We have so far discussed the repulsive potential due to the double layer. There is also an attractive contribution to the potential from the van der Waals interaction between the plates. We have in Section 5.12 derived an approximate expression for this interaction for the configuration in Figure 9.16.

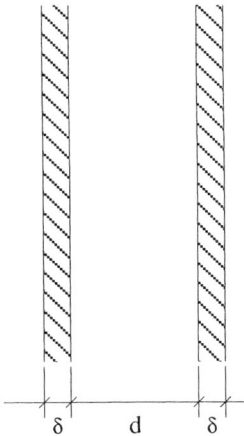

Figure 9.16: Plates studied in the text.

The potential per unit area is according to Equation (5.91)

$$V(d) = -\frac{A}{12\pi}\left[\frac{1}{d^2} + \frac{1}{(d+2\delta)^2} - \frac{2}{(d+\delta)^2}\right] \tag{9.66}$$

where

$$A = \frac{3\pi^2 n^2 \alpha(0)^2 \hbar\omega_0}{4} \tag{9.67}$$

is the so-called *Boer-Hamaker* constant of the material. When the plates are in a medium, the liquid in the present case, this constant consists of three terms:

$$A = A_{11} + A_{22} - 2A_{21} \tag{9.68}$$

where

$$A_{11} = \frac{3\pi^2 n^2 \alpha_1(0)^2 \hbar\omega_{0,1}}{4} \;;$$

$$A_{22} = \frac{3\pi^2 n^2 \alpha_2(0)^2 \hbar\omega_{0,2}}{4} \;; \tag{9.69}$$

$$A_{21} = \frac{3\pi^2 n^2 \alpha_1(0)\alpha_2(0)\hbar\omega_{0,1}\hbar\omega_{0,2}}{2(\hbar\omega_{0,1} + \hbar\omega_{0,2})}$$

The force per unit area is

$$F = -\frac{A}{6\pi}\left[(d)^{-3} + (d+2\delta)^{-3} - 2(d+\delta)^{-3}\right] \tag{9.70}$$

This force is always attractive so we will add an index A to this force and corresponding potential from now on. The two limits for the potential for thin and thick plates are

$$V_A \approx \begin{cases} -\dfrac{A\delta^2}{2\pi}\dfrac{1}{d^4} \;; d \gg \delta \\ -\dfrac{A}{12\pi}\dfrac{1}{d^2} \;; d \ll \delta \end{cases} \tag{9.71}$$

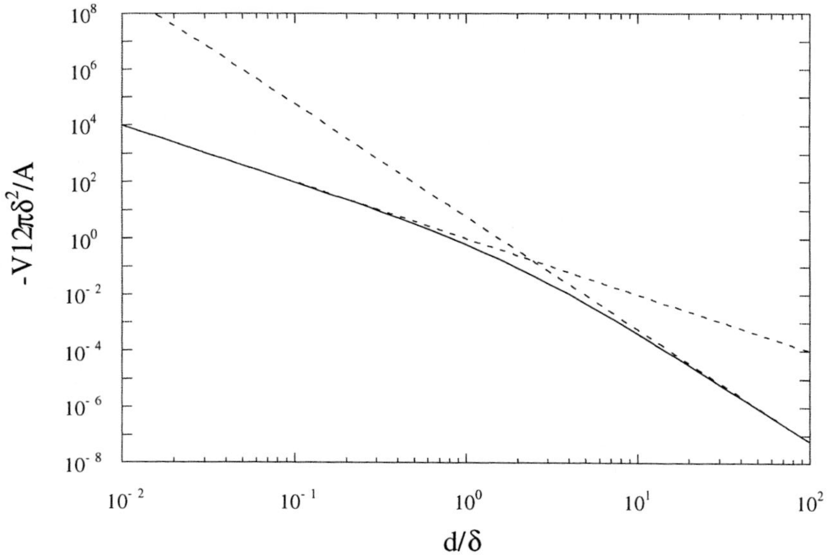

Figure 9.17: Attractive potential between plates and the asymptotic limits.

9.7 Interaction Energy and Force Between Objects with Double Layers

These limits (dashed curves) are given together with the full result in Figure 9.17. For very large separations we might need to take retardation into account. In the retarded or Casimir limit the potential is according to Equation (5.94)

$$V_A(d) = -\frac{23\hbar c n^2 \alpha(0)^2}{120}\left[(d)^{-3} - 2(d+\delta)^{-3} + (d+2\delta)^{-3}\right] \qquad (9.72)$$

The constant in front is in the presence of a medium modified accordingly, that is,

$$V_A(d) = -\frac{23\hbar c n^2 [\alpha_1(0) - \alpha_2(0)]^2}{120}\left[(d)^{-3} - 2(d+\delta)^{-3} + (d+2\delta)^{-3}\right] \qquad (9.73)$$

where the indices *1*, and *2* refer to the plates and the medium, respectively.

9.7.1.2 Stability conditions

In this section we study how the stability of the colloid depends on the parameters A, q_0, and y_0. Each figure is for a choice of the two first parameters. The curves in each figure are for the y_0-values 2, 4, 6, 8, and 10, counting from below. We have used the van der Waals interaction for the attractive part of the potential and the simplified double layer potential, given in Equation (9.65), for the repulsive part. In Figure 9.18, A has the value 2×10^{-12} ergs which is probably a most reasonable value.

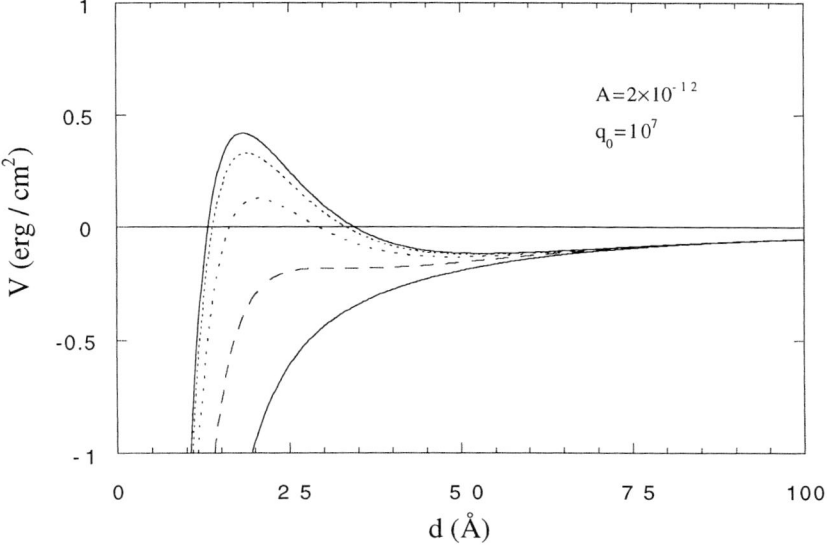

Figure 9.18: Total potential between plates for different potential strengths.

We have assumed T = 300K and that the liquid is water. The electrolyte ion density is given by the parameter q_0, which is chosen to be 1×10^7 cm^{-1}. We see that for a strong enough potential there is a shallow minimum for large distances and a barrier for smaller distances, preventing the particles to come into contact and coagulate. For increasing potential the minimum becomes more shallow and moves towards larger separations.

In Figure 9.19 we have kept the concentration unchanged but increased the Hamaker constant. We can not change this constant, but its value is experimentally uncertain, so we have to find out how sensitive the results are to its value.

We find that the barrier disappears completely and there is no stability. At smaller values for the Hamaker constant the barrier is more pronounced, as can be seen in Figure 9.20, but the outer minimum is now more shallow.

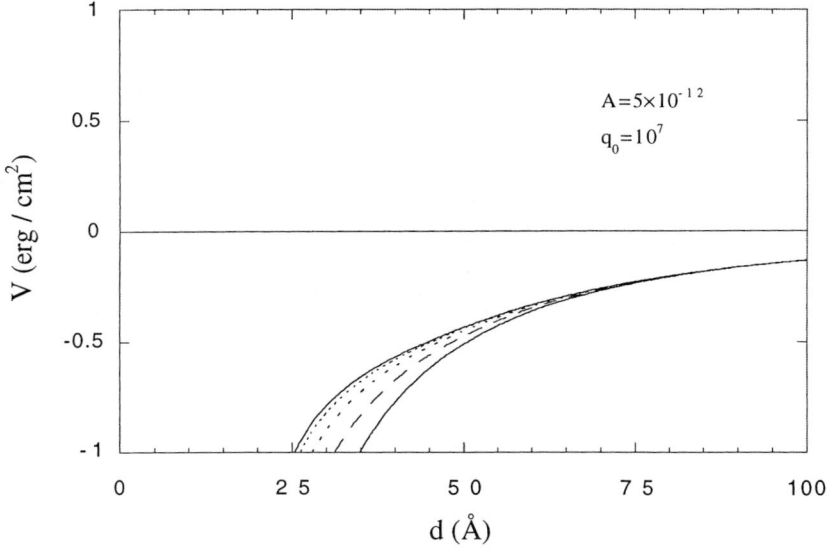

Figure 9.19: Same as Figure 9.18 but for a larger Hamaker constant.

9.7 Interaction Energy and Force Between Objects with Double Layers

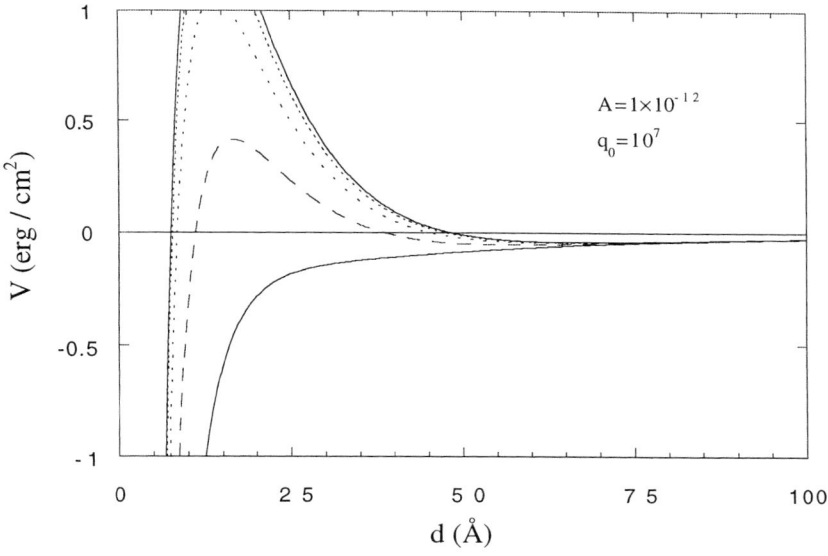

Figure 9.20: Same as previous two figures but for a smaller Hamaker constant.

Figures 9.21 and 9.22 show that decreasing the concentration leads to weaker barriers and more shallow outer minima that are shifted towards larger separations.

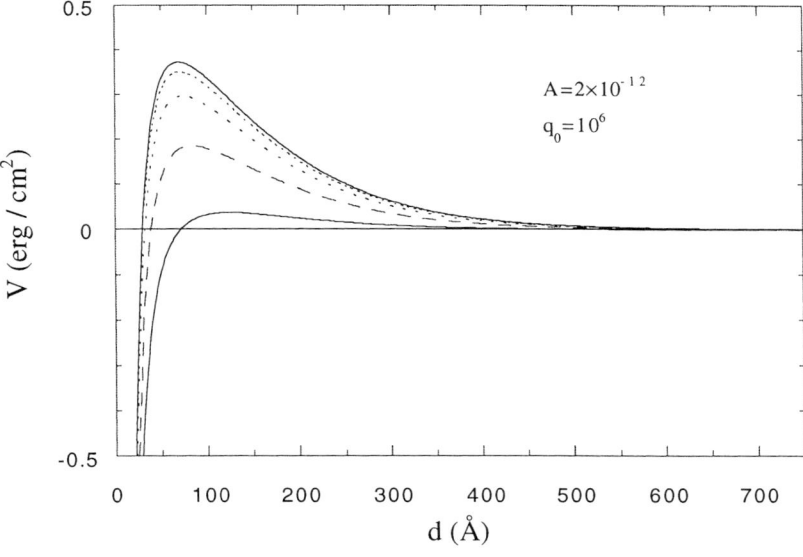

Figure 9.21: Same as Figure 9.18 but for a lower concentration.

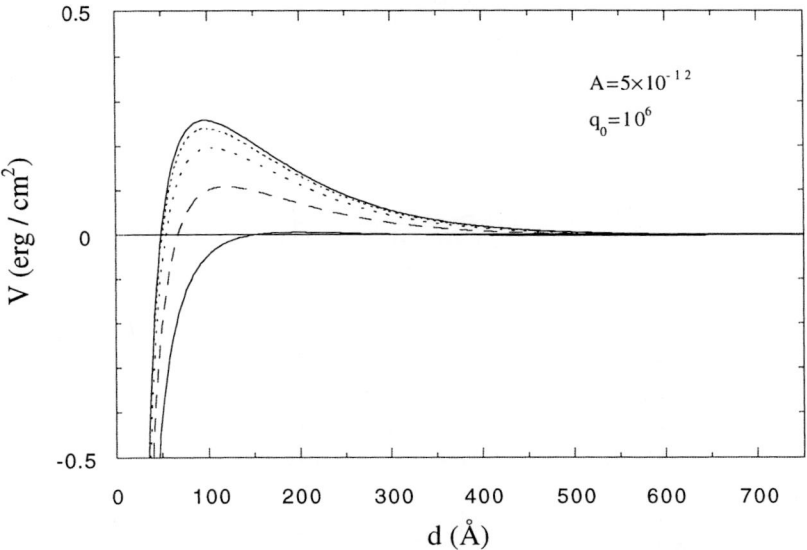

Figure 9.22: Same as Figure 9.19 but for a lower concentration.

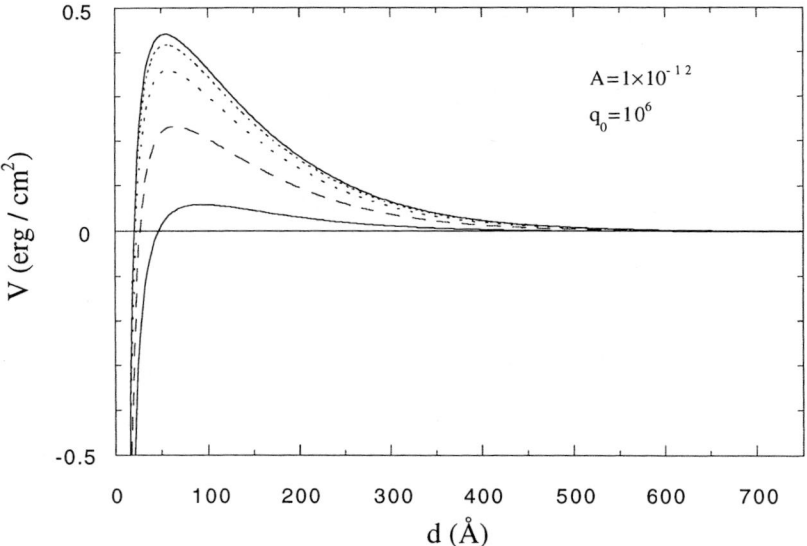

Figure 9.23: Same as Figure 9.20 but for a lower concentration.

One may define the stabilization criterion in the following way: The colloid is stable if there is a barrier with positive energy. Then the transition point is determined by:

9.7 Interaction Energy and Force Between Objects with Double Layers

$$\frac{dV}{dd} = 0 ;$$
$$V = 0$$
(9.74)

for the same d.

To be able to solve these equations analytically we need to chose a simpler approximation of the repulsive potential:

$$V_R \approx \frac{64n}{\beta q_0}\gamma^2 e^{-q_0 d} \; ; \; \gamma = \frac{e^{y_0/2}-1}{e^{y_0/2}+1}$$
(9.75)

For the attractive potential we choose:

$$V_A = -\frac{A}{12\pi}\frac{1}{d^2}$$
(9.76)

The solution is straight forward and one finds that this occurs for $d = 2/q_0$ and at the electrolyte concentration:

$$n = \frac{107.5\gamma^4 \kappa^3}{A^2 \beta^5 (Ze)^6}$$
(9.77)

We see that the valency has important effect on the concentration needed for flocculation. This valency dependence agrees quite well with experiments as can be seen in Figure 9.24, where we have plotted $n(Z)/n(1)$. The solid line is the theoretical prediction from Equation (9.77). The circles, squares and triangles are from column 1, 2, and 3, respectively of Tables 9.2 and 9.3. The solid symbols are the mean values and the open ones are maximum and minimum values.

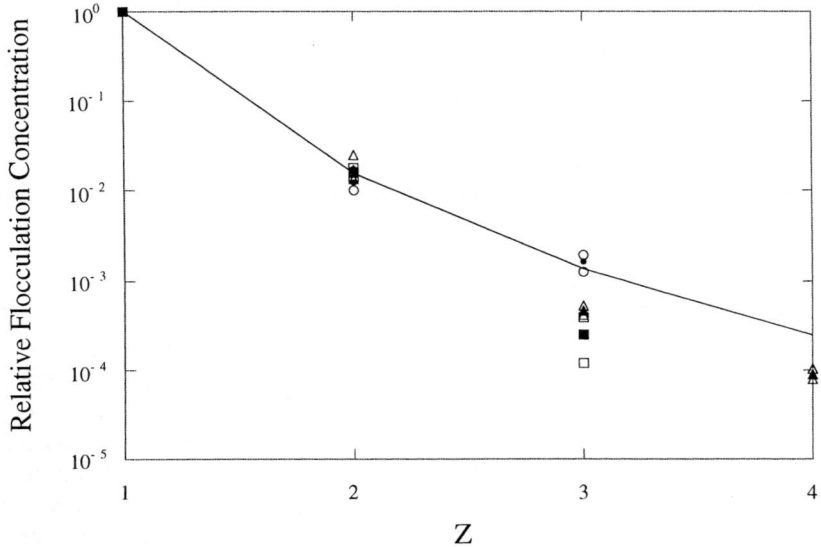

Figure 9.24: The dependence of relative flocculation concentration on the valency.

9.7.2 Interaction between spherical particles

For spherical particles we use the distance parameters defined in Figure 9.25.

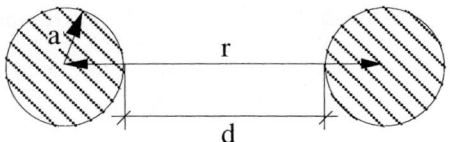

Figure 9.25: Parameters for two spherical particles.

The attractive van der Waals potential can according to Equation (5.96) be expressed as

$$V_A = -\frac{A}{6}\left[\frac{2}{s^2-4}+\frac{2}{s^2}+\ln\left(\frac{s^2-4}{s^2}\right)\right] \; ; \; s = r/a \tag{9.78}$$

or according to Equation (5.97) as

9.7 Interaction Energy and Force Between Objects with Double Layers 355

$$V_A = -\frac{A}{12}\left[\frac{1}{x(2+x)} + \frac{1}{(1+x)^2} + 2\ln\left(\frac{x(2+x)}{(1+x)^2}\right)\right] \quad ; \quad x = \frac{d}{2a} \tag{9.79}$$

For large separations we have:

$$V_A = -\frac{16A}{9}\left[\frac{1}{s^6} + \frac{6}{s^8} + \cdots\right] \tag{9.80}$$

and for small separations:

$$V_A \approx -\frac{A}{24}\frac{1}{x} = -\frac{Aa}{12}\frac{1}{d} \tag{9.81}$$

The repulsive part of the potential, used to generate the figures in this section, was obtained from interpolations of the numbers given in tables XV-XXI in VO [1]. These numbers were obtained from numerically solving the differential equation governing the double-layer potential.

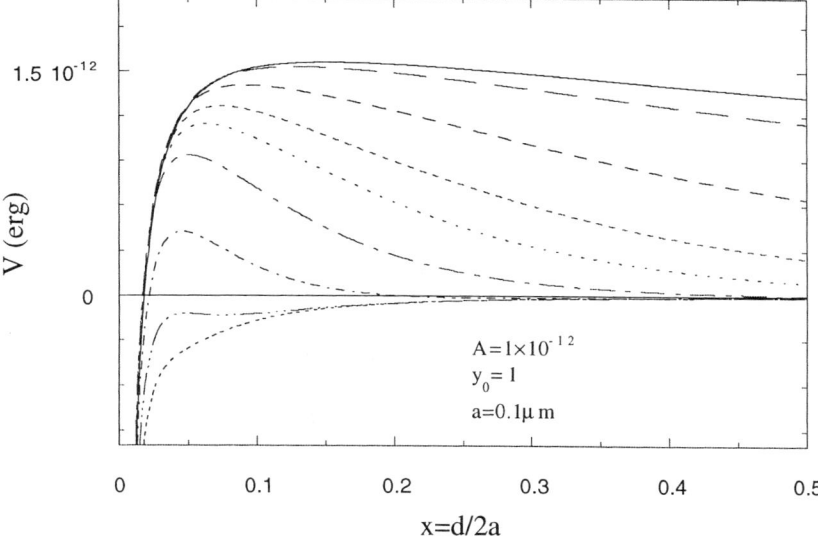

Figure 9.26: Potential between two spherical particles for different concentrations.

In Figure 9.26 we show the potential between two spherical particles for different concentrations defined by the q_0 values 10^4, 3×10^4, 10^5, 2×10^5, 3×10^5, 5×10^5, 10^6, 10^7, and 10^8 cm^{-1}, respectively, counted from above. We should note one very important difference with respect to the flat double layer. Here the size of the barrier does not decrease with decreasing electrolyte concentration. This means that the colloid is stable for very low concentrations. In the case of flat particles the concentration has to stay between upper and lower limits for the colloid to be stable. The value of the surface potential for the choice of $y_0 = 1$ is 26 mV. In Figure 9.27 we study how the potential barriers are influenced by the size of the surface potential and in Figure 9.28 we concentrate on how the second, or outer, shallow minimum depends on potential and concentration.

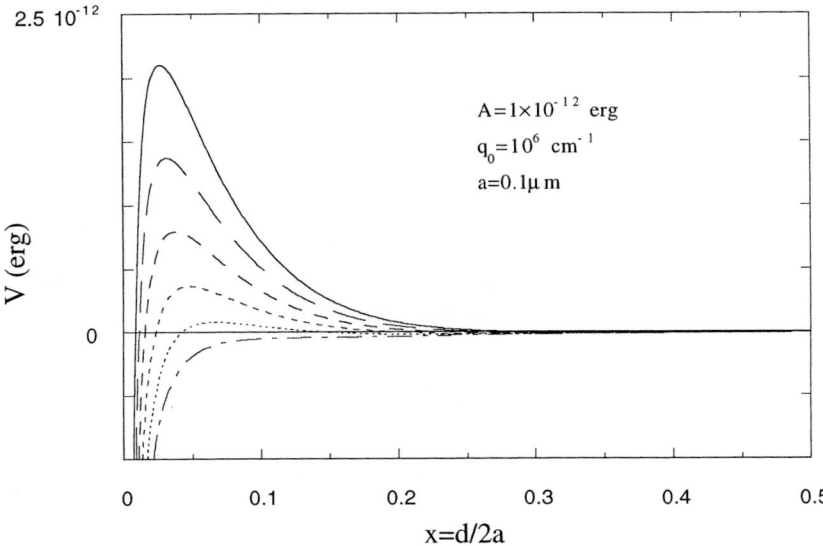

Figure 9.27: Potential between two spherical particles for different surface potentials.

The curves in Figure 9.27, counted from below, are for the surface potentials 15, 20, 25, 30, 35 and 40 mV, respectively.

The solid curves in Figure 9.28 correspond to $q_0 = 3 \times 10^6$ cm^{-1}, the dotted to $q_0 = 10^6$ cm^{-1}, and the dashed to 3×10^5 cm^{-1}, respectively. The solid ones are for y_0 equal to 2, 4 and 8, the dotted ones for y_0 equal to 1, 2, 4, 6, 8, and 10, and the dashed for y_0 equal to 1, 2, and 4, respectively, all counted from the left or bottom.

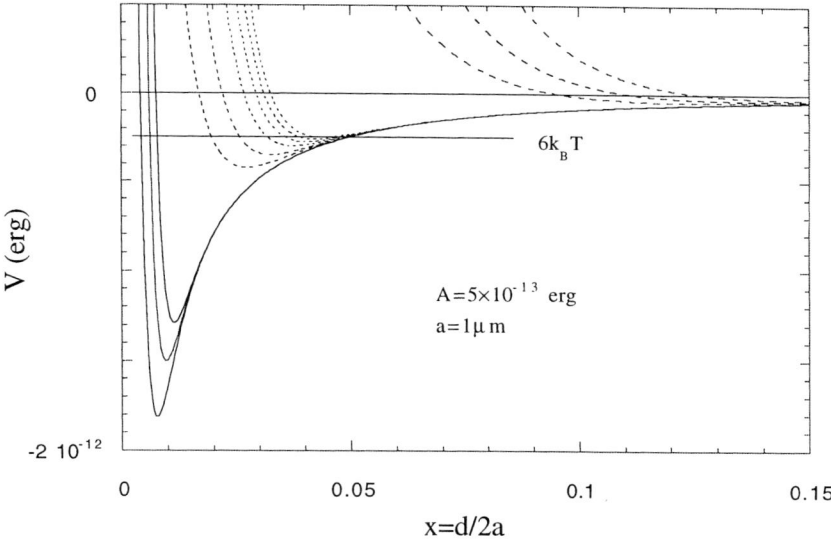

Figure 9.28: Potential between two spherical particles.

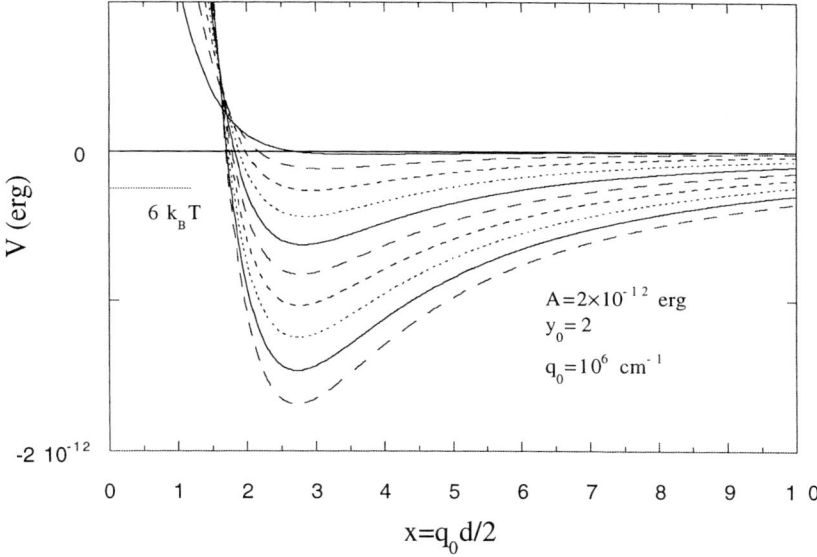

Figure 9.29: Potential between two spherical particles.

In Figure 9.29 we have kept the potential, the Hamaker constant, and the electrolyte concentration fixed and varied the size of the particles. The size of the particles has been given the values 0.1, 0.2, ..., 1.0 μm. The larger the particle the deeper the second minimum. If the minimum is more shallow than $6k_BT$ two particles bound at the second minimum will soon be thermally released. Thus, the larger the particles the more stable is the flocculation at the second minimum. The critical value is indicated in the figure by the short horizontal line.

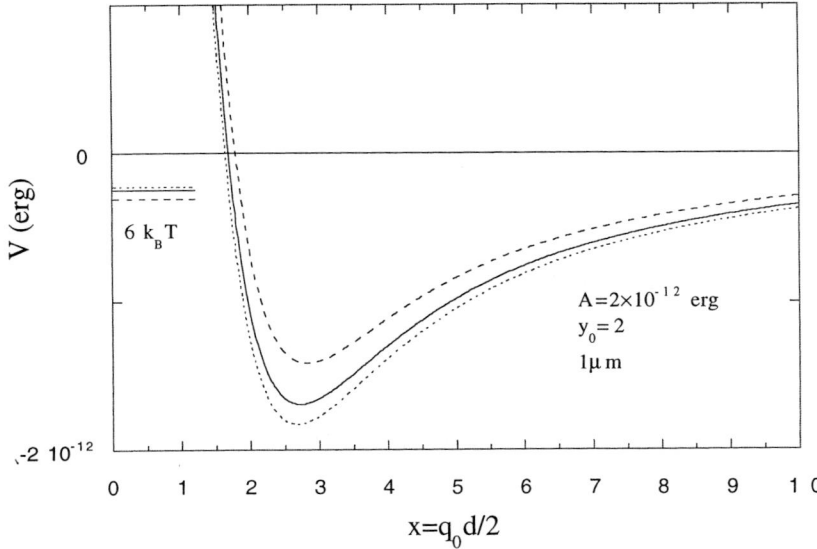

Figure 9.30: Potential between two spherical particles.

In Figure 9.30 we study the effects of varying the temperature. The system is the same as in Figure 9.29 for $1\mu m$ particle size. Apart from the room-temperature result we have determined the potential minimum at the freezing point, dotted curve, and at the boiling point, dashed curve. We have also indicated the $6k_BT$ values for these cases. We find that the depth of the minimum changes much more than the $6k_BT$ values. This means that a decrease in temperature favors the reversible, weak flocculation at the second minimum. What changes with temperature is q_0. In Figure 9.31 we plot the barrier. We find that the barrier preventing irreversible coagulation is reduced with decreasing temperature.

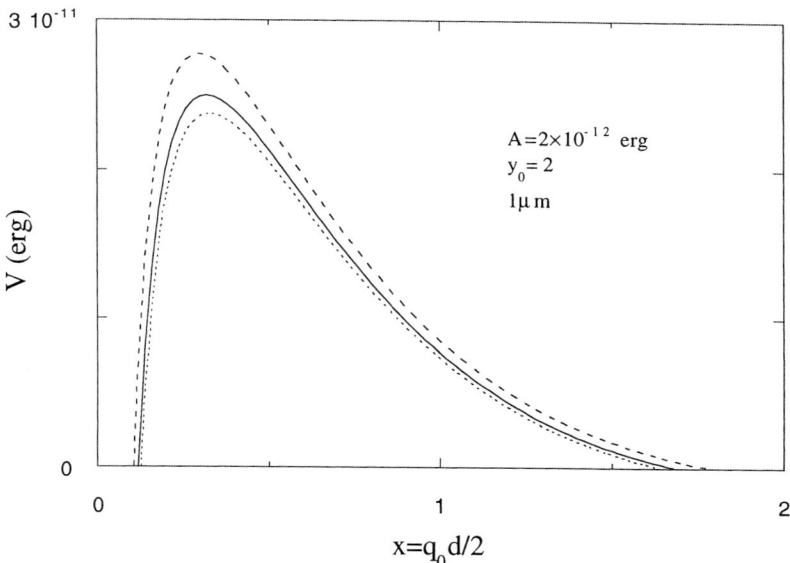

Figure 9.31: Barrier between two spherical particles.

This is a proper place to summarize our findings for flat and spherical double layers. For flat double layers we have seen that the coagulation-preventing barrier disappears in the diluted limit while for spherical particles the height of the barrier saturates and the barrier prevails. If we keep everything constant except the size of the spherical particles the second minimum in the potential becomes deeper with increasing particle size, but the barrier height is reduced. This is also consistent with the fact that for large particles one can treat the particles as flat. The larger the particle the smaller the saturation-value for the barrier at low concentration. The van der Waals interaction saturates at very small separations; of the order of the inter-atomic spacing in the particle. This is not included in our calculations. Besides there is a repulsive potential-contribution when the electron wave-functions in the two particles overlap. This has not been included either. If these effects are included there is a primary minimum in the potential at particle contact. This minimum is much deeper than the secondary minimum we have discussed here. It is furthermore always present. If the particles happen to reach this minimum it is much more difficult to bring them apart again. It is sometimes possible, though. The success is also depending on how long the particles have been stuck together. In the peptization process it is not enough to reduce the electrolyte concentration to a value where the colloid is stable; one really has to wash out the precipitate carefully before repeptization occurs.

Let us now expand on the secondary minimum. This minimum is much more shallow. This means that it is easier to reverse the flocculation process. The particles are furthermore not close enough for a chemical binding or a recrystallization to occur. This secondary minimum is probably responsible for some interesting effects found experimentally:

In sols with plate-shaped or rod-shaped particles one often encounters something called *tactoids*: A phase containing a high concentration of colloid in equilibrium with a low density phase. The high density phase is anisotropic; the plates or rods are aligned. This is probably due to the effect that larger particles have deeper secondary minima. Plates with their flat faces in contact correspond to larger particles and the minima are deeper in this configuration and can be deep enough for the particles to be trapped. The same holds for rod-shaped particles.

Another example is *thixotropy*. This is an easily reversible form of gelation. The high concentration phase can form a network throughout the whole system. This gelation is easily lifted by shaking the sample. This effect is sometimes used in paint. There are thixotropic paints where left alone in the can the paint forms a gel that does not flow. One may turn the pot upside down without spilling the paint. When stirred or shaken it flows easily.

An interesting experiment was performed by von Buzagh [5]. He let a dilute suspension of quartz particles settle on a glass plate. After reversing the glass plate under the liquid he observed that a smaller or larger part of the particles fell from the plate but another part adhered to it. The percentage of adhering particles increased strongly by increasing the electrolyte concentration. The particles while adhering to the glass still showed a lateral Brownian motion and particles of the size around 3 μm were adhered; smaller ones had a too shallow secondary minimum; larger ones were pulled off by gravitation. Similar or related experiments are performed today [6].

References

[1] E. J. W. Verwey, and J. T. G. Overbeek, *Theory of the stability of lyophobic colloids*, (Elsevier New York, 1948).
[2] G. Gouy, J. Physique **9**, 457 (1910).
[3] D. L. Chapman, Phil. Mag. **25**, 475 (1913).
[4] O. Stern, Z. Electrochem. **30**, 508 (1924).
[5] A. von Buzagh, Kolloid. Z., **47** 370 (1929); **51**, 105 (1930).
[6] A. E. Larsen, and D. G. Grier, Phys. Rev. Lett. **76**, 20 (1996); Nature **385**, 230 (1997).

Appendix 1

Table A1.1: Conversion from CGS to SI units; from : J. D. Jackson, in *Classical Electrodynamics*(John Wiley & Sons, Inc., New York 1962) p. 619. Reprinted by permission of John Wiley & Sons, Inc.

Quantity	CGS	SI
Velocity of Light	c	$\sqrt{1/\mu_0\varepsilon_0}$
Electric Field (potential, voltage)	$\mathbf{E}(\Phi, V)$	$\sqrt{4\pi\varepsilon_0}\,\mathbf{E}(\Phi, V)$
Displacement	$\mathbf{D}$	$\sqrt{4\pi/\varepsilon_0}\,\mathbf{D}$
Charge Density (charge, current density, current, polarization)	$\rho(q, \mathbf{j}, \mathbf{I}, \mathbf{P})$	$\sqrt{1/4\pi\varepsilon_0}\,\rho(q, \mathbf{j}, \mathbf{I}, \mathbf{P})$
Magnetic Induction	$\mathbf{B}$	$\sqrt{4\pi/\mu_0}\,\mathbf{B}$
Magnetic Field	$\mathbf{H}$	$\sqrt{4\pi\mu_0}\,\mathbf{H}$
Magnetization	$\mathbf{M}$	$\sqrt{\mu_0/4\pi}\,\mathbf{M}$
Conductivity (capacitance)	$\sigma(C)$	$\sigma(C)/4\pi\varepsilon_0$
Dielectric Function	ε	$\varepsilon/\varepsilon_0 = \varepsilon_r$
Permeability	μ	$\mu/\mu_0 = \mu_r$
Resistance (impedance, inductance)	$R(Z, L)$	$4\pi\varepsilon_0 R(Z, L)$

Appendix 2

The Fourier transform can be defined in different ways. We define the Fourier transforms with respect to position and time and their inverses in the following way:

$$f(\mathbf{q}) = \int d^3 r \, e^{-i\mathbf{q}\cdot\mathbf{r}} f(\mathbf{r})$$

$$f(\mathbf{r}) = \frac{1}{\Omega} \sum_{\mathbf{q}} e^{i\mathbf{q}\cdot\mathbf{r}} f(\mathbf{q}) = \int \frac{d^3 q}{(2\pi)^3} e^{i\mathbf{q}\cdot\mathbf{r}} f(\mathbf{q})$$

$$f(\omega) = \int_{-\infty}^{\infty} dt \, e^{i\omega t} f(t)$$

$$f(t) = \int_{-\infty}^{\infty} \frac{d\omega}{(2\pi)} e^{-i\omega t} f(\omega)$$

(A2.1)

With these sign conventions, Fourier transforming differential equations has the following substitutional effects:

$$\frac{\partial}{\partial t} \rightarrow -i\omega$$
$$\nabla \cdot \rightarrow i\mathbf{q} \cdot$$
$$\nabla \times \rightarrow i\mathbf{q} \times$$

(A2.2)

Index

2D (two-dimensional), 126, 154, 160

ABCs, 305
Absorption, 48
Absorption band, 44
Active side, 106
Additional boundary conditions, 305
Adhesion, 232, 236
    work of, 232, 235
Aerogel, 318
Aerosol, 317
Aluminum, 39, 40, 72
Analytic continuation, 205
Anions, 321, 324, 328
Arago, D. F., 315
Arakawa, E. T., 165
Argument principle, 198, 209, 210
Ashcroft, N. W., 77
ATR, 306, 307, 313
    angle scan, 311–313
    frequency scan, 310, 311, 313
Attenuated Total Reflection, 306, 307, 313
Axilrod, B. M., 196
Axilrod-Teller interaction, 177

Back-bending, 300, 313
Beryllium, 293
Bessel equation, 283
    modified, 272, 283
Bessel functions
    cylindrical, 273
    modified, 187, 272
    spherical, 261
Björk Patrik, 165
Boer, J. H., 196
Boer-Hamaker constant, 190, 223–225, 252, 253, 347, 350, 351, 357
Boltzmann constant, 53, 192

Boltzmann equation, generalized, 57
Boström, M., 165, 256
Brewster modes, 298, 300, 312
Brillouin zone, 290
Brownian motion, 57, 320, 321, 323, 328, 360
Bulk modes, 17

Cadmium Sulfide, 73, 114, 285
Capacitance of the double layer, 333, 335, 337
Capillary rise, 244, 245, 247
Carotenuto, L., 284
Casimir, H. B. G., 138, 165
Casimir force, 97, 138, 154, 157, 158, 190, 207, 208, 212, 220
    finite temperature, 142, 229, 230
    zero temperature, 139
Casimir-Polder interaction, see also Casimir force, 207, 208
Catalytic effects, 179
Cations, 321, 324, 328
Causality, 68, 76
CEM, 154, 158, 163, 165
Cerenkov radiation, 73
Chapman, D. L., 360
Characteristic integral, 186, 217
Charge density, 18
Chemical potential, 192, 193, 339
Chen, F., 165, 256
Chiao, R. Y., 77
Chu, S., 77
Classical electron radius, 80
Coagulation, 320
Coalescence, 269, 320
    non, 269
Cohesion, 232, 236
    work of, 232
Colloid, 317–321, 324, 329, 335, 336, 339, 349, 352, 356, 359, 360

stability of, 138, 317, 319–321, 324, 339, 345, 349, 350
Conductivity, 18
Constituent particles, 80
Constitutive relations, 18, 19
Continuous phase, 317
Contour
    integration in complex frequency plane, 198
        finite temperature, 210, 211
        zero temperature, 199, 201
Convolution integral, 19
Copper, 45
Correlation energy, 79, 93, 94, 97
Correlation energy method, 154, 158, 163, 165
Correlation functions
    retarded, 31, 204
    time ordered, 93, 204
Counter ion, 321, 323, 324, 328, 335, 337
Coupling constant, 92, 94, 96
Craig, R. A., 165
Current density, 18
Current drag, 154
Cylinder, 186, 187, 271
Cylindrical coordinates, 271
    parabolic, 277

Damping, 35
Debye, P., 77
Debye rotational relaxation, 56
Debye-Hückel
    screening, 323, 329
        length, 324
Dell'Aversana, P., 284
Density matrix, 82
Depeptization, 321
Determinant, 155, 174, 175, 204
    Slater, 83
Dielectric function, 18, 31, 34, 93, 95, 96
    for insulator, 42, 287
    for metal, 45, 151
    for plasma, 51
    for water, 60, 222–224
    relation to conductivity, 21
Dielectric matrix, 155
Diffuse layer, 321, 323, 328, 334, 335, 337
Dipole moment, 33
    permanent, 54, 168, 170, 171
    time dependent, 202

Dipole-dipole interaction, 175, 177, 181, 184
Dispersed phase, 317
Dispersion
    anomalous, 44, 57, 291
    exciton, 304
    of light, 24, 34, 35, 43, 44, 69, 70, 107, 108, 114, 116
    relations
        Kramers Kronig, 31, 221, 287
        spatial, 302–305
Dissipation, 57
Distribution function, 57
    Boltzmann, 193
Double layer, 317, 320, 321, 323, 333, 335, 336, 339–341, 359
    flat, 324, 329, 337, 342, 347, 349
    spherical, 326, 336, 355
Drude tail, 50

Earthquake, 285, 307, 314
Edge, 277
Effective area, 182, 187, 188
Effective separation, 225
Ehrenreich, H., 77
Eigenvalue equation, 175
Electric displacement, 18
Electric field, 18
Electrolytes, 320
Emulsion, 317
    solid, 317
Enders, W., 77
Energy density, 79, 109
Energy flux, 109
Enthalpy, 209
Equation of
    continuity, 18
        Fourier transformed, 21
    motion, 32, 51, 174, 202
    state, 173
        ideal gas, 173
        van der Waals, 192
Euler-Lagrange equation, 248
Euler-Maclaurin summation formula, 141
Evanescent modes, 297, 298, 300
Exchange energy, 79, 81, 85, 90, 93, 94, 97
Exchange hole, 84, 85, 87, 88
Exchange-correlation hole, 88
Exciton
    Frenkel, 303

Index 367

Mott, 303
Wannier, 303
Exner, F. M., 315

f-sum rule, 37–39, 45, 97, 161
Fano modes, 298, 300
Fermi
    sphere, 89
    temperature, 53
    wave vector, 37
Fermi-Dirac occupation number, 93
Flocculation, 320, 322, 323, 336, 353, 354, 358, 359
Foam, 317
    solid, 317
Force
    dispersion, 168
    electromagnetic, 167
    fundamental, 167
    gravitational, 167
    hydration, 320
    induction, 168
    nuclear, 167
        strong, 167
        weak, 167
    orientation, 168
    steric, 320
Fourier transform, 18, 363
Free energy
    Gibb's, 193, 209
    Helmholtz', 142, 170, 171, 209
Frequency summations, 211, 221, 222, 229, 230
Fresnel's equations, 305, 307–309
Friction, 32, 57
    coefficient of, 51, 53
    coefficient of mutual, 51
    inner, 57

Gallium Arsenide, 154, 285
Gauge
    Coulomb or transverse, 24, 26, 258
    Lorentz, 24, 28
        generalized, 28
Gelatin, 318, 320
Gold, 45
Gouy, G., 360
Gouy and Chapman theory, 328, 329, 336
Gouy layer, 337

Gradient
    in cylindrical coordinates, 271
    in parabolic cylindrical coordinates, 277
Grating, 297
Grigoriev, I. S., 256
Ground-state-energy-theorem, 92

Halevi, P., 315
Half-space, 186, 187
Halo, 314
Hamaker, H. C., 196
Hamaker constant, 190, 223–225, 252, 253, 347, 350, 351, 357
Hankel functions, 283
    cylindrical, 273
    spherical, 261
Harris, B. W., 165, 256
Hartman, T. E., 77
Hartree Fock approximation, 81, 84–88, 91, 92
Heaviside unit step function, 38
Heitmann, W., 77
Helium, liquid, 197, 253
Hermitian function, 278
Hertz, H., 17, 29
HF, 81, 84–88, 91, 92
Homogeneous phase, 317
Hooke's law, 32
Hopfield, J. J., 315
Hydrophilic, 319
Hydrophobic, 318, 319

Inagaki T., 165
Instantaneous interaction, 26
Insulators, 32
Interface
    flat, 99
    non-planar, 257
Internal energy, 209, 212, 221, 226
Ion, 170, 177
Israelachvili, J., 256

Jackson, J. D., 361
Jellium model, 81, 147

Khare, V., 315
Kleppner, D., 165
Kohn, W., 147, 165
Kretschmann configuration, 307

Kwiat, P. G., 77

Lamoreaux, S. K., 165, 256
Lang, N. D., 147, 165
Langbein, D., 196
Laplace's equation, 258
Laplacian
    in cylindrical coordinates, 271
    in parabolic cylindrical coordinates, 277
Larsen, A. E., 360
LC circuit, oscillations in, 145
Lifshitz, E. M., 7
Lipophilic, 319
Lipophobic, 319
Lithium, 208, 212
Local approximation, 18
London, F., 173, 196
London approximation, 176, 179, 252, 254
Longitudinal field component, 22
Lucas, A. A., 146, 165
Lyddane-Sachs-Teller relation, 42, 289
Lyophilic, 319
Lyophobic, 319

Macroscopic objects, 180
Magnetic field, 18
Magnetic induction, 18
Magnetic permeability, 18
Magnetization, 19
Mahan, G. D., 165, 196
Many-body theory, 79, 95
Maradudin, A. A., 306, 315
Marangoni convection, 270
Massless bosons, 79
Maxwell, J. C., 17, 29
Maxwell's equations, 18
    Fourier transformed, 21
    in vacuum, 64
Meilikhov, E. Z., 256
Meniscus, 244, 251
Mercury, 149, 151, 153, 233
Mermin, N. D., 77
Milk, 318
Mode summation method, 154, 164, 165
Models
    Drude's classical, 36, 151
        generalized, 36
    Lorentz' classical, 32, 173
        dielectric function, 34

Mohideen, U., 165, 256
Moment of inertia, 57
MSM, 154, 164, 165

Needle, 282
Neitzel, G. P., 284
Newton's second law, 32
Nimtz, G., 77
Nordling, C., 165

Octupole, 184
Ohm's law, 18, 19
Oleophilic, 319
Oleophobic, 319
Optical constants, 60
Optical glory, 285, 314
Oscillator strength, 38
Ostwald ripening, 244, 320
Oswald ripening, 244
Otto configuration, 307, 310
Overbeek, J. T. G., 138, 360

p-polarized, 121, 126, 158, 159, 308
Pair interactions, 188, 192, 252, 253
Pair-correlation function, 82, 84, 86–88
PAM, 307
Parabolic-cylinder functions, 279
Particle coarsening, 244
Partition function, 142
Passive side, 106
Pauli's exclusion principle, 81, 89
Peptization, 321, 359
Pertner, J. M., 315
Phillip, H. R., 77
Phonons
    acoustical, 286
    optical, 286
        longitudinal, 111
        transverse, 111
    surface, 116
Plasma frequency, 33
Plasmon
    bulk, 97, 108, 145
    dispersion, 40
    surface, 97, 107, 108, 145
Plasmon-pole approximation, 161, 163
PMA, 307
Point particle, 79
Poisson's equation, 26

Polariton, 35, 291
  exciton, 302
  surface, 108
    phonon, 116
    plasmon, 108
Polarizability
  atomic, 33
  molecular, 58
  multipole, 184
  of a sphere, 181, 259
  orientational, 55
Polarization, 19
Polder, D., 138, 165
Potential
  scalar, 24
  vector, 24
Poynting vector, 110

Quadropole, 184
Quantization
  first, 81
  second, 88
Quantum well, 154
Querry, M. R., 77

Radiative modes, 273, 297, 306, 314
Rainbow, 314
Random Phase Approximation, 37, 39, 93, 163
Ranfagni, A., 77
Rayleigh, B., 284
Reflectance, 48, 54
Reflectivity, 45
Refractive index, 44, 45, 62
Relaxation time, 51
Rest energy, 79
Retardation effects, 26, 65, 101, 107, 126, 138, 158
Reynolds, O., 284
Rotational operator, 20
Roy, Anushree, 165, 256
RPA, 37, 39, 93, 163
Ruppin, R., 284

s-polarized, 121, 158, 159, 308
Scharnhorst, K., 77
Schmit, J., 146, 165
Segelstein, D. J., 77
Self-energy, 79, 85–87, 90, 91

Self-interaction, 89, 90, 94
Self-sustained fields, 23
Self-sustained potentials, 27
Separation of colors, 62
Sernelius, Bo E., 77, 165, 256
Shchegrov, A. V., 306, 315
Shear plane, 338
Silicon oil, 269
Silver, 46
Single particle continuum, 96, 155, 293
Size quantization, 80, 126
Slab
  geometry, 117
  metal, 124
  semiconductor, 128
Sodium Cloride, 285
Soft order, 318
Sol, 317
Solid suspension, 317
Solution
  anti-symmetric, 125, 129
  symmetric, 125, 129
Sparnaay, M. J., 138, 165
Speed of light, 44
  in a medium, 35, 44, 73
  in vacuum, 35, 44, 64
Sphere, 181, 182, 185–187, 257
  layered, 265
Spherical harmonics, 258
Spurious modes, 24
Steinberg, A. M., 77
Stern
  layer, 337
  plane, 338
  potential, 338
Stern, O., 360
Stern's theory, 337
Structure factor, dynamical, 161
Strutt, J. W., 284
Super sonic speeds, 71
Superluminal speeds, 35, 64
Surface energy, 213–218, 232, 233, 239, 240, 247, 251, 269
  of metals, 145
Surface roughness, 261, 292, 297, 306
Surface tension, 233
  of metals, 153
Surfactants, 320
Susceptibility, multipole, 183, 186

Sánchez-Gil, J. A., 306, 315

Tachyons, 66
Tactoids, 360
TE modes, 121, 140, 215, 226, 229, 230, 260, 273, 300
Teller, E., 196
Tensor, dipole-dipole, 168, 178
Thermal wave length, 193
Thermodynamic potential, 209
Thixotropy, 360
Thomas Fermi
    screening, 323
        length, 160, 161, 214
Thunderstorms, 167, 170
TM modes, 121, 140, 215, 226, 229, 230, 260, 273, 300
Tontodonato, V., 284
Transitions
    dipolar, 31
    exciton, 31
    interband, 31, 38, 287
    intraband, 50, 97
Transmittance, 48
Transverse field component, 22
Tunneling, 74

Vacuum fluctuations, 79, 145
van der Waals, J. D., 173
van der Waals force, 97, 138, 154, 157, 158, 173, 190, 320
    finite temperature, 229, 230
    zero temperature, 229, 230
Velocity
    angular, 57
    group, 44, 65, 68
    phase, 44, 65, 68
Verwey, E. J. W., 138, 360
Vibration modes, 62
Virtual modes, 297
Viscosity, 56, 57, 269
von Buzagh, A., 360

Wang, S. Q., 165, 196
Water, 54, 60, 221–225, 227, 233
Wave packet, 35, 68
Wave vector, cutoff, 89, 140, 147, 214, 217
Wedge, 280
Wetting, 232, 234–236
    angle, 247
    suppression, 270
Whipping cream, 318
Wieliczka, D. M., 77
Wong, S., 77

Xerogels, 318

Zenneck modes, 298, 300
Zero-point energy, 79, 94, 95, 97, 154, 212, 216
Zerogels, 318
Zeta-potential, 338

Österman, J., 165